THE MK PROCESS AT 50 YEARS: A POWERFUL TOOL FOR ASTROPHYSICAL INSIGHT

A SERIES OF BOOKS ON RECENT DEVELOPMENTS IN ASTRONOMY AND ASTROPHYSICS

Printed by BookCrafters, Inc.

First published 1994

Library of Congress Catalog Card Number: 94-72082
ISBN 0-937707-79-1

D. Harold McNamara, Managing Editor of Conference Series
408 ESC Brigham Young University
Provo, UT 84602
801-378-2298

A SERIES OF BOOKS ON RECENT DEVELOPMENTS IN ASTRONOMY AND ASTROPHYSICS

Vol. 1-Progress and Opportunities in Southern Hemisphere Optical Astronomy: The CTIO 25th Anniversary Symposium
ed. V. M. Blanco and M. M. Phillips ISBN 0-937707-18-X

Vol. 2-Proceedings of a Workshop on Optical Surveys for Quasars
ed. P. S. Osmer, A. C. Porter, R. F. Green, and C. B. Foltz ISBN 0-937707-19-8

Vol. 3-Fiber Optics in Astronomy
ed. S. C. Barden ISBN 0-937707-20-1

Vol. 4-The Extragalactic Distance Scale: Proceedings of the ASP 100th Anniversary Symposium
ed. S. van den Bergh and C. J. Pritchet ISBN 0-937707-21-X

Vol. 5-The Minnesota Lectures on Clusters of Galaxies and Large-Scale Structure
ed. J. M. Dickey ISBN 0-937707-22-8

Vol. 6-Synthesis Imaging in Radio Astronomy: A Collection of Lectures from the Third NRAO Synthesis Imaging Summer School
ed. R. A. Perley, F. R. Schwab, and A. H. Bridle ISBN 0-937707-23-6

Vol. 7-Properties of Hot Luminous Stars: Boulder-Munich Workshop
ed. C. D. Garmany ISBN 0-937707-24-4

Vol. 8-CCDs in Astronomy
ed. G. H. Jacoby ISBN 0-937707-25-2

Vol. 9-Cool Stars, Stellar Systems, and the Sun. Sixth Cambridge Workshop
ed. G. Wallerstein ISBN 0-937707-27-9

Vol. 10-The Evolution of the Universe of Galaxies. The Edwin Hubble Centennial Symposium
ed. R. G. Kron ISBN 0-937707-28-7

Vol. 11-Confrontation Between Stellar Pulsation and Evolution
ed. C. Cacciari and G. Clementini ISBN 0-937707-30-9

Vol. 12-The Evolution of the Interstellar Medium
ed. L. Blitz ISBN 0-937707-31-7

Vol. 13-The Formation and Evolution of Star Clusters
ed. K. Janes ISBN 0-937707-32-5

Vol. 14-Astrophysics with Infrared Arrays
ed. R. Elston ISBN 0-937707-33-3

Vol. 15-Large-Scale Structures and Peculiar Motions in the Universe
ed. D. W. Latham and L. A. N. da Costa ISBN 0-937707-34-1

Vol. 16-Atoms, Ions and Molecules: New Results in Spectral Line Astrophysics
ed. A. D. Haschick and P. T. P. Ho ISBN 0-937707-35-X

Vol. 17-Light Pollution, Radio Interference, and Space Debris
ed. D. L. Crawford ISBN 0-937707-36-8

Vol. 18-The Interpretation of Modern Synthesis Observations of Spiral Galaxies
ed. N. Duric and P. C. Crane ISBN 0-937707-37-6

Vol. 19-Radio Interferometry: Theory, Techniques, and Applications, IAU Colloquium 131
ed. T. J. Cornwell and R. A. Perley ISBN 0-937707-38-4

Vol. 20-Frontiers of Stellar Evolution, celebrating the 50th Anniversary of McDonald Observatory
ed. D. L. Lambert ISBN 0-937707-39-2

Vol. 21-The Space Distribution of Quasars
ed. D. Crampton ISBN 0-937707-40-6

Vol. 22-Nonisotropic and Variable Outflows from Stars
ed. L. Drissen, C. Leitherer, and A. Nota ISBN 0-937707-41-4

Vol. 23-Astronomical CCD Observing and Reduction Techniques
ed. S. B. Howell ISBN 0-937707-42-4

Vol. 24-Cosmology and Large-Scale Structure in the Universe
ed. R. R. de Carvalho ISBN 0-937707-43-0

Vol. 25-Astronomical Data Analysis Software and Systems I
ed. D. M. Worrall, C. Biemesderfer, and J. Barnes ISBN 0-937707-44-9

Vol. 26-Cool Stars, Stellar Systems, and the Sun, Seventh Cambridge Workshop
ed. M. S. Giampapa and J. A. Bookbinder ISBN 0-937707-45-7

Vol. 27-The Solar Cycle
ed. K. L. Harvey ISBN 0-937707-46-5

Vol. 28-Automated Telescopes for Photometry and Imaging
ed. S. J. Adelman, R. J. Dukes, Jr., and C. J. Adelman ISBN 0-937707-47-3

Vol. 29-Workshop on Cataclysmic Variable Stars
ed. N. Vogt ISBN 0-937707-48-1

Vol. 30-Variable Stars and Galaxies, in honor of M. S. Feast on his retirement
ed. B. Warner ISBN 0-937707-49-X

Vol. 31-Relationships Between Active Galactic Nuclei and Starburst Galaxies
ed. A. V. Filippenko ISBN 0-937707-50-3

Vol. 32-Complementary Approaches to Double and Multiple Star Research, IAU Colloquium 135
ed. H. A. McAlister and W. I. Hartkopf ISBN 0-937707-51-1

Vol. 33-Research Amateur Astronomy
ed. S. J. Edberg ISBN 0-937707-52-X

Vol. 34-Robotic Telescopes in the 1990s
ed. A. V. Filippenko ISBN 0-937707-53-8

Vol. 35-Massive Stars: Their Lives in the Interstellar Medium
ed. J. P. Cassinelli and E. B. Churchwell ISBN 0-937707-54-6

Vol. 36-Planets and Pulsars
ed. J. A. Phillips, S. E. Thorsett, and S. R. Kulkarni ISBN 0-937707-55-4

Vol. 37-Fiber Optics in Astronomy II
ed. P. M. Gray ISBN 0-937707-56-2

Vol. 38-New Frontiers in Binary Star Research
ed. K. C. Leung and I. S. Nha ISBN 0-937707-57-0

Vol. 39-The Minnesota Lectures on the Structure and Dynamics of the Milky Way
ed. Roberta M. Humphreys ISBN 0-937707-58-9

Vol. 40-Inside the Stars, IAU Colloquium 137
ed. Werner W. Weiss and Annie Baglin ISBN 0-937707-59-7

Vol. 41-Astronomical Infrared Spectroscopy: Future Observational Directions
ed. Sun Kwok ISBN 0-937707-60-0

Vol. 42-GONG 1992: Seismic Investigation of the Sun and Stars
ed. Timothy M. Brown ISBN 0-937707-61-9

Vol. 43-Sky Surveys: Protostars to Protogalaxies
ed. B. T. Soifer ISBN 0-937707-62-7

Vol. 44-Peculiar Versus Normal Phenomena in A-Type and Related Stars
ed. M. M. Dworetsky, F. Castelli, and R. Faraggiana ISBN 0-937707-63-5

Vol. 45-Luminous High-Latitude Stars
ed. D. D. Sasselov ISBN 0-937707-64-3

Vol. 46-The Magnetic and Velocity Fields of Solar Active Regions, IAU Colloquium 141
ed. H. Zirin, G. Ai, and H. Wang ISBN 0-937707-65-1

Vol. 47-Third Decinnial US-USSR Conference on SETI
ed. G. Seth Shostak ISBN 0-937707-66-X

Vol. 48-The Globular Cluster-Galaxy Connection
ed. Graeme H. Smith and Jean P. Brodie ISBN 0-937707-67-8

Vol. 49-Galaxy Evolution: The Milky Way Perspective
ed. Steven R. Majewski ISBN 0-937707-68-6

Vol. 50-Structure and Dynamics of Globular Clusters
ed. S. G. Djorgovski and G. Meylan ISBN 0-937707-69-4

Vol. 51-Observational Cosmology
ed. G. Chincarini, A. Iovino, T. Maccacaro, and D. Maccagni ISBN 0-937707-70-8

Vol. 52-Astronomical Data Analysis Software and Systems II
ed. R. J. Hanisch, J. V. Brissenden, and Jeannette Barnes ISBN 0-937707-71-6

Vol. 53-Blue Stragglers
ed. Rex A. Saffer ISBN 0-937707-72-4

Vol. 54-The First Stromlo Symposium: The Physics of Active Galaxies
ed. Geoffrey V. Bicknell, Michael A. Dopita, and Peter J. Quinn ISBN 0-937707-73-2

Vol. 55-Optical Astronomy from the Earth and Moon,
ed. Diane M. Pyper and Ronald J. Angione ISBN 0-937707-74-0

Vol. 56-Interacting Binary Stars
ed. Allen W. Shafter ISBN 0-937707-75-9

Vol. 57-Stellar and Circumstellar Astrophysics
ed. George Wallerstein and Alberto Noriega-Crespo ISBN 0-937707-76-7

Vol. 58-The First Symposium on the Infrared Cirrus and Diffuse Interstellar Clouds
ed. Roc M. Cutri and William B. Latter ISBN 0-937707-77-5

Vol. 59-Astronomy with Millimeter and Submillimeter Wave Interferometry
ed. M. Ishiguro and Wm. J. Welch ISBN 0-937707-78-3

Vol. 60-The MK Process at 50 Years: A Powerful Tool for Astrophysical Insight
ed. C. J. Corbally, R. O. Gray, and R. F. Garrison ISBN 0-937707-79-1

Inquiries concerning these volumes should be directed to the:
Astronomical Society of the Pacific
CONFERENCE SERIES
390 Ashton Avenue
San Francisco, CA 94112-1722
415-337-1100

ASTRONOMICAL SOCIETY OF THE PACIFIC
CONFERENCE SERIES

Volume 60

THE MK PROCESS AT 50 YEARS:
A POWERFUL TOOL FOR ASTROPHYSICAL INSIGHT

A Workshop of the Vatican Observatory

Tucson, Arizona, U.S.A. September 1993

Edited by

C. J. Corbally, R. O. Gray, and R. F. Garrison

TABLE OF CONTENTS

PART VII. SURVEYS AND AUTOMATED CLASSIFICATION

PART VIII. GENERAL DISCUSSION

GENERAL DISCUSSION (continued)

PREFACE

This workshop was just the obvious thing to do in 1993, at least for spectral classifiers, for this was the year of the 50th anniversary of the publication of the MKK *Atlas of Stellar Spectra*[1]. That inauguration of the MK System and its underlying MK Process cried out for some kind of a celebration, since a great deal of our present understanding of the physics of stars and of galaxies has flowed from the accurate two-dimensional classification of stars in terms of their temperature types and luminosity classes.

The format of the workshop was conceived while Chris Corbally and Richard Gray were eating an excellent pizza in Trieste, and they hope you will find that the fare of this workshop turns out to be as tasty and nourishing (scientifically) as that pizza. Our starting point, as is appropriate for a workshop marking the anniversary of an Atlas, is with the techniques now available to spectral classification. These techniques include the advent of digital spectra, the continued opening of spectroscopy into the ultraviolet and infrared wavelength regions, and the development of automated spectral classification. The techniques also include that foundation of the MK Process, the grid of standard spectra and current work on that grid.

The kind of astronomy that can be done through the spectral classification of stars flows from the techniques. These astrophysical insights are well represented in the workshop but, since spectral classification is at the stage of integrating the newly available techniques, those insights can appear rather buried in this volume. However, precisely in emphasizing the techniques, so this workshop prepares for a blossoming future of astronomy through spectral classification. It was a desire to communicate our excitement over spectral classification that led us to invite Janet Rountree to give a public lecture in downtown Tucson during the workshop, a summary of which lecture is reproduced in this volume.

The Contents pages of this volume will give you an overview of the topics. The review papers generally start each part and are a good way to put the other papers into context. The papers here were carefully edited, rather than strictly refereed, and so the discussions that follow (taken from written sheets at the time) are an important part of recording the workshop process in action and of including clarifications. The hope is that, from these and from the General Discussion, the reader will gain a sense of being present at the workshop, even if not having been so physically.

---

[1] *An Atlas of Stellar Spectra*, W.W. Morgan, P.C. Keenan, E. Kellman, (Yerkes Observatory, Univ. of Chicago) 1943.

The knowledge that, in both the techniques and in the science flowing from these, we "stand on the shoulders of giants" prompted us to put a significant section of history into the workshop. These papers are entertaining for their portrayal of the characters involved in spectral classification, but, as they are read during cloudy nights at the telescope, we hope that they also convey how good science is done. It was a problem whether to put Mihalas's and Olsen's papers in the science or in the history section, so well did they both blend historical narrative with the science that flows from classification work.

From the readers of this volume the editors would ask that they keep some historical elements of MK classification in mind. These elements can well be gathered from the Preface to *The MK Process and Stellar Classification*[2], the proceedings of the meeting of which this is the successor, and they need not be repeated here.

The Sponsor

The sponsorship and the venue of the workshop fell as obviously into place as did the meeting itself. The Vatican through its Observatory had organized a Colloquium in 1978 to honor Father Angelo Secchi and his work. Three years later, the Vatican established a research base for the Observatory in Tucson, Arizona, U.S.A., seeking darker skies than Rome for its observations. The partnership with the Steward Observatory led to the joint project of the Vatican Advanced Technology Telescope (VATT) which was built on Mount Graham, Arizona. In 1993 it was appropriate to bring together celebrations of the Vatican Observatory Research Group's presence in Arizona, as embodied in the inauguration of the VATT, with anniversary celebrations of a technique in which the Vatican has been historically involved. The participants would agree, we gather, that the conjunction was fruitful for science and delightful as an experience.

The Honorees

If so far we have not mentioned William W. Morgan and Philip C. Keenan, it is only because their presence suffused the workshop: Morgan through his former graduate students and co-workers; Keenan by actually coming to Tucson. The workshop honored their contributions to science and was designed to show them our appreciation and affection. During the workshop, we presented them both with medallions and scrolls commissioned by the Vatican at the bequest of Cardinal Castillo Lara, under whom the Vatican Observatory operates. The scrolls are

---

[2] *The MK Process and Stellar Classification*, ed. R.F. Garrison (David Dunlap Observatory, Toronto) 1984.

illustrated (p. xxi) following the message of Dr. Morgan to the conferees (p. xvii), and the words are sincerely meant by the sponsors and similarly echoed by the participants.

Organization and Acknowledgements

The members of the scientific organizing committee were Chris Corbally, S.J., (Chairperson), Robert Garrison, Richard Gray, Hugo Levato, Dimitri Mihalas, and Erik Olsen, with regular help from Helmut Abt and Jack MacConnell.

The local organizing committee, to whom the thanks from the participants can well be repeated here, were George V. Coyne, S.J., and Joseph E. Henry, S.J. Father Henry also helped as an eagle-eyed copy-editor.

We thank sincerely Michelle Santos (Steward Observatory) for her cheerful coping with registration and many other details. Her skill in transcribing the final discussion from audiotapes was invaluable. The staff of the Arizona Inn, Tucson, where the workshop was held, made us all feel most welcome by their helpfulness. Ray White graciously undertook the photography during the workshop.

Christopher J. Corbally, S.J.
Richard O. Gray
Robert F. Garrison

Tucson, 1994

A Message from W. W. Morgan to the Participants

31 August 1993

From: William Wilson Morgan

To: The Participants at the Celebration of the Fiftieth Anniversary of the Publication of the MKK Atlas of Stellar Spectra (1943):

[Statement to be read by Bob Garrison]

I am very deeply moved by the preparations for the conference in celebration of the fiftieth anniversary of the publication of the MKK Atlas. I only wish that my health were good enough that I could be there.

The publication of the ATLAS in 1943 was for me one of the most moving occasions of my life in astronomy. Happenings since that time have only strengthened that first feeling. There is something very moving that the two words "ATLAS" and "PROCESS" mean to me. They are in effect a short summary of all that I have loved to be working on during my career in astronomy.

It is particularly moving to me to know that the 1943 ATLAS was so fundamental that it is still useful today and that it is remembered well enough for a celebration to be appropriate. I am most thankful that at this time a conference has been organized by Dr. Corbally, under the auspices of the Vatican Observatory, to provide the opportunity for younger people to discuss some of the developments in modern spectral classification, most of which have evolved from those two simple words. The MK Process is one of the most beautiful developments of early thoughts that I have encountered in astronomy.

I wish you well and will look forward to reading about the celebration. May God bless and keep you all.

Dr. W. W. Morgan
Yerkes Observatory
Box 258
Williams Bay, Wisconsin
USA 53191

"MK Process at 50 Years" Workshop Participants

Abt, Helmut A., Kitt Peak National Obs., USA (14)
Boeshaar, Patricia C., Drew U., USA (2)
Boyle, S.J., Richard P., Vatican Obs. (36)
Catalano, Francesco A., Istituto di Astronomia, Italy
Chavez-Dagostino, Miguel, S.I.S.S.A.-I.N.A.O.E./Trieste, Italy (32)
Chargeishvili, B. Ketevan, Abastumani Astrophy. Obs., Rep. of Georgia
Comeron Tejero, Fernando, Univ. de Barcelona, Spain (6)
Corbally, S.J., Christopher J., Vatican Obs. (44)
Corral, Luis J., I.N.A.O.E., Mexico (9)
Coyne, S.J., George V., Vatican Obs.
Craine, Eric R., Western Research Co., USA (23)
Cramer, Noel, Obs. de Genève, Switzerland (37)
Crawford, David L., KPNO/NOAO, USA (15)
Culver, Roger B., Colorado State U., USA (29)
Drilling, John S., Louisiana State U., USA (45)
Faraggiana, Rosanna, Dept. of Astronomy/Trieste, Italy
Garrison, Robert F., David Dunlap Obs./Toronto, Canada (30)
Gerbaldi, Michèle, Inst. Astrophysique/Paris, France
Gieren, Wolfgang P., Univ. Catolica de Chile, Chile (22)
Golay, Marcel, Obs. de Genève, Switzerland
Gray, Richard O., Appalachian State U., USA (21)
Grosso, Monica G., Complejo Astro. el Leoncito, Argentina (18)
Hauck, Bernard J., Inst. d'Astro./Univ. Lausanne, Switzerland (28)
Hensberge, Herman, Royal Obs., Belgium
Hoffleit, E. Dorrit, Yale Univ., USA (42)
Houk, Nancy, Univ. of Michigan, USA (8)
Keenan, Philip C., Ohio State Univ., USA (1)
LaSala, Jerry, Univ. of Southern Maine, USA (7)
Levato, Hugo, Complejo Astro. el Leoncito, Argentina (41)

Liebert, James W., Steward Obs./Univ. of AZ, USA (40)
Lloyd Evans, Thomas H. H., South African Astron. Obs., South Africa (10)
MacConnell, D. Jack, CSC/STSI, USA (27)
Malaroda, Stella M., Complejo Astro. el Leoncito, Argentina (19)
McCarthy, S.J., Martin F., Vatican Obs. (43)
Meyers-Rice, Barry A., Steward Obs., USA
Mihalas, Dimitri, Univ. of Illinois, USA (16)
Mujica, Raul, I.N.A.O.E., Mexico (3)
Navarro-Corral, Silvana G., I.N.A.O.E., Mexico (17)
Nicolet, Bernard, Obs. de Genève, Switzerland (39)
Olsen, Erik H., Copenhagen Univ. Obs., Denmark (12)
Osterbrock, Donald E., Lick Obs./Univ. of Cal., USA (24)
Park, Hong Suh., Korea Astronomy Obs., Korea (34)
Parthasarathy, M., Indian Inst. of Astrophysics, India (5)
Penprase, Bryan E., Cal Tech/Pomona College, USA
Philip, A. G. Davis, VVO/Union College, USA (26)
Rao, Sreedhar S., Japal-Rangapur Obs., India (38)
Roman, Nancy G., USA (20)
Rountree, Janet C., Univ. of Arizona, USA

Workshop Participants (continued)

Ruiz, Milagros, Steward Obs., USA
Saucedo-Morales, Julio (4)
Shvelidze, Teimuraz D., Abastumani Astrophy. Obs., Rep. of Georgia (11)
Sion, Edward M., Villanova Univ., USA
Torres-Dodgen, Ana V., Monterey Inst., USA
Vieira, Eduardo F., ESA-IUE Tracking Station, Spain (33)
von Hippel, Ted A., Inst. of Astron./Cambridge, UK (35)
Walborn, Nolan R., Space Telescope Science Inst., USA (13)
Warren, Jr., Wayne H., Hughes STX Corp./NASA, USA (25)
Weaver, Wm. Bruce, Monterey Inst., USA (31)
White, Raymond E., Steward Obs., USA (behind camera)

Commemorative Scrolls Presented to W.W. Morgan and P.C. Keenan

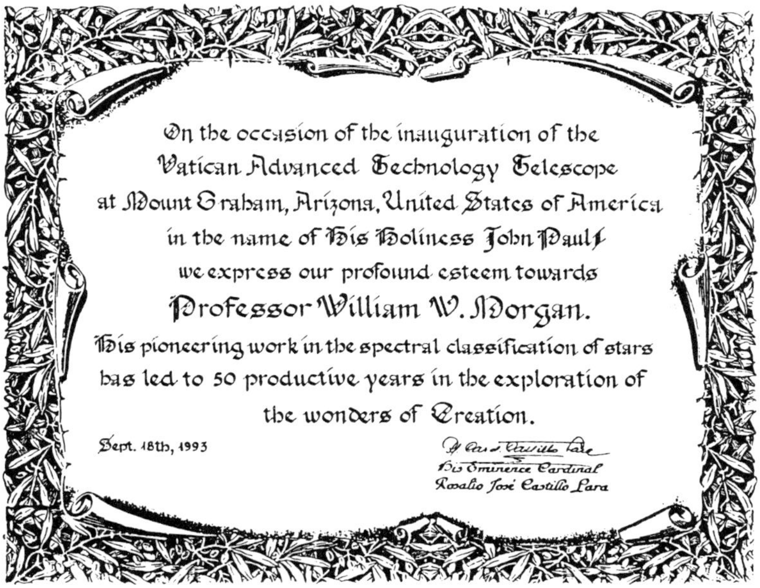

On the occasion of the inauguration of the
Vatican Advanced Technology Telescope
at Mount Graham, Arizona, United States of America
in the name of His Holiness John Paul II
we express our profound esteem towards
Professor William W. Morgan.
His pioneering work in the spectral classification of stars
has led to 50 productive years in the exploration of
the wonders of Creation.

Sept. 18th, 1993

His Eminence Cardinal
Rosalio José Castillo Lara

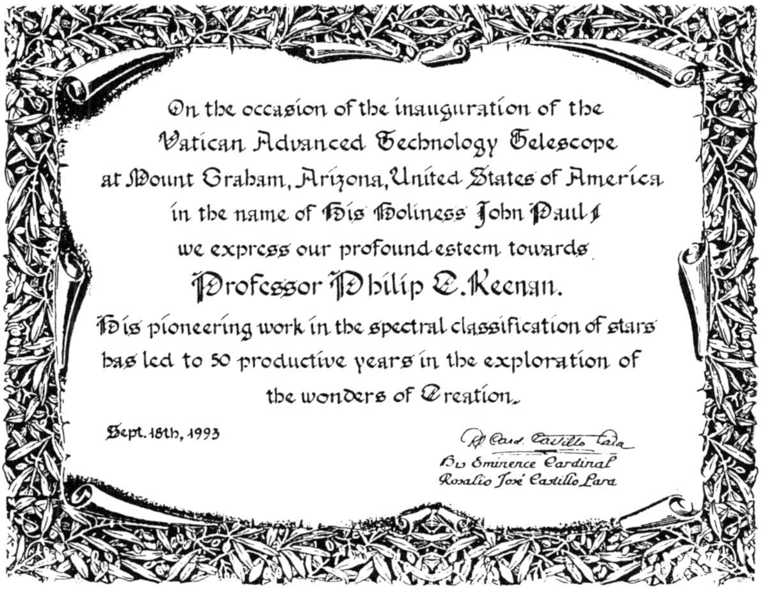

On the occasion of the inauguration of the
Vatican Advanced Technology Telescope
at Mount Graham, Arizona, United States of America
in the name of His Holiness John Paul II
we express our profound esteem towards
Professor Philip C. Keenan.
His pioneering work in the spectral classification of stars
has led to 50 productive years in the exploration of
the wonders of Creation.

Sept. 18th, 1993

His Eminence Cardinal
Rosalio José Castillo Lara

PART I.

THE MK PROCESS AND STANDARDS

The MK Process at 50 Years
ASP Conference Series, Vol. 60, 1994
C.J. Corbally, R.O. Gray, and R.F. Garrison (eds.)

A Hierarchy of Standards for the MK Process

R. F. Garrison

David Dunlap Observatory, University of Toronto
Box 360, Richmond Hill, Ontario, Canada L4C 4Y6

Abstract. "An Atlas of Stellar Spectra" was produced fifty years ago at the Yerkes Observatory, University of Chicago, by W.W. Morgan, P.C. Keenan, and E. Kellman. A brief review is presented, stressing that the MK Process, with its emphasis on standards, is the main reason for the long-term success of the MK System. Sets of standard stars are analyzed and a hierarchy is proposed.

1. Introduction

Fifty years ago, in 1943, "An Atlas of Stellar Spectra" was produced at the Yerkes Observatory, University of Chicago, by W.W. Morgan, with the aid of P.C. Keenan, and E. Kellman. The original ("MKK") atlas represented a new approach to the classification of stars, one that depends on a set of standard "specimens." The System was refined by Morgan in 1953 (Johnson and Morgan 1953, referred to as "MK"), and again in 1973 by Morgan and Keenan (referred to as the "revised MK"). New atlases appeared in 1976 (Keenan, for the stars cooler than the Sun) and in 1978 (Morgan, for the stars hotter than the Sun).

In 1983, at a workshop held in Toronto, the philosophical underpinnings of the MK System were generalized to include extensions to the MK System as well as other areas of classification, not just those in stellar spectroscopy or even in astronomy. The MK Process (Mihalas 1984, Morgan 1984) is a methodology which uses the objects themselves as standards. This approach gives the MK System a richness and depth which cannot be achieved by any mere description in words or numbers (e.g. temperature, surface gravity, abundance, etc.). A star is given a certain MK type because it is "like" the standard specimen. A detailed discussion of "The Use and Abuse of Standard Stars" is given in Garrison (1985). The MK Process, with its emphasis on standards and specimens, is one of the main reasons for the long-term success of the MK System.

Another key to the success of the MK System and to its continuing usefulness into the future is that the classifications are independent of the calibration, which can be given in a separate table. Thus, as the temperature and luminosity calibrations change from year to year, observer to observer, instrumentation to instrumentation, reduction to reduction, and theory to theory, the more fundamental MK classification remains the same. A normal star which was "like" the Sun at MK resolution (1-3 Angstroms) twenty years ago will still be "like" the Sun at the same resolution now and in another twenty years, whereas the calibration of the type in terms of temperature, luminosity or chemical composition will almost certainly have changed. Without this advantage, the MK System would have gone the way of the dinosaurs long ago.

In this first paper of the conference, I want to talk about standards, because they not only represent the MK System, they *are* the MK System.

2. The Hierarchy

As mentioned above, the MK System has been refined since 1943, and several lists of standards have been produced by Morgan and/or Keenan. The types of some of the original standard stars have been modified slightly and other types refined, for a variety of reasons, but mostly because of the availability of better data. For those observers not familiar with the stars or with the MK System, this can be confusing.

In order to reduce the level of confusion about standards, a new hierarchy is being introduced herein. "Anchor Points" is a set of standards which have survived through the years without change, and which represent the most stable points in the system. There is no attempt, with the Anchor Points, to cover the complete grid of possible classifications.

"Primary Standards" are those which have been thoroughly tested over many years and which best represent the types, but which may not have such a long and consistent history. The set of Primary Standards include the Anchor Points and they do cover a complete grid of classification boxes.

"Secondary Standards" are those which are used as surrogates when the Primary Standards cannot be observed, by reason of position or brightness. For example, there is a need for standards fainter than tenth magnitude, because some of the modern electronic detectors on large telescopes cannot be used to observe the bright standards.

There are some stars which have good data and good classifications by Morgan, Keenan or one of their close associates, but which do not have the status of standards. In some situations where high precision is not required, these can be used to obtain approximate types, but this is not recommended as a general practice. It is better to return to the standards. The list of good classifications by Keenan and McNeil (1989), which also includes most of the standard stars cooler than the Sun, is an example of this category.

Finally, there are some very peculiar stars, which don't fit into the standard star scheme, and which do not form a large enough or stable enough group to become standards. Individual members of this group can be used as prime examples of a given peculiarity (for example, the metallic-line or magnetic-A stars). Stars like P Cygni (a very luminous early-B type supergiant with very peculiar hydrogen-line profiles) have become prototypes for their peculiarities. Such stars are considered to be good specimens of the peculiarity and can be used as guides, but are not considered to be standards.

3. Anchor Points

As an example of one layer of the hierarchy, the Anchor Points are presented in Table 1, selected from a complete list of all standards. The table is a list of the stars whose classifications have not changed since 1943. These stars can be thought of as the "Anchor Points" of the MK System. Additional criteria applied to this list of stars are that they must also have a reliable modern published type (by Morgan, Keenan, Richard Gray for A stars, Nolan Walborn for O stars, and/or myself) and that the types also must be consistent with my own classifications of new spectra. Thus these stars represent the MK System as it was then and is now. The stability and consistency of the MK System, as represented by the Anchor Points, is one of its greatest advantages.

The first three columns of Table I are self explanatory. Column four is a list of classifications from the tables in the introductory booklet of the MKK Atlas (1943). In parentheses are listed types from the Atlas which were not listed in the tables (only one star in the final list: HR 617, K2 III), or which differed slightly from the types in the tables (only one star in the final list: HR 5191, B3 V, to which an "n" was added in the Atlas).

Column five is a list of classifications from Morgan's table of MK standard stars given in the UBV definition paper by Johnson and Morgan (1953). This revision of types is usually referred to in the literature as the definition of "MK" standards (rather than "MKK", which refers to the 1943 Atlas). Only the latest O stars are listed in the 1953 paper; the best source for O-star types from that era is Michigan Publications No.X (1951), which contains a list of OB-star types. These additional types are in parentheses and are identical for stars in the final list, except that S Mon is not listed in the 1953 paper.

Column six is a list of types from the Annual Reviews article by Morgan and Keenan (1973). In their table 1, Morgan gives the best standards for stars earlier than the Sun, all marked with "daggers"; in their table 2, Keenan gives a list of his best classifications, with "daggers" to indicate the ones with the most reliable standard types. The types from these two tables have become known as "dagger types." Also included parenthetically in column six are types from the atlases produced in the late 1970's by Morgan (1978) and Keenan (1976).

The types listed on the second page are taken from several modern lists to make sure that the MKK and MK types are still considered valid. The second and third columns include types for stars cooler than the Sun from various publications by Keenan and his associates (Keenan and Pitts 1980, Keenan and Yorka 1985, Keenan and Yorka 1988, Keenan and McNeil 1989). Morgan (circa 1986, unpublished) gave me a list of what he called "strong stars," which he reclassified from a series of new plates taken by Helmut Abt. (These are not to be confused with strong-lined stars, so I have left out the word strong in the column heading.) He considered them to be the most reliable standard types for the GK stars; hence I have included those classifications here in column four. Also included in column four are lists of new types by Walborn (1973) for the O stars and Gray (Gray 1986; Gray and Garrison 1987, 1989a,b; Garrison and Gray 1994) for the B8-F2 stars.

Finally, I have my own spectra of all of the listed standard stars, taken at both 1.2 A and 2.4 A resolutions. I have checked these for consistency with the listed type, and personally have confirmed all of them independently.

Thus, these stars represent the MK System as it was in 1943, as it was in 1953, 1973, 1978, 1989, and as it is now. The Anchor Points are the *de facto* standards, though they may not be the best ones or the ones we wish were standards. However, because of its long history, the MK System now has a life of its own. Instead of trying to manipulate it and distort it, we must listen to it.

4. Primary Standards

In the case of the Anchor Points no attempt has been made to fill in the grid of standard stars. A set of Primary Standards, equally reliable and representing the best known specimen of each spectral type and luminosity class, is needed to fill in the gaps. The set of Primary Standards includes the set of Anchor Points. Work on this set is not yet complete and will be published separately.

5. Secondary Standards

The Anchor Points and Primary Standards are distributed randomly over the sky, with most located north of the equator. Thus, in some cases, a primary standard cannot be observed, by virtue of its position relative to the observer. For those types whose Primary Standards are unobservable from some key locations, we are in the process of setting up Secondary Standards located near the equator, so that observers in both hemispheres easily can use the same standards. Ideally, these all would be within 10 degrees of the equator, but we find that for some of the rarer types it is necessary to extend the limit to 20 degrees.

In case the Sun is located too near one of the standard stars, it is essential to have more than one Secondary Standard for each type. Ideally, these would be placed 12 hours apart, but in practice at least 6-8 hours apart, so one of them always can be reached when the Sun is too near the other.

Another problem is the brightness of the standards. Most of the Primary Standards are brighter than fifth magnitude and some are among the brightest stars in the sky. Modern detectors are very efficient and most spectrographs, especially those on large telescopes, are being built to observe the faintest stars. In some cases, there is no provision for a neutral density filter or fine mesh screen to decrease the light for standards, so stars brighter than tenth magnitude cannot be observed. Thus we need at least a skeletal grid of Secondary Standards at tenth and fifteenth magnitudes to tie the observations with large telescopes to the standard system.

This is not an easy task! There is no problem with the lower main-sequence stars and the common red giant stars, but the rarer and more luminous the type, the more difficult it is to find a relatively unreddened specimen at faint magnitudes. For example, how many blue supergiants with small reddening can we expect to find at fifteenth magnitude? There are ways around the problem, but most are not very easy or practical.

For example, we could use the same spectrograph with a small telescope of the same f/ratio. Or we could diaphragm down the large telescope to build a library of digital electronic spectra which can be used by any observer for a given spectrograph. In practice, we may even have to use spectra taken with another telescope and spectrograph, even though the scattering properties of the two spectrographs may differ slightly.

However, these are problems we need to think about and worry about. Otherwise, work done with the brightest stars in other galaxies may not be well calibrated against stars in our own galaxy, and much subtle information will be lost or misinterpreted.

As an interim measure, we may publish faint Secondary Standards for the more common and less luminous stars, leaving the larger problem of the rare, luminous stars until later.

6. Peculiar Stars

Even though peculiar stars, by definition, lie outside the system, it is useful to have specimens of them to show the classifier what to expect. Examples are the Ap stars, the Am stars, and the red giants.

Some peculiar stars are so numerous and well understood that they eventually work their way into the standard system, one way or another, and the system is expanded to include them. An example of these is the class of rapidly rotating stars, which were incorporated into the MKK (1943) system using the n or nn notation (meaning rapidly rotating and very rapidly rotating). However there were many gaps in the set of standards for rapidly rotating stars, and the differences in line ratios introduced by rapid rotation made it necessary to be very careful in classifying rapid rotators, and especially in setting up standards. Richard Gray (1986), in his thesis at the University of Toronto, carefully studied the problem, set up additional high and low rotation standards for the A-type stars, based on Morgan's Primary Standards, and used them to classify over a thousand stars with high precision. His work was very successful and resulted in an increase in the understanding of a number of astrophysical problems, such as the λ Bootis stars and the rotation of Vega.

For most of the peculiar stars, it is good to have a specimen of the peculiarity, a prototype to show how the peculiarity manifests itself. In this case, we don't expect the star being classified to be exactly "like" the specimen, but only to exhibit the same peculiarity.

An example might be the Ap(Si) stars. They have a range of spectral types, luminosity classes, and silicon-line strengths, but they share the characteristic of having silicon lines stronger than in any of the normal stars. In modern classification, they are given a type which is the best match to a standard along with the silicon designation to characterize the type of peculiarity, say B9 IVp (Si). A logical extension of this system would be to add a silicon-line strength indicator on a scale of 1-5. The problem with this is that the strength of silicon in normal stars varies with temperature and luminosity, so a silicon star with a certain strength at B8 V might be peculiar, while the same silicon strength might be observed in an A2 II star and be perfectly normal. Besides this, while the temperature and luminosity of a silicon star are reasonably stable, the silicon-line strength for a given star is often variable on a scale of days (rotation) or decades (spot evolution). Therefore, it is better to leave the classification more general, signaling the user that this star is among those with peculiar and variable silicon-line strength. Similar arguements apply to the class of Be stars and others. Nevertheless, it is useful to have specimens or prototypes of the particular kind of peculiarity.

In the case of the metallic-line (Am) stars, the range of metals strengths is greater than that in normal stars, and normal criteria are markedly inconsistent.

An Am star may have a Ca II K line similar to that of an A2 star, hydrogen lines similar to those in an A5 star, and metal lines similar in strength to those in an A7 star. In this case, it is useful to give estimates of the extent of these discrepancies. Thus, in the example, we might classify a star as kA2hA5mA7 V:, which signals the user that it is an Am star and gives an indication of how extreme it is.

For metal-weak or metal-strong stars, Keenan has developed a very workable system for estimating the degree of peculiarity. He gives the best match with standards for temperature-dependent, but abundance-independent, lines and features, then gives an estimate of the excess or deficiency. In most cases, when iron is weak or strong, so are chromium, titanium, vanadium, CN, and CH, so a classification "K0 III Fe-1" indicates that the pattern matches K0 III except that all the line strengths are weakened with respect to the K0 III standards. A classification of "K0 III Fe-1, CN+1" indicates that while the iron is weak, the CN is strong, making the spectrum unusually peculiar. This nomenclature is so useful that it eventually will become part of the standard system.

Other peculiar stars are sufficiently unique or variable that the most we can do is offer a description of how they *don't* fit into the system.

7. Other Dimensions and Other Wavelength Regions

There will be much discussion of these extensions during this workshop, so I will be brief. There is no "third dimension," involving abundance or age, which will apply all over the HR diagram. Rather, there are as many other dimensions as there are chemical elements. Luckily for us, lines of many of the elements are not visible at classification dispersion, and indeed don't play a major role in the determination of the general properties of the star. Also, as mentioned above in the discussion of Keenan's nomenclature for red giants, there are certain groups of elements, such as the iron group, whose abundances seem to be coupled. Thus, in practice, the number of dimensions is not as large as might be feared.

The setting up of parallel standard systems for weak-lined and strong-lined stars is a very delicate process and must be approached with extreme care. Unfortunately, not very many attempts to date have been successful. A good example of a careful piece of work is that of Richard Gray, who, as a result of his work with me and with Olsen in Denmark, has carefully set up a preliminary set of standards for the weak-lined, F-type stars (Gray 1989).

8. Conclusion

The conclusion is simple and can be summed up in a sentence. Yesterday, at a Mexican restaurant here in Tucson, we were given Chinese fortune cookies (an interesting example of multiculturalism!) and, believe it or not, mine said: "Success starts with the STANDARDS you set for yourself."

Table 1. MK Standard Stars: Anchor Points September 1993

HD	HR or other	Star Name	MKK43 Table (Atlas)	MK53 Table 1 (Michpub1951)	MK73 Table 1,2 (K76M78 Atlases)
47839	HR 2456	15 S Mon	O7	(O7)	O7 (V)
214680	HR 8622	10 Lac	O9 V	O9 V	O9 V
36512	HR 1855	upsilon Ori	B0 V	B0 V	B0 V
37128	HR 1903	epsilon Ori	B0 I	B0 Ia	B0 Ia (B0 Ia)
41117	HR 2135	chi-2 Ori	B2 I	B2 Ia	B2 Ia (B2 Ia)
206165	HR 8279	9 Cep	B2 I	B2 Ib	(B2 Ib)
120315	HR 5191	eta UMa	B3 V (n)	B3 V	B3 V
32630	HR 1641	eta Aur	B3 V	B3 V	B3 V
53138	HR 2653	omicron2 CMa	B3 I	B3 Ia	B3 Ia (B3 Iab)
58350	HR 2827	eta CMa	B5 I	B5 Ia	B5 Ia (B5 Ia)
34085	HR 1713	beta Ori	B8 Ia	B8 Ia	B8 Ia
103287	HR 4554	gamma UMa	A0 V	A0 V	A0 V
172167	HR 7001	alpha Lyr	A0 V	A0 V	A0 V (A0 Va)
87737	HR 3975	eta Leo	A0 Ib	A0 Ib	(A0 Ib)
21389	HR 1040	+58 607	A0 Ia	A0 Ia	(A0 Ia)
197345	HR 7924	alpha Cyg	A2 Ia	A2 Ia	A2 Ia (A2 Ia)
216956	HR 8728	alpha PsA	A3 V	A3 V	A3 V
89025	HR 4031	zeta Leo	F0 III	F0 III	F0 III (F0 III)
36673	HR 1865	alpha Lep	F0 Ib	F0 Ib	F0 Ib (F0 Ib)
113139	HR 4931	78 UMa	F2 V	F2 V	F2 V (F2 V)
20902	HR 1017	alpha Per	F5 Ib	F5 Ib	F5 Ib (F5 Ib)
30652	HR 1543	pi-3 Ori	F6 V	F6 V	F6 V
194093	HR 7796	gamma Cyg	F8 Ib	F8 Ib	(F8 Ib)(WWM)
54605	HR 2693	delta CMa	F8 Ia	F8 Ia	(F8 Ia)(WWM)
109358	HR 4785	beta CVn	G0 V	G0 V	G0 V (PCK)
121370	HR 5235	eta Boo	G0 IV	G0 IV	(G0 IV) PCK
204867	HR 8232	beta Aqr	G0 Ib	G0 Ib	G0Ib(PCK) (G0Ib)
Sun	Sun	ASTEROIDS		G2 V	G2 V
20630	HR 996	kappa Cet	G5 V	G5 V	(G5 V)
161797	HR 6623	mu Her	G5 IV	G5 IV	G5 IV(G5 IV)note
206859	HR 8313	9 Peg	G5 Ib	G5 Ib	G5 Ib
101501	HR 4496	61 UMa	G8 V	G8 V	(G8 V)
188512	HR 7602	beta Aql	G8 IV	G8 IV	G8 IV
62345	HR 2985	kappa Gem	G8 III	G8 III	G8 IIIa
113226	HR 4932	epsilon Vir	G8 III	(G8 III)	G8 IIIab (G8III)
48329	HR 2473	epsilon Gem	G8 Ib	G8 Ib	G8 Ib
185144	HR 7462	sigma Dra	K0 V	K0 V	K0 V
62509	HR 2990	beta Gem	K0 III	K0 III	K0 IIIb
197989	HR 7949	epsilon Cyg	K0 III	K0 III	K0- III
222404	HR 8974	gamma Cep	K1 IV	K1 IV	K1 IV see notes
22049	HR 1084	epsilon Eri	K2 V	K2 V	(K2 V)
12929	HR 617	alpha Ari	(K2 III)	K2 III	K2 IIIab
153210	HR 6299	kappa Oph	K2 III	K2 III	K2 III
127665	HR 5429	rho Boo	K3 III	K3 III	K3 III
31398	HR 1577	iota Aur	K3 II	K3 II	K3 II
201091	HR 8085	61 Cyg A	K5 V	K5 V	K5 V
164058	HR 6705	gamma Dra	K5 III	K5 III	K5 III
6860	HR 337	beta And	M0 III	M0 III	M0 IIIa
1013	HR 45	chi Peg	M2 III	M2 III	M2+ III(M2 III)
39801	HR 2061	alpha Ori	M2 Ib	M2 Iab	M1-M2 Ia-Ib
206936	HR 8316	mu Cep	M2 Ia	M2 Ia	M2 Ia

HD 161797: G5 IV in atlas illustration, but IV-V in booklet table.
HD 222404: K1 IV in atlas, but K1 IV CN1 in booklet table.

Table 1. MK Standard Stars: Anchor Points September 1993 (Continued)

HD	KY88 (KM89)	KY85 (KP80)	WWM GK STD 86 Walborn OB, GRAY BAF
47839			O7 V((f))
214680			O9 V
36512			B0 V
37128			B0 Ia
41117			B2 Ia
206165			B2 Ib
120315			B3 V
32630			B3 V
53138			B3 Ia
58350			B5 Ia
34085			B8 Ia GRAYSTD/Wb
103287			A0 Van GRAYSTD
172167			A0 Va GRAYSTD
87737			A0 Ib GRAYSTD
21389			A0 Ia
197345			A2 Ia
216956			A3 Va GRAYSTD
89025			F0 IIIa
36673			F0 Ib GRAYSTD
113139			F2 V
20902			
30652			
194093			
54605			
109358	G0 V	G0 V	G0 V
121370			G0 IV
204867	G0 Ib	G0 Ib	
	G2 V	G2 V	
20630	G5 V	G5 V	
161797	G5 IV	G5 IV	G5 IV
206859	G5 Ib	G5 Ib	
101501	G8 V	G8 V	G8 V
188512	G8 IV	G8 IV	G8 IV
62345	G8 III	G8 IIIa	G8 III
113226	G8 IIIab	G8 IIIab	
48329	G8 Ib	G8 Ib	
185144	K0 V	K0 V	K0 V
62509			K0 III
197989	K0 III	K0- III	K0 III
222404			
22049	K2 V	K2 V	
12929			
153210	K2 III	K2 III	
127665			K3 III
31398			
201091	K5 V	K5 V	K5 V
164058	K5 III	K5 III	K5 III
6860	M0+ IIIa	M0+ IIIa	
1013	M2+ III	M2+ III	
39801	M1-M2 Ia-Iab var	M1-M2 Ia-Iab	
206936	M2- Ia	M2- Ia	

References

Garrison, R.F. 1985, in *The Calibration of Fundamental Quantities*, eds. D.S. Hayes *et al.* (Kluwer: Dordrecht), 17

Garrison, R.F. and Gray, R.O. 1994 AJ(in press)

Gray, R.O. 1986, Thesis, University of Toronto

Gray, R.O. 1989, AJ, **98**, 1049

Gray, R.O. and Garrison, R.F. 1987, ApJS, **65**, 581

Gray, R.O. and Garrison, R.F. 1989a, ApJS, **69**, 301

Gray, R.O. and Garrison, R.F. 1989b, ApJS, **70**, 623

Johnson, H.M. and Morgan, W.W. 1953 ApJ, **117**, 313

Keenan, P.C. and McNeil, R.C. 1976, *An Atlas of Spectra of the Cooler Stars: Types G,K,M,S and C* (Columbus: Ohio State Univ. Press)

Keenan, P.C. and McNeil, R.C. 1989, ApJS, **71**, 245

Keenan, P.C. and Pitts, R.E. 1980, ApJS, **42**, 541

Keenan, P.C. and Yorka, S.B. 1985, Bull d'Inf CDS No29, p199

Keenan, P.C. and Yorka, S.B. 1988, Bull d'Inf CDS No35, p37

Mihalas, D. 1984, in *The MK Process and Stellar Classification*, ed. R.F. Garrison (Toronto: David Dunlap Obs.), 4

Morgan, W.W. 1951, Univ. of Michigan Publ., No X, 33

Morgan, W.W. 1984, in *The MK Process and Stellar Classification*, ed. R.F. Garrison (Toronto: David Dunlap Obs.), 18

Morgan, W.W. Abt, H.A. and Tapscott, J. 1976, *Revised MK Spectral Atlas for Stars Earlier than the Sun* (Chicago: Yerkes Observatory)

Morgan, W.W. and Keenan, P.C. 1973, ARA&A, **11**, 29

Morgan, W.W., Keenan, P.C., and Kellman, E. 1943, *An Atlas of Stellar Spectra* (Chicago: Yerkes Obs)

Walborn, N.R. 1973, AJ, **78**, 1067

Discussion

Keenan: Why consider the best standards to be those which have not had their types changed? Actually just those stars for which the most improved spectra are available are those that have had their types changed. On the other hand, many of the stars with unchanged types have merely been kept the same because better spectra have not been obtained.

Garrison: Your point is well taken, and I worried about that problem. Therefore, I also required that the "Anchor Points" be stars which also have been carefully considered by Walborn, Gray or myself from modern spectra of the best quality, so I don't think any of the Anchor Points suffer from the lack of good data.

There are, of course, stars which don't have a history, but have been introduced recently and have been found to be exceptionally good standards. These are what I call primary standards. Primary standards include the Anchor Points as a subset.

Abt: The MK system applies primarily to population I stars. Morgan wanted to extend the system for intermediate (FG) stars to population II. Keenan has faced that partially for later types. Do you feel that some slightly metal-poor stars have crept into the standards?

Garrison: Obviously there is a continuum of spectral types between any two population types, so in some sense, at the highest resolution and with the best fine analyses, the stars will not all belong to particular populations. However, in my experience, it is possible to distinguish fairly small differences in the line strengths, yet I have not found any problem with assigning most of the bright stars to a single population. I doubt that this is much of a problem for the primary standards.

Abt: A related question is that if people wish to apply the MK system to 20th mag stars, perhaps by using automated spectral classification, they will encounter a much larger fraction of population II stars. Do you think that there are adequate standards for them?

Garrison: Certainly not yet! We have really only begun to set up the system for metal weak stars, but we have made a good start.

Olsen: In connection with your reply to Helmut Abt, I have a question. I have noted that Richard Gray has defined an *extension* to the MK system by setting up new, independent standards in the range F5-G5 V-III for metal weak stars. Does your reply to Abt mean that you don't approve of his approach?

Garrison: Not at all. Richard has made a very good attempt at a first step for that small range of types.

Crawford: Relative to the skeleton structure, you may not need them all. I remember very well many years ago, when a cataloger was criticising Morgan for not having standards for some spectral classes, Bill replied that if God made some more stars, he'd classify them. I think that if they are there, observers will find candidates, and then you can critique them.

Garrison: I've heard the same comment and it is inevitable that when new stars are classified from massive surveys, there are bound to be stars which fall between the cracks among the standards, yet are not extrapolations from the system, but are merely easy interpolations.

Osterbrock: In the beginning of your talk you emphasized correctly, that the MK system is *defined* by its standards. Yet at the end of your talk you spoke of setting up a network of faint secondary standards, and seemed to imply that not all types would be included, but only enough to interpolate between. This seems to me to be inconsistent with your earlier statement. No matter how you arrive at the approximate type by interpolation, in the end the acid test of the type must be that it matches the standard stars (perhaps secondary) of that type more closely than it matches any other type.

Garrison: What you say is correct as far as it goes. A star of a given type is, in principle, exactly like the standard at the resolution of the MK system. However, from a practical point of view, the secondary standards do not in themselves define the system. They are surrogates which approximate the system of primary standards and "Anchor Points".

In practice, it is necessary at times to use interpolation, perhaps because we don't happen to have been able to observe some particular standard. Also, sometimes it is possible to see a smaller difference in a smoothly varying feature than is given in the grid of standards. In principle, it may not be pure to interpolate, but in practice, we sometimes have to do so.

Roman: In answer to Don Osterbrock: if I understand him, he said that a GO V must match a GO V standard. This is impossible. No two stars look identical. The interpolation is necessary. The GO V must look *more* like a GO V standard than like any other type.

Has the line strength parameter method also been applied to the metallic line stars?

Garrison: No, the line strength parameter has not been applied to the Am stars yet, but I believe that some variation of it should be applied to them now that there are enough of them to treat as a significant sample set.

McCarthy: Regarding the use of the word "strong" speaking of stellar standards, it will be important to avoid misunderstanding "strong" to mean "prominent" or "principal" and "strong" to mean "intense" or "metal rich". This is a linguistic item, but deserves to be clearly distinguished.

Garrison: Your point is very well taken and it didn't occur to me that this would be a problem. The word "strong" in describing what Morgan meant as "really reliable" was a direct quote from the handwritten list of 1986 from Morgan. I will try to avoid that in my list of previous types. (Changed in the manuscript.)

Parthasarathy: It may be difficult to set up fainter (10-16 mag range) secondary MK standards for the complete MK types of stars. However, setting up fainter secondary MK standards for low luminosity stars, dwarf stars, carbon stars and metal rich stars is useful for galactic and extragalactic work.

For setting up of metal poor MK standard stars, the results of the analysis of high resolution spectra of large number of metal poor stars are now available covering the range in $T_{\rm eff}$, $\log g$, and [Fe/H].

Garrison: As I said, in setting up the faint secondary standards, we will accept suggested "candidates" from *any* technique, but we will then examine them critically using MK process techniques before calling them secondary standards.

For the faint secondary standards, we are having no problem setting up standards at faint magnitudes for cool main sequence stars. The problem is with the rare, luminous stars.

Weaver: Automatic classification techniques are starting to distinguish classifications in the sub-subclass region. Should the standards/Anchor Points be taken as defining a discrete bin or as an exact class, e.g., F2.0?

Garrison: The sub-subtypes, if they really can be distinguished, should be based on the Anchor Points as defining the exact zero point. For example, γ Dra is K5.000 III.

Gray: You have suggested diaphragming down a large telescope to build a library of bright standard stars. This would not necessarily be a good idea. As the incoming beam into the spectrograph would have a higher f/ratio, the collimator would be underfilled, and the resolution (as fewer lines on the grating would be illuminated) might suffer.

The MK Process at 50 Years
ASP Conference Series, Vol. 60, 1994
C.J. Corbally, R.O. Gray, and R.F. Garrison (eds.)

Some Problems and Developments of Current Classification

Philip C. Keenan

The Ohio State University, Department of Astronomy, 174 West Eighteenth Avenue, Columbus, OH 43210

Abstract. Several major groups of stars have received new or improved spectral types on the Revised MK system, and will be included in a new edition of the Perkins Catalog. These include:
1. 14 marginal barium stars reclassified at 0.3 Å resolution. These can serve as standards for future surveys at lower resolution.
2. Small-amplitude variables of type S, with the 1980 notation of Boeshaar and Keenan slightly modified.
3. A selection of the red carbon stars that have been brought into the MK system in 1993. The new types replace the C system.

One group of carbon stars, those originally called late R stars or early N stars, still pose a serious problem.

1. Introduction

Future development of the MK system of spectral classification, as that of any other technique of research, is in the hands of those who are active in the field. All that we old fossils should try to do is to clean up a few unfinished jobs to make it easier for the rest of you to go ahead. Today I just want to mention a few groups of stars whose classification continues to give difficulties.

The first problem that needs attention concerns the marginal barium stars listed in the Perkins Catalog. For those who are not familiar with the catalog, it is simply a list of all the stars classified definitely at the Perkins Observatory – hence the name.

Just a few of the stars were classified by the author at the David Dunlap Observatory on Toronto spectrograms, through the courtesy of Bob Garrison. The catalog is, of course, not complete to any apparent magnitude, and cannot be used for statistical studies. What it *is* designed for is to provide as many standards of spectral type as possible, distributed over the whole sky. The primary standards, distinguished by a plus mark in front of the star's name, are those based on several good spectrograms, preferably taken on several dates. The remaining secondary standards are not as accurate, but I think that they are good enough to be useful. The advantage of all the types in the catalog is that they are at least consistent, since they are repeatedly cross-checked up to the time of publication.

As time goes on it is sometimes necessary to make slight changes in even the best standards. The changes result simply from the gradual improvement in classification spectrograms over the past twenty years or so. This point was discussed in Keenan (1991), though I am sorry that the illustration for that paper was somehow lost before publication. It can be understood easily, however, that as we narrow the bins into which the stars are sorted, any one star may find itself outside the bounds of its original type. This happened to α Tau, which was originally defined as K5 III, though it always seemed just a shade later than the average K5 giant. With the introduction of bins of half-type or quarter-type width, it now falls into type K5.5 III.

2. Marginal Barium Stars

The history of our attempts to detect and classify marginal barium stars has been given in Griffin and Keenan, 1992. On spectrograms having the relatively high resolution of 0.3 Å (Figure 1), Ba II 4554.0 is easily resolved from the neighboring titanium and other lines. On low-dispersion classification spectrograms of 2 to 3 Å resolution, however, the whole group of strong lines from at least 4549 Å to about 4556 Å are blended together (Figure 2). It is evident that the barium line must be considerably stronger than its neighbors in order to have its enhancement reliably estimated on the low dispersion spectrograms. Experience eventually led to the conclusion that a barium index of almost one was required before it could be assigned with confidence in the course of classification. Consequently, many stars previously suspected of being marginal barium stars had their barium indices omitted from the Perkins Catalog. Undoubtedly, some of these would be restored if spectrograms of sufficient resolution were available. If we are ever to have reliable statistics on the frequency of marginal barium stars, it seems necessary that surveys be carried out at a resolution not worse than about one angstrom.

In the years since the appearance of the Perkins Catalog, intensity plots have been made of nearly all of the coudé spectrograms used by Keenan and Wilson (1977) in establishing a group of confirmed marginal barium stars. On these plots central intensities can be compared more accurately than by inspection of the original spectrograms. A list of 14 marginal barium stars, with revised indices, is given in Table 1. These can serve as standards of classification for stars observed in surveys with lower resolution.

3. S Type Stars

The spectra of stars of type S are complicated by the presence of more than one abundance variable. The computations of Scalo (1973), and the observations of Ake (1978, 1979), showed that the ratio C/O was the primary variable in determining the sequence M-MS-S-SC-C. The mean ratio of s-process elements to iron-peak elements is only partially correlated with C/O, however. This complication provided the reason for the classification introduced by Keenan and Boeshaar (1980), based on Ake's work, in which the temperature type was followed by a slant line and a C/O index, followed by indices for Zr (from the ZrO bands) and Ti (from the TiO bands).

This scheme has seemed sufficient to describe the spectra of the known S stars, but the notation was more cumbersome than that of most Revised MK

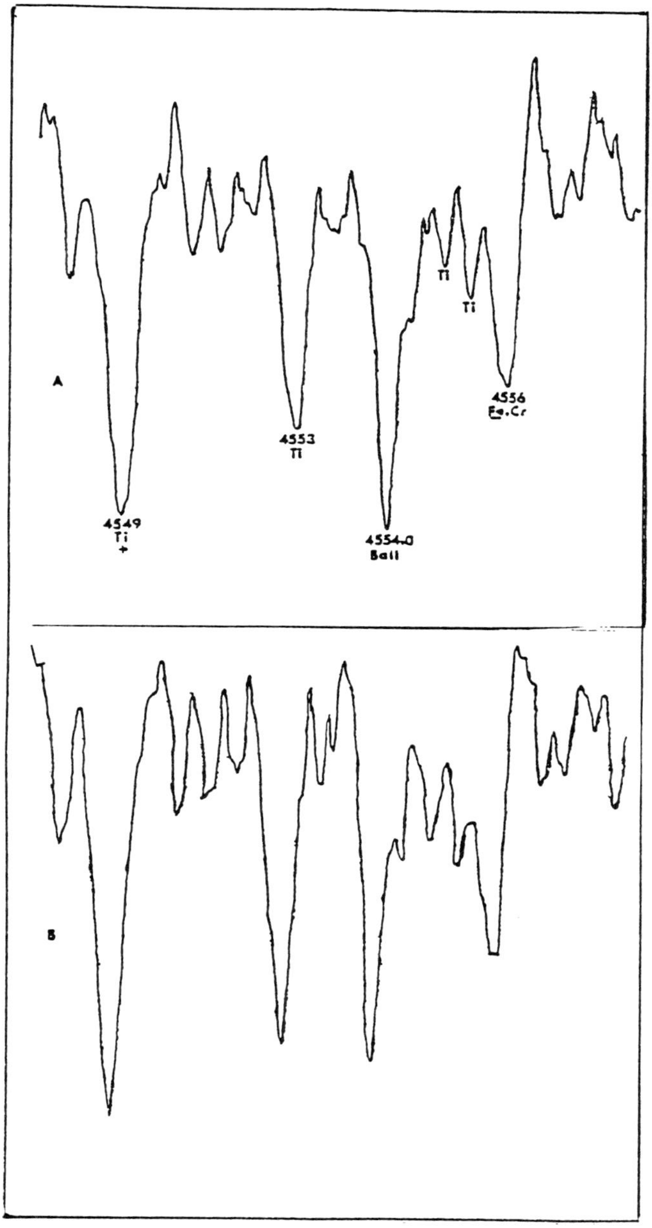

Figure 1. Palomar coude spectrograms at 9 Å/mm.
A. 16 Ser G9.5 IIIa: Ba 0.6. B. η Cyg K0 III

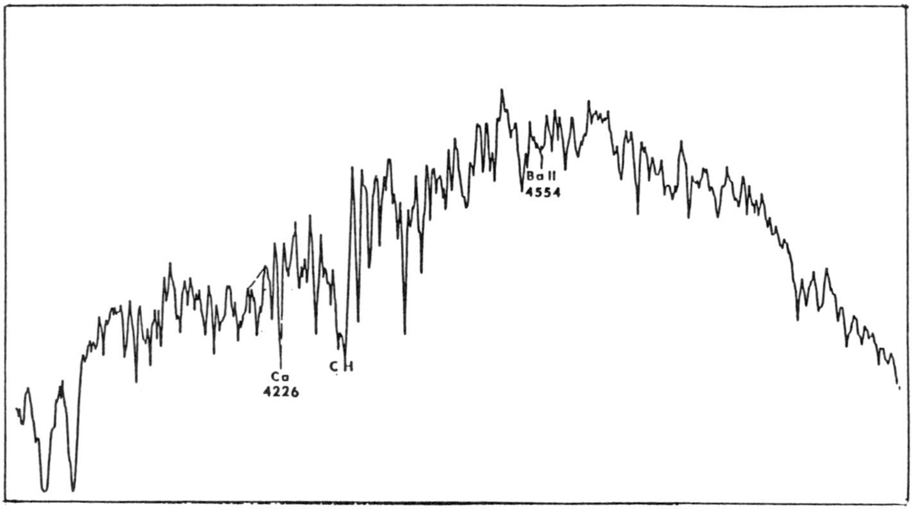

Figure 2. Perkins classification spectrogram at 78 A/mm of η Cyg K0 III.

types of other kinds of stars. It would be convenient if the S stars (excluding, of course, Mira variables) could be included in general catalogs of spectral type. It has since occurred to us that the desired simplification could be accomplished by dropping the C/O index, and giving only the Zr and Ti abundance indices. After the temperature type of any spectrum has been assigned, the value of C/O can be inferred by comparing the difference and the sum of these indices. For example, HD 63733 is now classified as S3.5 Zr2.5 Ti2, which is reasonably self-explanatory, and short enough to fit easily into any catalog.

4. Carbon Stars

Of the two main reasons that the C system of classification of the red carbon stars has proved unsatisfactory, the one most often cited (Tsuji, 1981) is perhaps the least serious. That one is the failure of the sequence of types to correlate well with the currently accepted temperature sequence for the cooler carbon stars (N stars). That could be remedied by shifting some types around if only we could find more reliable criteria of temperature in these heavily banded spectra.

An objection of greater weight against the attempt to arrange all the carbon stars in one temperature sequence is that it ignores the fundamental population differences between R stars, N stars, CH stars, and Hd (hydrogen-deficient) carbon stars. Yamashita (1972, 1975) made some improvement for two of these four population groups by adding "CH" to the types for the Population II CH stars, and Hd for the hydrogen-deficient carbon stars, but without any numerical scales.

With the added information that has since become available, this finally seemed the right time to develop a classification that would bring the carbon

Table 1. Marginal Barium Stars

	Name	HD	Type
1	ξ Cet	13611	G7 III Fe-1 Ba 0.4
2	67 Cet	14129	G8.5 III Ba 0.1:
3	97 Cet	20791	G8.5 III Ba 0.2
4	ι Cnc A	34739	G8 II-III Ba 0.2
5	27 Ori	35410	G9.5 III-IV Ba 0.5
6	5 CVn	107950	G7 III Ba 0.3
7	16 Ser	139195	G9.5 IIIa: Ba 0.6
8	BS 6196	150416	G8 III Ba 0.4
9	——	179870	K0 IIb: Ba 0.2
10	24 Aql	181053	G9.5 III: Ba 0.6
11	60 Sgr	189005	G6 IIIa: Ba 0.3
12	——	196673 A	G9 II-III Ba 0.7
13	31 Vul	198809	G7 IIIa Fe-1 Ba 0.2
14	ζ Cyg	202109	G8+ IIIa Ba 0.5

stars into the Revised MK system. The notation is described in Keenan (1993). That paper pointed out also the growing need for a compact classification that would be flexible enough to describe the spectral differences between carbon stars in the solar neighborhood and those in the galactic bulge, or in other systems. For example, a difference in metallicity can probably be represented by a positive CN index.

The details of the revised classification will not be repeated here, but Table 2 gives examples of the revised types of representative members of the four presently recognized population groups of carbon stars.

Table 2. Revised MK Types of Typical Carbon Stars

Population Group	Sample Type	
R Stars	HD 223392	C-R3 III: C2 4
N Stars	Z Psc = HD 7561	C-N5 C2 4
CH Stars	HD 209621	C-H3 III: C2 4
Hd Stars	HD 182040	C-Hd2 II: C2 3.5 CH 0

Unusual intensities of any other features can be indicated by additional abundance indices, such as Li 7 for WZ Cas.

The ratio $^{12}C/^{13}C$ is especially important in view of its large variation from star to star. Yamashita (1972, 1975) retained Bouigue's (1952) designation of stars with excess isotopic carbon as J stars, merely adding J to their types. In the Revised MK system, the abundance index is Jx, with x running from 0 to about 7. Since R stars are systematically richer in ^{13}C than are N stars, for which J is normally about 2, there is no need to add a J index to a type unless the strength of the isotopic features in that spectrum is abnormal for that population group.

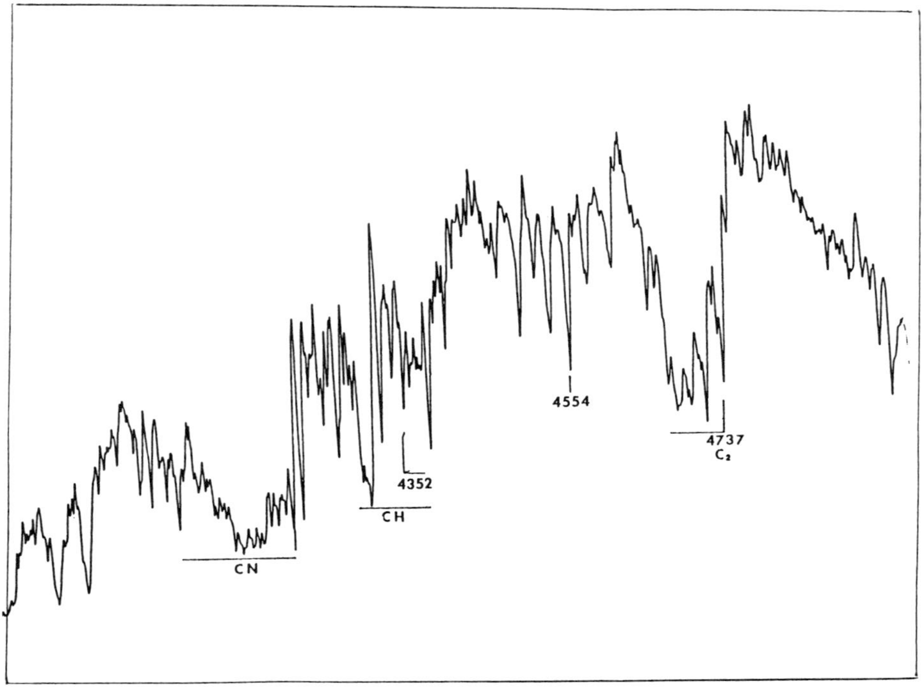

Figure 3. Moderate CH star, HD 24281 C-H 4 IIIa: C_2 3. V = 8.6: CH 4352 present, Ba II 4554 enhanced. Cerro Tololo spectrogram at 78 Å/mm.

If abundance indices are added only when they are needed, the notation can be kept compact enough to allow carbon stars to be included in any catalog of spectral types, and a limited number of them will be added to the Perkins Catalog if the planned second edition is completed.

5. Recognition of CH Stars

There are still many unsolved problems presented by the spectra of carbon stars. One of these concerns the recognition of CH stars in surveys carried out at high galactic latitudes, or in the halo surrounding the galactic bulge. In the well-known paper of Hartwick and Cowley (1985), in which they pointed out the value of CH stars as tracers of galactic halos, they had adequate resolution (1 Å on their reticon scanner spectra) to recognize the CH bands and to measure radial velocities. Yet one must be sure that all the suspected objects really have the spectra of CH stars, for if it is merely assumed that they must be CH stars because of their location or apparent motions, the argument can become circular.

To review briefly the conditions necessary for the identification of CH stars consider Figure 3. The blue system of CH bands is dominated by the well-known

G-band, due to R branches of the $\Delta v = 0$ system, but weaker sequences and branches extend shortward beyond 4200 Å and longward to about 4384 Å. The G-band itself is so strong that although it can be seen at quite low resolutions, its appearance alone is not always sufficient to define a CH star. The much weaker P branches can be seen in Figure 3 to form an apparent head near 4352 Å. There is an overlapping C_2 band extending shortward from 4383 Å, but even when the C_2 band is strong, the characteristic shelf at 4352 Å is the signature of a CH star. In the figure this is confirmed by the strengthening of Ba II 4554 Å line, even though the barium line is blended with one of a group of CN bands.

What is the lowest resolution that will permit the identification of a CH star? The problem is complicated by the steep spectral gradient of the cooler carbon stars in the region shortward of about 4400 Å – a gradient made to appear steeper by the falling off in sensitivity of CCD detectors in the blue.

The answer depends upon the amount of the excess of CH. When the CH bands are so strong that Ca 4226 and the blue CN bands are obviously weakened, CH stars can be recognized even on survey plates of very low resolution. For moderate CH stars like HD 24284, however, I judge that a resolution better than 5 Å is necessary for their certain identification. Hence, it is not surprising that many of the low-dispersion surveys list a number of stars without being certain whether they are CH or R stars.

6. R Stars or N Stars?

The revised classification of carbon stars that I have ventured to suggest is not perfect, and is far from complete. It needs more than refining, for there is at least one important group of stars that have resisted classification. These are the objects that were originally called late R stars or early N stars (Shane, 1920). Many of them have such strong bands of carbon compounds that from spectra of low dispersion, it is not certain whether the lines of s-process elements are enhanced, though it seems likely that they are slightly stronger than in early R stars, but not as strong as in typical N stars. These stars cannot be thought of as transitional objects between such completely different populations as R and N stars. They include such well-known problem stars as Y Hya and RY Dra, as well as HD 19553, and probably HD 37212. Additional members of the group have been listed by Lloyd Evans (1986).

Most stars with strong Merrill-Sanford bands of SiC_2 fall into this group, including the remarkable stars VX And in the northern sky and T Mus in the far south. Although both Gordon (1968) and Yamashita (1972) found that there is no general correlation between the strengths of the SiC_2 bands and those of the isotopic carbon bands, it is noteworthy that these two stars have both the SiC_2 and ^{13}C features stronger than in almost any other known star.

The presence of the SiC_2 bands may not imply anything unusual about the composition, for they are extremely sensitive to temperature. Figure 4 shows a rough sketch based on the equilibrium computations of Tsuji (1981) and Johnson *et al.* (1975). Since the scale of the figure is logarithmic, it is not surprising that the SiC_2 bands are usually either invisible or very strong. When well calibrated they should prove accurate temperature indicators, but in the meantime they

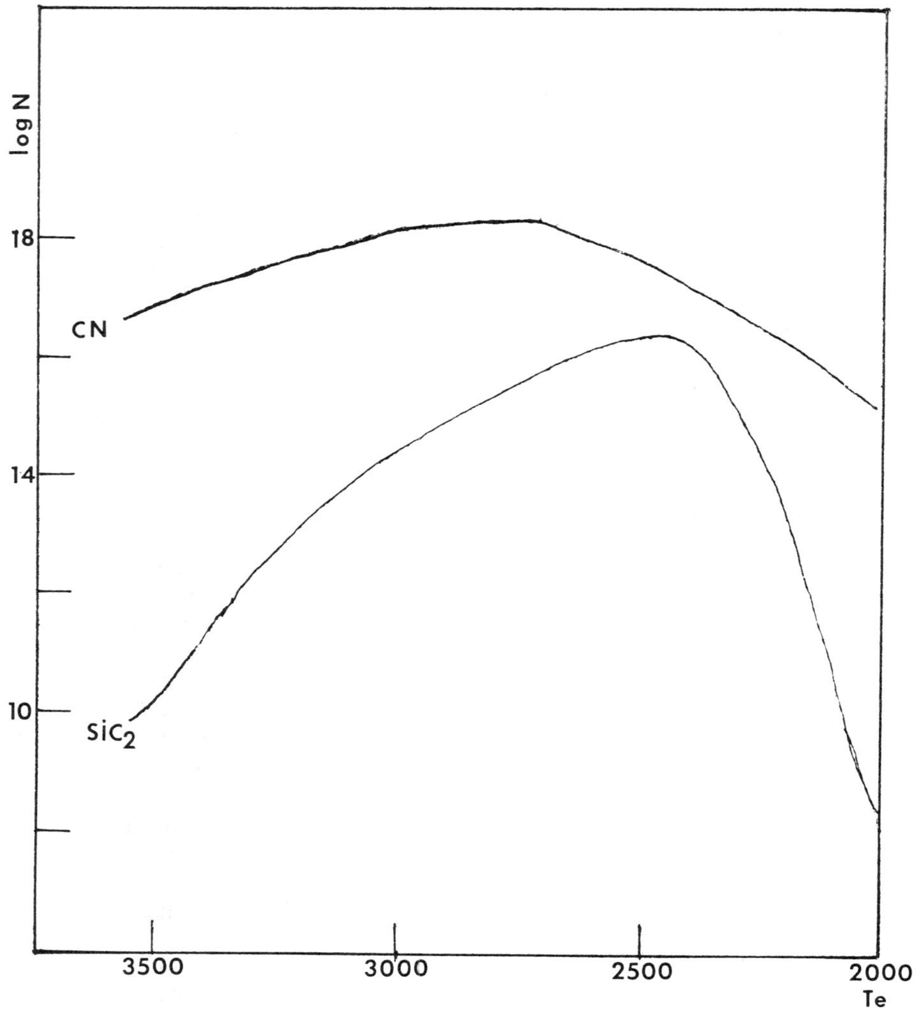

Figure 4. Sketch of the column densities of SiC_2 and CN, based on the calculations of Tsuji (1981) and of Johnson *et al.* (1975)

suggest that stars like RY Dra and T Mus may have temperatures approaching 2500 K or even 2000 K.

To return to the question of the nature of these stars of ambiguous type, it is clearly needful to study enough of them to establish their distribution, motions, and frequency of binarity, as well as their luminosities. One clue that may be important is that at least three of the stars mentioned have strong isotopic carbon bands – more typical of R stars than of N stars. When this is coupled with their low temperatures, which probably put them to the right of the Hayashi lines, the possibility suggests itself that they might represent later stages in the evolution of R stars – stages at which heavy elements are just beginning to be dredged up.

That suggestion will probably be shot down, but does not this whole group of stars present a problem which challenges the attention of this Workshop?

References

Ake, T. 1978, Thesis, Case Western Reserve University.

Ake, T. 1979, ApJ, **234**, 538.

Gordon, C. P. 1968, PASP, **80**, 597.

Griffin, R. F., Keenan, P. C. 1992, Obs., **112**, 168.

Hartwick, F. D. A., Cowley, A. P. 1985, AJ, **90**, 2244.

Johnson, H. R., Beebe, R. F., Sneden, C. 1975, ApJS, **29**, 123.

Keenan, P. C. 1991, Van Vleck Obs. Contr., No. 12, 127.

Keenan, P. C. 1993, AJ, **105**, in press.

Keenan, P. C., Boeshaar, P. C. 1980, ApJS, **43**, 379.

Keenan, P. C., Wilson, O. C. 1977, ApJ, **214**, 399.

Lloyd Evans, T. 1986, MNRAS, **220**, 723.

Scalo, J. M. 1973, ApJ, **194**, 361.

Shane, C. D. 1920, LOB 10, No. 329, 29.

Tsuji, T. 1981, JAp&A, **2**, 95.

Yamashita, Y. 1972, Ann. Tokyo Ast. Obs., (2), **13**, 169.

Yamashita, Y. 1975, Ann. Tokyo Ast. Obs., (2), **15**, 47.

Discussion

Parthasarathy: The carbon stars with oxygen rich dust shells have been found to be J-type carbon stars. It is interpreted that they have turned into J-type carbon stars very recently from M star phase. Are there signatures on the classification dispersion spectra to distinguish these stars or to detect them?

Keenan: You raise an important point, but I have not looked into it yet.

Lloyd Evans: All have enhanced ^{13}C so are J stars. Some have strong SiC_2 bands so must be cool – N stars not R stars.

Keenan: Why should R stars not be cool?

Warren: We have noticed that in successive editions of the Perkins catalog, spectral types have changed slightly, particularly with regard to the abundance indices. Are these "standard" types expected to stabilize or are differences really seen in various spectra of the same stars?

Keenan: Yes, the types are convincing as a result of extensive cross comparisons. Most of the "changes" have been only slight refinements.

Garrison: With the existence of carbon-star types on more than one "system" nomenclature, how do we know which era or which system a type is on?

Keenan: The only time that I changed from one system to another (for carbon stars), the notation was made different enough to make the system plain.

Garrison: Do you have a table of correspondence between one system and the other? That is, if a star has a type in the old system, can we translate it to the new system without looking at a spectrum?

Keenan: The C-types correspond only to temperatures as they were supposed to be in 1943. There is no one-to-one correspondence between them and the revised MK types.

Houk: When the stars are classified into populations, is this done totally on the basis of the spectra?

Keenan: Yes, the types assigned to the carbon stars in the new classification are based entirely on the spectra. It may turn out later that the separation into populations is sometimes ambiguous. That is the reason that I call them "population groups," for their boundaries may not coincide with those of the classical Population I, Population II, etc.

Houk: Do you consider these types to be MK types even when they are derived from high dispersion spectra (16 Å/mm, etc.)?

Keenan: Yes.

The MK Process at 50 Years
ASP Conference Series, Vol. 60, 1994
C.J. Corbally, R.O. Gray, and R.F. Garrison (eds.)

Spectroscopy of Standard Stars in the Region 4000–9000 Å

L. J. Corral, S. G. Navarro, J. H. Peña[1], R. Peniche[1], I. Hirata.
I.N.A.O.E., Tonantzintla, Pue., 72000 México

Abstract. We present low resolution spectra of MK standard stars in the spectral region 4000–9000 Å.

1. Introduction

The LFOSC (Landessternwarte [LSW, Heidelberg] Faint Object Spectroscopic Camera) is a wide field, multi-object, low dispersion spectrograph and is attached to the 2.1 m telescope of the Guillermo Haro Observatory at Cananea, México. The principal use of the instrument is the identification of the ROSAT sources' optical counterparts, but is also suitable for making other studies with it. (c.f. Peña et al. 1993)

The LFOSC has the following characteristics: wide field of view (10×6 arc minutes); Johnson's V and B and Cousin's R and I filters for direct imaging; the possibility of isolating the interesting object from the surrounding sources using a mask with holes that is made at the observatory, in order to provide the spectra without light contamination; a focal reducer with an effective focal number of 2.4 and two grisms that are used as dispersive elements (5.5 Å/pixel ["blue" grism] and 7.5 Å/pixel ["red" grism]).

As the masks are made using a direct image of the field, a star near the edge of the field may have some of its spectrum fall off the CCD and so end up with a reduced spectrum. This problem of getting different spectral coverages is only significant when the size of the spectral images is comparable with the size of the detector, as is the case for CCDs at present.

Another characteristic of LFOSC is that the CCD used has a "normal" spectral response, which means that the spectral response falls to zero around 4000 Å. As the MK system was defined in the 3800–5000 Å region, we could not rely on atlases of normal stars made with an MK dispersion (e.g., Morgan *et al.* 1978, Houk *et al.* 1974) for classification of the spectra taken with the LFOSC. We observed MK standard stars to make our own comparison grid of spectra.

2. Observations

The stars were chosen from the list of Johnson and Morgan (1953) in order to have a consistent classification list. The stars were observed with a long slit,

---

[1]Instituto de Astronomía, U.N.A.M., México, D.F., 04510,

0.3 mm wide ($\sim$ 2 arcsec.), centered in the field, in order to have the maximum spectral coverage. The stars were observed from Nov 20 - 25, 1991, by two of us (L.C and I.H.).

The first night we took comparison spectra (xenon, neon and Halogen lamp) three times during the night, but, in order to minimize a possible error in the wavelength transformation function, on the following nights each star's spectrum had its own comparison spectrum. We found during the reduction this was not necessary.

3. Reduction

For the data reduction we used a software package made by F. J. Zickgraf, of LSW, based on MIDAS. The spectral images were cut to take away bad columns, the bias was subtracted, and the spectra were extracted from the star and the comparison lamps images. Both the star and the neon and xenon spectra were divided by the normalized Halogen lamp spectra. This was done in order to correct the pixel to pixel spectral response.

Table 1. Observed Stars.

HD	Name	MK Type	V	HD	Name	MK Type	V
30614	α Cam	O9.5 Ia	4.29	36486	δ Ori	O9.5 II	2.23
37128	ϵ Ori	B0 Ia	1.7	194779		B3 II	7.77
194839		B0.5 Ia	7.49	34578	19 Aur	A5 II	5.03
216411		B1 Ia	7.2	193443		O9 III	7.24
14143		B2 Ia	6.65	214993	12 Lac	B2 III	5.25
198478	55 Cyg	B3 Ia	4.84	22928	δ Per	B5 III	3.01
34085	β Ori	B8 Ia	0.12	13161	β Tri	A5 III	3.0
199478		B8 Ia	5.67	89025	ζ Leo	F0 III	3.44
223960		A0 Ia	6.9	13174	14 Ari	F2 III	4.98
17378		A5 Ia	6.25	62345	κ Gem	G8 III	3.57
7927	ϕ Cas	F0 Ia	4.98	62509	β Gem	K0 III	1.14
10494		F5 Ia	7.31	12929	α Ari	K2 III	2.0
217476	HR 8752	G0 Ia	5.0	9927	51 And	K3 III	3.57
206936	μ Cep	M2 Ia	4.08	1013	χ Peg	M2 III	4.8
210809		O9 Ib	7.56	211336	ϵ Cep	F0 IV	4.19
37742	ζ Ori	O9.5 Ib	1.77	216385	σ Peg	F7 IV	5.16
204172	69 Cyg	B0 Ib	5.94	214680	10 Lac	O9 V	4.88
9311		B5 Ib	7.15	207538		B0 V	7.31
35600		B9 Ib	5.74	3360	ζ Cas	B2 V	3.76
31910	β Cam	G0 Ib	4.03	198183	λ Cyg	B5 V	4.53
206859	9 Peg	G5 Ib	4.34	196867	α Del	B9 V	3.77
210745	ζ Cep	K1 Ib	3.35	31647	4 Aur	A0 V	4.94
200905	ξ Cyg	K5 Ib	3.72		HR 7784	A1 V	6.23
36389	119 Tau	M2 Ib	4.38	1280	θ And	A2 V	4.61
36371	χ Aur	B5 Iab	4.76	8538	δ Cas	A5 V	2.68
202850	σ Cyg	B9 Iab	4.23	58946	ρ Gem	F0 V	4.18
212593	4 Lac	B9 Iab	4.57	210027	ι Peg	F5 V	3.76
195593	44 Cyg	F5 Iab	6.19	9826	ν And	F8 V	4.09
44537	ψ(1) Aur	M0 Iab	4.91	4614	η Cas	G0 V	3.44
39801	α Ori	M2 Iab	0.5	3651	54 Psc	K0 V	5.87
				201091	61 CygA	K5 V	5.22
				201092	61 CygB	K7 V	6.04

The neon lamp had all the bright lines above 5000 Å and the xenon lamp has a group of lines around 4500 Å. A "quick" reduction was made for the comparison spectra and a composite image was created using the red part ($\lambda >$ 5000 Å) of the neon spectrum and the blue part of the xenon spectrum. A final wavelength tranformation function was calculated with this composite image and it was applied to the star spectra. The star spectra were later normalized with a function chosen by hand.

Usually two spectra of each star were obtained, one with each grism. The observed stars appear in Table 1.

Some spectra are shown in figures 1 (Class I), 2 (Class III), and 3 (Class V). They were taken with the "red" grism and cover from 4500 to 9000 Å. The complete Atlas will be published later. Also an electronic copy of it will be available elsewhere.

References

Houk, N., Irvine, N., and Rosenbush, D. 1974, in *An Atlas of Objective-Prism Spectra*, The University of Michigan.

Johnson, H. L., and Morgan, W. W. 1953, ApJ, **117**, 313.

Morgan, W. W., Abt, H. A., and Tapscott, J. W. 1978, in *Revised MK Spectral Atlas for Stars Earlier than the Sun*, Yerkes Observatory, University of Chicago and Kitt Peak National Observatory.

Peña, J. H., Peniche, R., Bravo, H., Corral, L., Hirata, I., and Yam, O. 1993, *Rev. Mex. Astron. Astrof.*, submitted.

Discussion

MacConnell: To what uses will you put your atlas?

Corral: The atlas was made for use in a project to classify open cluster stars which have Strömgen photometry available. We have plans to complete the spectral type coverage in the near future, and copies will be available upon request.

McCarthy: What are limiting magnitudes for the stars you have observed?

Corral: We want to go to 13th magnitude.

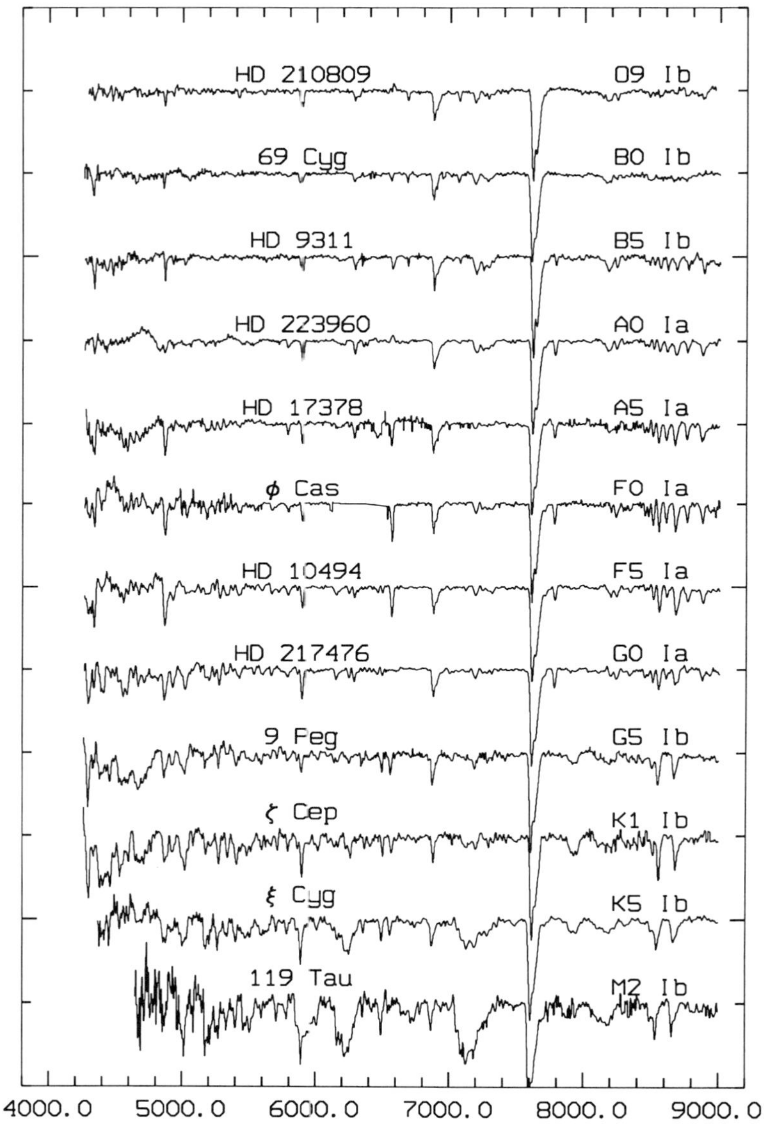

Figure 1. Luminosity Class I stars.

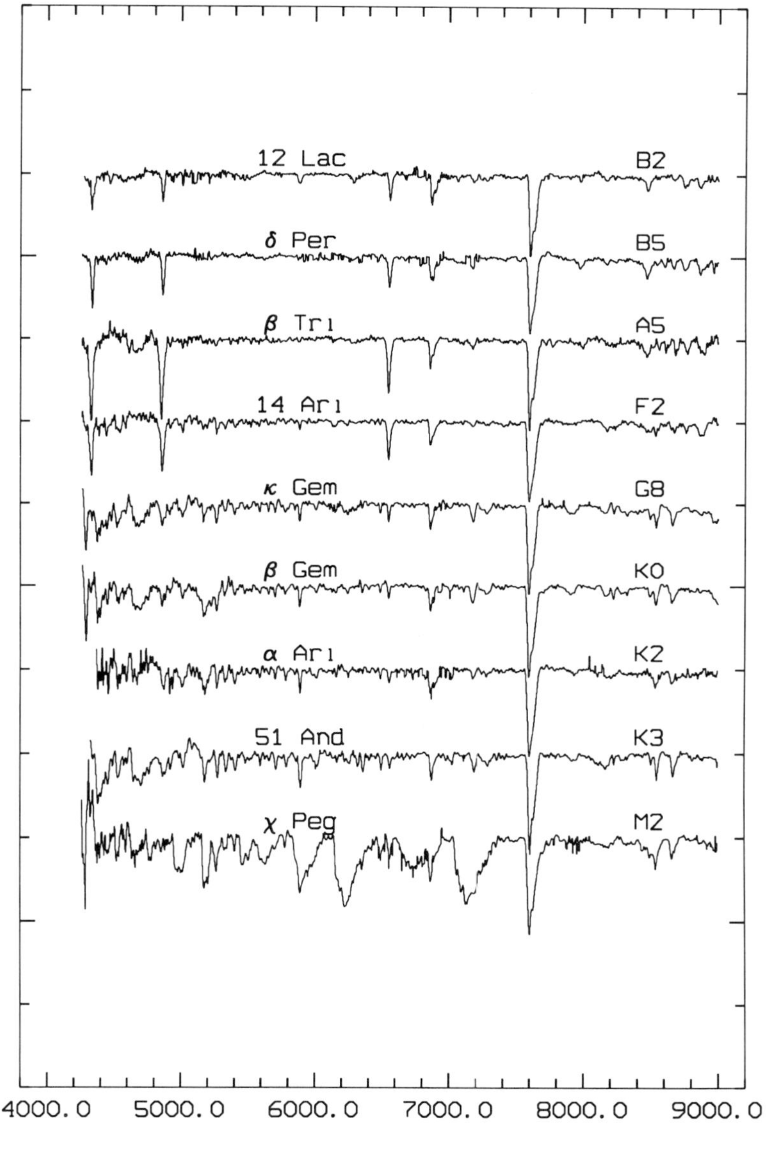

Figure 2. Luminosity Class III stars.

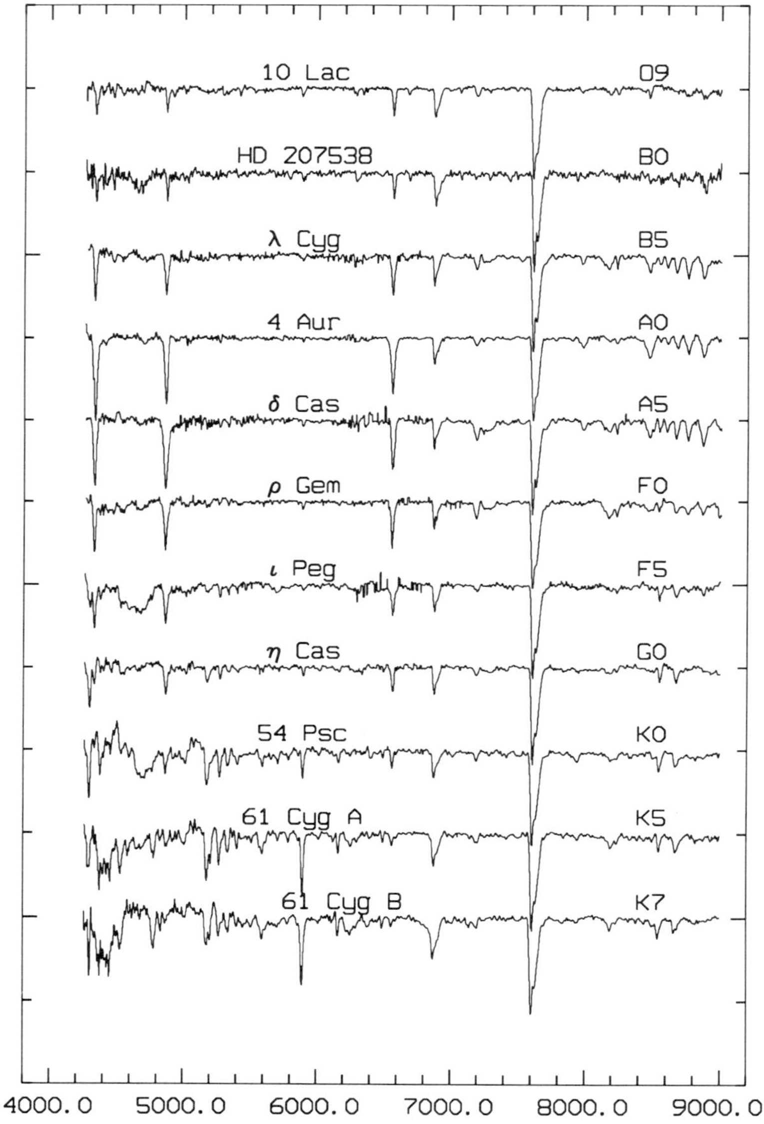

Figure 3. Luminosity Class V stars.

The MK Process at 50 Years
ASP Conference Series, Vol. 60, 1994
C.J. Corbally, R.O. Gray, and R.F. Garrison (eds.)

An Analysis of MK Spectral Types in the New Bright Star Catalogue (BSC)

Wayne H. Warren Jr.

Hughes STX Corporation, NASA Goddard Space Flight Center, Greenbelt, MD 20771

Dorrit Hoffleit

Department of Astronomy, Yale University, New Haven, CT 06511

Abstract. MK spectral types are among the most important data in the BSC, since they can be used to estimate distances, effective temperatures, and luminosities for the brightest stars. The third edition of the BSC (1964) was the first to include MK types, but only 40 percent of the stars had been classified in the MK system at the time. This percentage has steadily increased with every succeeding version and is presently at 93. This paper reports statistics of MK types in the present (1993) version of the catalog and discusses prospects for completion of the two-dimensional classification work for the "bright" stars.

1. Introduction

MK spectral types first appeared in the third edition of the *Bright Star Catalogue* (Hoffleit 1964), the first edition to be published after the introduction of the MK system. The following table shows how the number of stars having MK classifications has increased through the years.

Table 1. History of MK Types in the BSC

Edition	3	4	5	5
Year	1964	1982	1988	1993
% MK	40	87	89	93

Even the last figure (93%) is found to be somewhat misleading when a more detailed examination is made. There are actually 999 primary stars without MK luminosity classes, but most are for good reason. The following breakdown is found:

Table 2. Special Types without Luminosity Classes

Type	Number
Wolf-Rayet	5
S stars	8
C stars	20
Peculiar spectra	191
Metallic-line stars	183
Mt. Wilson L classes	264
Total	671

where stars with peculiar (Ap, pec, n, nn, s, etc.) and metallic-line spectra are only those having no luminosity classifications. Thus, if we discount the above stars, only 328 (4%) are left that have only Henry Draper-type temperature classes. Among these, the following breakdown is found:

Table 3. Stars having only Henry Draper Types

Type	Number	Type	Number
B8	6	G0	4
B9	5	G5	26
A0	10	G8	2
A1	4	K0	156
A2	8	K1	1
A3	5	K2	35
F0	10	K3	3
F2	5	K5	36
F5	5	K6	1
F7	2	M1	4

Of course, many of the unclassified stars with peculiar spectra, particularly the broad-line (n, nn), emission (e), and pec. stars may be classifiable, so the above (4%) is an upper limit.

2. Remarks

Since the BSC defines a magnitude-limited sample, the observed distribution of spectral types merely reflects the stellar luminosity function. As will be seen in the more detailed tables below, the great majority of BSC stars earlier than spectral type G2 are dwarfs, while those from G2 through G5 are distributed rather uniformly in luminosity, and those later than G5 are mostly giants. Thus, most of the unclassified stars (238 of the 328) are expected to be giants in the range G8 through M1.

The following tables provide detailed counts of temperature and luminosity types in the current (August 1993) version of the BSC. The complexity of the

MK classification system and the occurrence of many ranges in temperature subclass and intermediate luminosity classes required that certain types be grouped together, as described in the following notes. The spectral types of companions, when included in the spectral classification field, are separated in the columns by a slash (/).

Although no detailed analysis will be attempted with this still preliminary edition of the BSC5, the following are interesting characteristics of the tables. We can probably assume that most of these characteristics hold in general (with the exception of abundance parameters) for the entire stellar population of the Galaxy.

1. The chemically peculiar stars of the upper main sequence appear at B6 or B7 and abruptly disappear at early F, while the metallic-lined stars don't appear until A0 and carry through to F5 in parallel with the δ Del stars (if indeed they exist as an independent class), many of which are "m".
2. The rather uniform distribution of stars within each luminosity class while moving in temperature, even for companions, is rather striking, although not surprising. Although the number of giants increases significantly at G8 and peaks at K0 among earlier types, the distribution is rather uniform in high-luminosity types.

The statistics will be redone following completion of the BSC5, especially because the tabulated numbers add up to about 90 less than the number of stars in the BSC (9096; of the 9110 objects in the BSC, 14 are either novae or nonstellar), a disagreement attributable to the fact that it was not possible to group all similar spectral types together by machine sorting, so many counts had to be done manually. Following the completion of BSC5, a more complete analysis will be made.

3. Tables

NOTES TO TABLES

- Stars having variable spectra or where a range in temperature subclass is given are included with the hottest type.
- Metallic-line stars having luminosity classes are grouped with those having no luminosity classes under the "m" column.
- For later types (F-M), luminosity class Ib-II is grouped under Ib and Ia-Iab and Iab-Ib are grouped under Iab, while I-II is under I.
- λ3955 stars are grouped with classical Ap, while λ4200 and λ Boo stars are included under "pec".
- δ Sct and δ Del stars are combined under δ Sct.
- Shell stars are counted only where indicated in the spectral type. Many shell stars are only indicated by the remarks to the BSC.

Acknowledgments. The help of Martha Warren with preparation of the large tables is greatly appreciated.

References

Hoffleit, D. 1964, The *Bright Star Catalogue*, 3rd revised edition (New Haven, Yale University Observatory)

Hoffleit, D. (with the collaboration of Jaschek, C.) 1982, The *Bright Star Catalogue*, 4th revised edition (New Haven, Yale University Observatory)

Hoffleit, D., & Warren Jr., W. H. 1991, The *Bright Star Catalogue*, 5th revised preliminary edition, Selected Astronomical Catalogs, Vol. 1 CDROM (Greenbelt, MD, Astronomical Data Center, NASA/GSFC) (this is the most recent version available in machine-readable form)

Discussion

Houk: What is the distribution of the remaining missing MK types? Are they mostly in the north? Would they be mostly fainter than Garrison has done and those I haven't yet done between 4.8 and 6.5 mag?

Warren: Yes to all three questions. They are mostly K0 and later in the north, but there are quite a few stars between +02°and -12°as well. Thus, we expect to fill in a number of types as soon as your fifth volume of the Michigan Spectral Survey becomes available.

Garrison: If you give me a list of the missing stars by December 1, I will observe whatever I can reach from Chile (and Toronto).

Warren: We are in the process of preparing a list and will be pleased to send it to you. It may be the MK classifications already exist for some of these stars and that we have just not encountered them as yet.

Keenan: Of the twenty carbon stars in the *Bright Star Catalogue*, how many are R stars?

Warren: Although the types in the current catalogue, which are on the older C classification system, do not identify a star as originally type N or R, we expect that there are no R stars in the catalogue because they are too faint. The new C classification system described by Keenan at this meeting should solve this problem, since the types include an N, R, CH, or Hd identifier.

MacConnell: How are the statistics for the stars $6.5 < V < 7.0$, and what is being done to get them classified?

Hoffleit: Nothing as yet has been done about updating *The Supplement* (pub. 1983). A new edition should be approximately three times as large because a great deal more *UBV* photometry is now available. The supplement is supposed to include all stars with known V-magnitudes to 7.1 that are not already in the *Bright Star Catalogue*, 4th edition. This I hope will become mainly Wayne Warren's project as, at age 86.5, I still have a lot of unfinished other projects – but I shall help him to whatever extent I can.

SPECTRAL-TYPE STATISTICS FOR THE BRIGHT STAR CATALOGUE
W, O, B STARS

Standard Classifications | Peculiarities

L T	None	sd	V	IV-V	IV	III-IV	III	II-III	II	I	Ib	Iab	Ia		classical p	p or pec	n, nn	s	shell
WC	2/1																		
WN	3																		
Totals	5/1																		
O																			
4			1																
5	1						1												
6	1		1							1						2			
6.5			3																
7			2/1																
7.5			2				3		1				1				3		
8	1						1						2			2			
8.5			1																
9			4/1				2			1	4		1				2		
9.5	1		5						4		1	1	2			2	1		
9.7											1								
Totals	4		19/2				7		5	2	6	1	6			6	6		
B	1/5	1	/1												1	1			1
0			6/1		3		10		1		3		5			4	4		1
0.3					1														
0.5			9		6		6		3		4		7			1	2		
1	1		31		7		10	1	4		6	2	6			5	2		
1.5			21/1		9		9						3			3	6		
2	1/2		66/4	52	44	3	17	1	2		3		2			10/1	16		
2.5			49		25		3				2		1			1	9		
3	2		89/7	1	44/2		17	2	1		1	1	3			10	12/1		2
4			36		18	1	7						1			6	10		
5			84/1	1	31		21/1	2	2		3	1	4			5	20		
6			48	1	23		18		1		1	1	2		1	4	6		
6.5			1																
7			45		20	5	36	2	2				1		3	5	11		
8	12		134/5	3	19	5	77/1	11	6		5		4		23	11	37	1	
8.5			1																
9	64/1		150/8	10	33/1	5	64	5	2		3	3	4		81/1	6	34	2	
9.5	6		74/5	3	7	3	17								10	3	14	1	
Totals	87/8	1	844/33	71	290/3	22	312/2	24	24		31	8	43		119/1	75/1	183/1	4	4

SPECTRAL-TYPE STATISTICS FOR THE BRIGHT STAR CATALOGUE
A, F STARS

T \ L	Standard Classifications																	Peculiarities						
	None	sd	V	d	IV-V	IV	sg	III-IV	III	g	II-III	II	I	Ib	Iab	Ia		m	Classical P	p or pec	n, nn	s	δ Sct	Shell
A	54/8			/1														34/2	20				1	1
0	55/1		250/8		3	27/2		4	30		1	2		3	1	6		1	42/5	10/3	42/2	14	1	
1	33/1		226/5		2	12		2	4					2		2		16	11/1	5/2	43	1		2
2	44/3		238/11		5	39		4	15		1	1		3	4	4/1		38	13	9/1	34	19		3/1
3	38/4		173/13		3	36/1		2	14		1	1		4	1	2		34/1	2	4	26	6	1	1
4	10		66/7		2	13/2			12									10		1	13	1	1	
4.5			1																					
5	44/2		79/4		9	20/1		5	20			4		2		1		42	3	2	20	4	2	
6	6		15/5	1	1	5			/2		1	1		1		1		4	3		2			
7	12/2		55/5		5/1	22		6	37					1	1			14	1	2	17	2	1	
8	7/1		39/3			12		2	13		1		1	1				8	1		11	2	1	
9	1		26	1	4	13	1	3	12/1			2				1		3		2	6/1	1	2	
9.5									1															
Totals	304/22		1168/61	2/1	34/1	199/6	1	28	158/3		5	11	1	17	7	17/1		204/3	96/6	35/6	214/3	50	10	7/1
F	/4																						1	
0	19		94/7	2	13	59/2		6	41	3		5		1	1	1		5	6	2	20/1	3	5	
1			25/2		1	16		9	14			4				1				1	1			
2	5		76/9	1	4	27/2		10	31	1	4	8	1	6	2	1		1		2	4		2	
3	/2		52/4		7/1	23		6	15		1	1		1		2		1		2	2	5	1	
4			48/2	2	2	18/1		3	14		1			1		1			1	1				
5	8		84/9	3	11	36		4	22/1		3	6		6	2	1		2			1	2	4	
6			81/7	5	10	24/2		4	12		3	6		3		2						4	1	
7	2		82/7	1	8	14/1			8/2		1	2		10	1						2	1		
7.5														1										
8			70/3		4	8		2	5			1	1	9		1						1		
8.5			1		1																			
9			30/3		1	4/1			2			1				1					1			
9.5			1																					
Totals	34/6		644/53	14	62/1	229/9	1	44	164/3	4	13	34	3	38	6	11		9	7	8	31/1	16	14	

SPECTRAL-TYPE STATISTICS FOR THE BRIGHT STAR CATALOGUE
G, K STARS

Standard Classifications / Peculiarities

T \ L	None	V	d	IV-V	IV	sg	III-IV	III	g	II-III	II	I	Ib	Iab	Ia	0 or 0-Ia		p or pec	s
G	/5							1											
0	4/1	55/4		7	11/1			16/2		1	1	2	8			2		3	2
0.5		2			1														
1		23/1		1	4			3/1			2		4						
1.5		4		2				1					1						
2		22/3			11/2			13/1		2	1		12			2			
2.5		3		1	1			1			1								
3		20		2	7			8	1		3		9		1	1			
4		14/1		1	1			12	3	1	1		2			2			
4.5		1																	
5		22/2		3	22/2		7	50	12	1	11	1	9	2	2				
6		5	1	1	6	1		47	14	2	5	2	3	1					
6.5		1						1				1							
7		3		1	4			36/1	16	4	2		2						
7.5		1		1				5		2									
8		12/1		1	22/1		23	291	17	19	17		9					3	
8.5					1			19											
9		6		1	5	1	5	102	19	1	1		2						
9.5							1	11											
Totals	4/6	194/12	1	22	96/6	2	36	617/5	82	33	45	6	61	3	3	7		6	2
K	/1								1										
0	117	18			24/1		21	357/5	28	24	10	1	5			2		2	
0.5		2						16		2	3								
1	1	8		1	18	1	14	268/1	31	11	13	1	3		1			1	
1.5					1		1	17			2		2						
2	35	7		1	8	1	5	264/1	22	9	11/1	1	5	1					
2.5								26		1	2		2						
3	3	8			1			212/2	23	7	17	1	14						
3.5								10		1				1					
4		3/1						152	24	3	4	3	4					1	
4.5								15					3						
5	36	3/1						165	22		2		2	1					
5.5								2											
6	1							7	2				1						
7		1						8											
8								1											
Totals	193/1	50/2		2	52/1	2	41	1520/9	153	58	64/1	7	41	2	1	2		4	

SPECTRAL-TYPE STATISTICS FOR THE BRIGHT STAR CATALOGUE
M, S, C STARS

Standard Classifications — Peculiarities

T \ L	None	V	III	g	II-III	II	I	Ib	Iab	Ia		p or pec	MS	J
M														
0		1/1	75	1	1	2		1						
0.5			7			1								
1	4		87	2	2	3		4						
1.5			7		1	1			1					
2			78		3	2			3	2		2	2	
2.5			14		1									
3			69	1	3	2		1					1	
3.5			10											
4			49			2		1					1	
4.5			13			1							2	
5			18			3	1	1					1	
5.5			2											
6			12									1		
6.5			2									1		
7	1		11										1	
8			2											
Totals	5	1/1	456	4	11	17	1	8	4	2		4	8	
S														
3	1													
3.5	2													
3.9	1													
4	1													
5	1													
6	1													
6.5			1											
7.3	1													
Totals	8		1											
C														
5	9													1
6	5													
7	6													
Totals	20													1

PART II.

THE PHYSICAL BASIS OF CLASSIFICATION

The MK Process at 50 Years
ASP Conference Series, Vol. 60, 1994
C.J. Corbally, R.O. Gray, and R.F. Garrison (eds.)

Stellar Classification and the Theory of Stellar Atmospheres

Dimitri Mihalas

Department of Astronomy, University of Illinois, 1002 W. Green St., Urbana IL 61801

Abstract. This paper sketches some aspects of the theory of stellar atmospheres over the past 70 years or so, and describes some of the interactions that have occurred between the theory and stellar classification.

1. Introduction

It is a great pleasure to participate in this celebration of the 50th anniversary of the publication of the MKK Spectral Atlas. The MK System and the MK Process have had a great influence on astrophysics in general, and on my own work in particular. I have known Bill Morgan for over 25 years, and during that time he has been a constant source of inspiration. I have learned a great deal from him, not only about stellar spectra, but also about how one goes about doing science.

The title of this talk as given in the preliminary program for the conference was "The Physical Basis of Stellar Classification". As I actually prepared the talk I realized that that was not satisfactory, because it implies that it is somehow possible to choose a unique physical structure that defines what we see in stellar spectra. That picture is too one-sided, like what one gets hearing only one side of a telephone conversation. It may be possible someday to "reduce" stellar classification to a few physical principles, but we are nowhere near doing that at the present time; indeed I certainly don't expect it to happen in my lifetime. Rather, I see the process at the present time as one of direct strong interaction back and forth between theory and observation. More like a good duet, or a pas de deux, when it works well.

The simple fact is, that at any given moment in time the observations are 30 to 50 years ahead of the theory in the sense that one finds observationally phenomena which the theory won't be able to cope with fully for at least that many years. That is why spectral classification is so important to theorists: it shows us what *actually* exists in the spectra of real stars, whether or not we are "ready" to understand it now.

I believe that the view expressed above is not very widely held by stellar atmospheres' theoreticians or by astrophysical theorists in general. I think I acquired this orientation from three distinct episodes in my training. First, as a graduate student at Caltech in the early '60s, I wrote my thesis under the direction of J. B. Oke, a card-carrying observer if there ever was one. Initially I set out to do an observational thesis on O stars, but thanks to Oke's wizardry with instrumentation, the data were relatively quickly acquired. Then I had to expend a huge effort on their theoretical/computational interpretation. Even then I knew that what I had done was a long way from providing an adequate explanation of what I had observed. In fact, I have been after that "adequate explanation" ever since. Then, in the late '60s when I was at Yerkes Observatory, Morgan routinely confronted me with real spectra, and helped me overcome the notion that modeling of spectra could be done with only a few simple free parameters. Finally, I spent a few years at Sacramento Peak Observatory in the early '80s, where I was suitably humbled by learning what I could about the detailed radiative, hydrodynamic, and magnetohydrodynamic structure of the solar atmosphere from the magnificent data various observers obtained there. This latter experience has made me wary ever since about accepting highly simplified models as adequate representations of real stellar atmospheres.

What I would like to do here this morning is to give a brief sketch of the story of how stellar atmospheres theory has grown, and to mention some of the fruitful interactions between theory and observation that have occurred.

2. Where We Came From

2.1. Foundations (1900) - (1950)

Looking back at the beginning of this century, we see that in 1900 there was essentially no theoretical basis for calculating stellar models or stellar spectra. But happily, there were true giants at hand. In the early 1900s, K. Schwarzschild and A. Schuster "invented" the equation of transfer, the basic theoretical tool needed in all studies of stellar atmospheres and spectra. Then began a long series of papers by A. S. Eddington and E. A. Milne, elucidating the implications and consequences of radiative transfer in stellar atmospheres. It is really a thrilling experience to re-read their papers today, and one is left with deep admiration at how far they were able to penetrate with so little information (both observational and theoretical) and no computational capability at their disposal.

Brilliant work was also done by Eddington's and Milne's colleagues and students. For example, W. McCrea invented the method of model atmospheres, and demonstrated the power of numerical methods. But I would like to mention in particular the work of Cecilia Payne (later C. Payne-Gaposchkin), whom most of you here would happily claim as an observer. Actually, she started out as more of a theoretician! Cecilia Payne, as will be described below, was influenced to become an astrophysicist by Eddington, who recognized her genius. She wrote her thesis at Cambridge under the supervision of Ralph Fowler, an eminent figure in the then-rapidly developing field of statistical mechanics. Her goal was to provide a physically sound broad-brush explanation for the progression of stellar spectra along the Main Sequence, and for the additional changes observed in what we would now call giants and supergiants. Her thesis was brilliant, and was subsequently published as the book *Stellar Atmospheres*. She was the first woman to receive a Ph.D. from Cambridge. Her analysis, using only statistical mechanics as developed by Fowler and Saha (all of this *before* the development of quantum mechanics!), showed compellingly that the appearance of the observed spectral sequence is basically the result of the variation in the average atmospheric temperature and density from one spectral type to another. In addition, she concluded that stellar atmospheres must be *hydrogen (and helium) rich*, and that the Russell mix, generally accepted as the cosmic mix, could apply only as a description of the relative abundances of the heavier elements. Thus she began her incredible career, making at one stroke a breakthrough in the study of stellar spectra, and also annoying the most distinguished astrophysicist in the U.S. (but recall she was English).

I would like to digress briefly here to tell one story about Cecilia. I first met her while she was on a sabbatical at Caltech, analyzing, if I remember correctly, the spectrum of R CrB. One evening she invited me to dinner with Sergei and their son Peter at their apartment. It was a good meal and an interesting evening. Several years later I was at an AAS meeting at Michigan State University in East Lansing. Morgan was there too, and he invited me to have dinner with him, mentioning that Cecilia would join us. We met in the dining room and after Morgan introduced me to her I remarked "I actually have met you once before at Caltech, but I was just another graduate student then, and you surely won't remember me." She immediately responded "My dear boy [I was younger then], *of course* I remember you. Why it was on ... (she quoted the date), and then on ... (she quoted another date) you came to our house for dinner, and we served ... (she quoted the menu)." Later in the meal, after listening to her and Morgan discussing the spectra of individual stars, which usually entailed her quoting from memory detailed references to the entire (?) bibliography of papers written about each one of them, I commented that she seemed to have a remarkable memory. To which she replied that she always had, and that it was that very asset which brought her into astronomy. She said "I was in sixth form in school in Cambridge, when one evening Professor Eddington gave a public talk on relativity. I went to it, and it was wonderful. I was so thrilled, that I went straight back to my room, sat down, and copied it out." I wasn't sure what she meant, so I asked for clarification. "I sat down, and wrote it out from memory. And I got it right too! I know that, because the next day I took it to Professor Eddington; he read it, and he said 'My dear girl, this is *exactly* what I said! Do you suppose I might have this so I could give it to my publishers to be printed?' " Eddington encouraged her to become a professional astronomer, and while she was a student, he took an active interest in her work and provided valuable guidance.

From the mid twenties to the mid fifties, there was a virtual explosion of work on stellar atmospheres, much of it stimulated by the invention of wave mechanics, which permitted, for the first time, reliable calculation of line strengths and continuum cross sections. Many great names in our field are prominent in this era. For example, Donald Menzel and Lawrence H. Aller worked hard to put results from quantum mechanics into quantitative analyses of the solar and stellar spectra. Albrecht Unsöld in Germany and E.R. Mustel in Russia were leaders in developing model atmospheres and quantitative computations of line strengths and curves of growth.

During the Second World War, astronomical work was greatly slowed. Yet great strides were made even at this difficult time. For example, S. Chandrasekhar carried out (by hand!) a monumental calculation of the continuum cross section of H^-, and from an analysis of the solar spectrum he and G. Münch showed that H^- is the dominant opacity source in the solar atmosphere. In Copenhagen Bengt Strömgren laid out a systematic program for constructing model atmospheres. He refined existing methods, and produced many tables of basic physical quantities (e.g. opacities and equation of state) which speeded model construction. In addition, he gathered around him a small "school" trained in these methods; after the War these individuals provided the "seed corn" for a rebirth of the subject in several countries. Likewise, sequestered from the advance of the German Army, a small number of Russian scientists worked in safety, and in isolation. It was then that V. Ambartsumian developed his powerful invariance method for solving transfer problems. And also when V. V. Sobolev had his astonishing insight that under certain conditions, actually met in many astrophysical flows, such as stellar winds, it is possible to solve the transfer problem in an expanding medium *analytically*, and that the solution is actually *easier* than for a static medium. These were remarkable times!

In this era, there were 30 years of slow, but steady, improvement in atomic data, opacities, and basic transfer theory. Yet even the best model atmospheres were crude: all them were static, assumed grey opacities, and invoked local thermodynamic equilibrium (LTE). It was not yet possible to enforce strict energy equilibrium (radiative equilibrium). Actually, I personally think that it is easily argued that despite the great increase in *quantitative* results, the *conceptual* basis of the theory was exactly the same as in Cecilia Payne's work; it was on the foundations she provided that so much had been built.

2.2. The Dam Breaks ('50s and '60s)

The situation changed suddenly and radically with the advent of digital computers after WW II. The first machines were so primitive and so difficult to code and operate that they remained mainly experimental devices. The watershed was when IBM introduced a series of large, fast computers in the late '50s, culminating in the IBM 7090. That machine showed, for the first time what could be done with numerical modelling. And today we have machines that are a million times faster and bigger. From today's perspective it is interesting to note that the PC I am using to write this manuscript is *much* larger and faster than the IBM 7090. It is essentially impossible to overestimate the importance of such a huge increase in computational power in such a short time. The only comparison that seems reasonable is the effect replacing photographic plates with CCD detectors has had on observational work. So far I have mentioned only hardware. But another breakthrough of equal importance was the release of the FORTRAN compiler by IBM in the late '50s. FORTRAN is often the object of scorn and complaint by computer scientists, and few words of praise are said about it. But those who can remember back to the pre-FORTRAN era will vouch for the tremendous increase in efficiency and saving of their own personal time that resulted from being able to write code for a machine in essentially mathematical notation instead of unforgiving machine language.

With powerful computers, it became possible for the first time to construct nongrey models, which yielded reasonably accurate monochromatic fluxes for comparison with observation. Towards the last half of the decade of the '60s it even became possible to include some spectral lines in the computation, and thus to address the problem of line-blanketing, which Milne had first explored some 40 years earlier. Further, it became possible to enforce the condition of radiative equilibrium accurately thanks to new temperature-correction procedures developed by E. H. Avrett and M. Krook at Harvard. Later, these methods were extended to enforce constancy of the sum of the radiative and convective fluxes.

All of this activity led to the construction of large numbers of model atmospheres for a wide range of effective temperature and surface gravity, and a wide selection of chemical abundances. The *Astrophysical Journal Supplements* started publishing telephone-book size collections of models. I must confess to having done my share of filling journal pages with models. Today, of course, one would keep all the data in machine-readable form, and simply mail a colleague a floppy disk, or send it over the network.

With the development of quantum theories of line broadening by H. Griem and his associates, it became possible to calculate accurate line profiles of H and He lines for comparison with observed profiles. Detailed fits of computed model fluxes with observed absolute energy distributions (becoming available for the first time thanks to the development of monochromatic spectrum scanners carefully calibrated against laboratory black bodies by A. D. Code, J. B. Oke, D. Hayes, and R. Schild) were finally possible. Combining the information from continuum fluxes with line profiles yielded pretty good estimates of the effective temperatures and surface gravities of individual stars. With enough computer power it was possible to do abundance analyses by computing the strength, as a function of abundance, of each line included in the spectrum analysis, varying the assumed abundance until a match with the observed equivalent width was obtained. This approach obviates all of the restrictive assumptions made in the usual curve-of-growth analysis, and leads to more reliable results.

Rapid progress was made in this period as a result of the collective efforts of many workers; so many that it would be difficult to mention all the relevant names. Let it suffice to say that the Harvard-Smithsonian Center for Astrophysics was (indeed still is) a major center of research on stellar atmospheres.

2.3. NonLTE ('70s)

One of the most troubling assumptions made by the classical theory of stellar atmospheres was that the material is in local thermodynamic equilibrium (LTE). Simple estimates show that in the line-forming regions of an atmosphere, radiative rates generally exceed collisional rates, both for excitation and ionization. But the radiation at small optical depth in the atmosphere cannot be in equilibrium (except by pure accident) because (1) it is dilute (an atom at the surface sees radiation only from one hemisphere and blackness in the other), and (2) it cannot be characterized by a single temperature (radiation at more transparent wavelengths arises from deeper, hotter layers of the star). Thus the possibility arises that the radiation field may drive the material away from LTE.

Actually nonLTE analyses of nebular spectra and the solar coronal spectrum had already been made many years before by Menzel and his collaborators (including L. H. Aller). These studies were made possible because one can assume that the subordinate lines (e.g. Balmer lines) are transparent, whereas the resonance lines (e.g. Lyman lines) are so optically thick that they can be taken to be in radiative detailed balance (photoexcitations exactly balance photodexcitations), thus evading the transfer problem in the lines (in both cases). This work showed that in these rarefied media departures from LTE are indeed major; for example, the ionization balance in the solar corona is different by about a factor of 10^6 from the LTE prediction. Nevertheless, the importance of departures from LTE in typical stellar spectra remained unproven, and was, in fact, controversial. Remarkable progress in understanding many of the physical aspects of line transfer when departures from LTE are permitted were made by a number of workers, mostly in Boulder. For example, Richard Thomas derived expressions for the source function in the line of a two-level atom, both without and with continuum, and distinguished the differences between collision-dominated and photoionization-dominated lines. He showed these two classes of lines behaved quite differently from one another, and gave results often quite at variance with the classical predictions of LTE. John Jefferies studied the difficult problems of photon switching and degradation in multilevel atoms, and introduced the very fundamental concept of the *photon pool*, i.e., the notion that a photon in any one line does not simply "belong" to that particular line, but actually can produce effects in *all* other lines in the complete transition array of the atom. Grant Athay was busy applying these ideas to numerical modelling of lines in the solar spectrum, particularly in the solar chromosphere. David Hummer carried out very precise solutions of the two-level atom problem, and was able to show how the *thermalization depth* of the line, i.e., the depth below which the line source function finally returns essentially to LTE, depended on collision parameters and the presence or absence of a background continuum. A number of important related studies were carried out by George Rybicki and Wolfgang Kalkofen.

Despite this progress, and the general acceptance of the need for nonLTE calculations by solar astronomers, stellar astrophysicists remained unconvinced because no one had been able to solve a realistic multilevel line-transfer problem in a realistic (one hoped) model atmosphere. I think that by then it was obvious to a number of us that to meet this challenge would require new, powerful, numerical methods. There were many important developments at this time, but at the risk of playing favorites I will mention only two. First, a way of recasting the transfer equation into an extremely nice and very general form which promotes very efficient solutions was developed by Paul Feautrier in Paris. Second, about the same time a powerful method of coupling transfer equations to the equations of statistical equilibrium, which allows the solution of the full nonlinear multilevel problem was developed by Lawrence Auer and me. Using this method we were able to show that departures from LTE were very important in the hydrogen/helium spectra of the O and early B stars. When the new theoretical profiles were compared with observation, we found much better fits than had been obtained previously, at a helium abundance in agreement with nebular values, and at surface gravities consistent with measured masses and radii; previously both the helium abundance and the surface gravity were systematically overestimated using LTE calculations.

Thus opened an era in which we finally had a *greatly* improved theoretical/conceptual picture of spectrum formation along with a capability of computing particular spectra for application to the analysis of real stars. During this decade nonLTE studies were made of the spectra of H, He I, He II, C III, N III, Ne I, Mg II, Si II - Si IV, and Ca II. The work on Ne I in B-star spectra showed that a factor of ten discrepancy in neon abundances from nebular spectra and analyses of B-star spectra could be ascribed to departures from LTE in the latter. Similarly, the work on Mg II showed that a factor of four discrepancy between the magnesium abundance obtained for O-stars and stars of spectral class B and later was caused by departures from LTE in the O-star spectra. And studies of the He I spectrum in B-stars yielded much better fits in the yellow-red part of the spectrum, although to this day, attempts to fit the He I λ 5876 line ($2p\,^3P - 3d\,^3D$) are not completely successful.

The studies of N III and C III had a completely different thrust, and I would like to say a few words about this work because it illustrates how spectral classification can provide critical input, key ideas, into astrophysical interpretation. As you all know, there is a subclass of the O-stars, the Of stars, which show emission at He II λ 4686 ($n = 3 - n = 4$), and at N III $\lambda\lambda$4634-40-41 ($3p\,^2P - 3d\,^2D$). The N III emission poses quite a puzzle for astrophysical analysis because only that one multiplet out of the entire N III spectrum is in emission; even the multiplet immediately below it in a cascade chain, ($3s\,^2S - 3p\,^2P$) at λ 4097 is always in *absorption*. For the theoretician the situation was further muddied because all of the classification work up to the early '70s spoke of these stars as if *both* emission phenomena occurred in all Of stars. Because of that, and because we know that many of these stars are extremely bright, we had simply assumed that the observed emission must be the result of extended, and probably rapidly expanding, envelopes, and we knew quite well that the then-existing theoretical tools were inadequate to the task of explaining such objects.

A breakthrough came when Nolan Walborn, at the time Morgan's graduate student, who was sitting just across the hall from me at Yerkes Observatory, pointed out to me that the phenomenology of Of-stars is more complex and interesting. He had discovered that actually there are three subclasses, which can be described briefly by the following little table, using Walborn's classification notation.

Spectral Type	He II λ4686	N III $\lambda\lambda$4634 − 41
O	absorption	absorption
O((f))	absorption	weak emission
O(f)	absent	emission
Of	strong emission	strong emission

We spent many long evenings discussing the possible significance of his results. The stars called Of by Walborn are the "classical" Of stars: both the He II and N III lines are in emission. We agreed that these stars undoubtedly have extensive, expanding envelopes (subsequently confirmed by space observations). But there are also Of spectra in which the N III lines are in emission, but the He II λ4686 line is *absent*, even though the Pickering series lines (e.g. $\lambda\lambda$ 5412, 4542, 4200, ...) are clearly present at normal strength; for such stars it appears that the λ 4686 line has been *filled in*. These stars probably have less extensive envelopes, with less dense winds. But the most interesting category for me was the O((f)) stars, because these stars seemed to be perfectly ordinary, compact, O-stars, with the only anomaly that a single multiplet of N III is in emission. It seemed to us that the O((f)) stars are showing us that there is some *intrinsic* emission mechanism (i.e. not dependent upon an extended envelope) at work in the N III lines, and that to the extent that their atmospheres really are compact and in hydrostatic equilibrium, it ought to be possible to explain the data with existing theory.

One is then faced with two problems: (1) why is the $(2s^23p\ ^2P^o - 2s^23d\ ^2D)$ transition in emission; and (2) when it is, why is the next transition downward $(2s^23s\ ^2S\ -\ 2s^23p\ ^2P^o)$ always in absorption? The answers to these questions have to do with atomic physics, not stellar atmospheres. The answer to the first is that by an accident of nature there is an autoionizing state $2s2p(^1P^o)3d(^2P^o, ^2D^o, ^2F^o)$ just a fraction of an e.V. above the continuum, so that free electrons can efficiently recombine with N IV ions into this state; then the doubly excited state can be *stabilized* by the very fast $2p \rightarrow 2s$ transition of the excited inner-shell electron, which leaves us with an atom in our $2s^23d$ state. Once the mechanism had been identified, it was easy to show that it would yield sufficient electrons to bring the $\lambda\lambda 4634 - 41$ lines into emission. The second question seems more mysterious, because the only reasonable answer is that *somehow* the $2s^23p$ state must be drained efficiently, so that electrons cascading $(3d \rightarrow 3p)$ would produce emission, but then diverted into some channel other than $(3p \rightarrow 3s)$ so that no emission would result at $\lambda 4097$. The rub is that there are no lower states into which $3p$ can drain: the ground state is $2s^22p\ ^2P^o$, and a $p \rightarrow p$ transition is forbidden; likewise, to reach the three $2s2p^2$ states below $2s^23p$ would require a *two-electron jump* which is normally strongly forbidden. However, as noticed by M. J. Seaton, nature has played another little joke on us. It just so happens that there is a $2p^3\ ^2P^o$ state at almost exactly the same energy as the $2s^22p\ ^2P^o$ state of interest to us; these two states are almost totally mixed quantum-mechanically, so that an electron in our $2s^23p$ state is equally an electron in the $2p^3$ state. The latter can jump to the three $2s2p^2$ states with a single (fast!) electron jump $(2p \rightarrow 2s)$.

Thus excess electrons flowing from the dielectronically recombining state $2s2p3d$ into $2s^23d$ jump $(3d \rightarrow 3p)$, become completely mixed up about whether they are $2s^23p$ or $2p^3$, then drain very quickly into the three lower-lying $2s2p^2$ states, thus preventing emission in the $(2s^23p \rightarrow 2s^23s)$ line. When the detailed calculation was done, quantitative agreement with the observations of the O((f)) stars was achieved, which validated the basic atomic physics invoked. If Walborn had not distinguished the three types of Of stars by his excellent classification scheme, we might not have understood this mechanism even today. To my mind, this is a really good example of how careful stellar classification can have direct impact on the theory.

Once the N III emission mechanism was understood, one could analyze the spectrum of C III, which shows similar effects. From visible-light data it was known that in the singlets $\lambda 5696$ $(3p^1P\ -\ 3d^1D)$ is in emission in Of stars, while in the triplets $\lambda 4650$ $(3s\ ^3S\ -\ 3p\ ^3P)$ is in absorption . This information is too fragmentary, and needs to be supplemented with data about the singlet $\lambda 8500$ $(3s\ ^1S - 3p\ ^1P)$ and the triplet $\lambda 9700$ $(3p\ ^3P - 3d\ ^3D)$, both in the infrared. Observations by Stuart Frost, Wes Lockwood, and me showed that $\lambda 8500$ is in absorption, while (much more difficult) observations by Dennis Ebbets showed that $\lambda 9700$ is in emission. Thus the C III lines show the same behavior as the N III doublets, with the $(3s - 3p)$ transitions in absorption and $(3p - 3d)$ transitions in emission for both singlets and triplets, which points to similar atomic physics. A nonLTE analysis by N. Sakhibullin, L. Auer, and K. van der Hucht yielded quantitative agreement with the data.

The kinds of NLTE computations described above finally ran out of steam in the late 1970s because (1) the computations, using existing machines and then-current algorithms (see below) simply became too costly, and (2) it was not possible to handle (nonLTE) line blanketing, which affects photoionization rates in the model atmospheres. Recently both of these problems have been surmounted: new algorithms yield tremendous speedups, and the nonLTE line-blanketing problem is yielding to new methods, such as the one devised by Lawrence Anderson at Toledo. We can reasonably expect much more progress in these areas in this decade.

2.4. Large-Scale Velocity Fields ('80s)

With the advent of space observation we learned conclusively that virtually all early-type stars have massive stellar winds. Mass-loss rates for extreme Of stars range up to 10^{-6} solar masses/year, and for WR stars up to 10^{-5} or even higher. In the ultraviolet spectrum of these stars one finds huge P-Cygni lines, which inherently contain information about the outflow velocity of the material as a function of distance from the star. In the '80s new theoretical methods for calculating NLTE spectra in rapidly expanding flows were developed. John Castor incorporated Sobolev's escape probability method into a complete formalism for determining the nonLTE statistical equilibrium of excited states and ionization stages, and developed a powerful iteration scheme to solve the resulting system. When applied to WR spectra, this method yielded good diagnostics of the temperature and density structure of the envelope, and its velocity profile. Further, good abundance diagnostics were obtained, and the old question was answered of whether the observed differences in WC and WN spectra are attributable to differences in chemical composition of these two groups (yes) or to some peculiarity in the excitation mechanisms leading to prominence of carbon in some stars and nitrogen in others (no).

About the same time a number of studies were made by David Hummer, Paul Kunasz, Peter Noerdlinger, George Rybicki, and me of how to solve the line transfer equation in the *comoving frame*, that is, in a set of frames, each moving with the fluid velocity at each point in the flow. This approach was first suggested by W. McCrea and K. Mitra way back in the '30s, but, of course, they lacked the computational power to carry it very far. The advantage of this method is that one follows the line profile in frequency through the flow, and can thus focus on the physics of the scattering and redistribution mechanisms in the lines without needing a huge number of frequency sample points. Another advantage of the method is that it works in both the supersonic and subsonic regimes (indeed, even in static material), and does not depend upon the assumption of large velocity gradients as does the Sobolev method. Several effective numerical methods for treating variants of the basic problem were found.

In the mid '80s a tremendous leap forward in the modelling of radiatively driven winds was made in Munich, under the leadership of Rolf Kudritzki. Back in the mid '70s, Castor, David Abbott, and Richard Klein had shown that stellar winds of the observed mass flux for early type stars will result from the force radiation exerts on thousands (millions?) of spectrum lines in the the Balmer continuum of these stars. This pioneering work was a real breakthrough, but it still had to make some restrictive assumptions, among them that the excitation/ionization state of the material could be calculated using LTE. Kudritzki realized that with the existence of the CRAY 1A computer, it would be possible to do much better. Thus the Munich group undertook to make truly *self-consistent* models of radiatively driven winds, allowing for a complete description of both the nonLTE and dynamical coupling between radiation and matter. Thus departures from LTE affect line strengths, hence the ability of a line to scatter photons, and thus the radiation force on the material; whereas the atmospheric structure resulting from these forces (i.e., the velocity distribution and density and temperature structure of the flow), determine the (nonLTE) excitation and ionization state of the material. This is a very complex problem, but they found efficient methods of solution. In their work they used the comoving-frame method so that they could handle accurately the transitions from the static atmosphere, to subsonic flow, to trans-sonic flow, to extremely supersonic flow with a single formalism.

The calculation was carried out for thousands of lines in the UV spectrum. which provided a definitive estimate of the amount of line-blanketing by the expanding envelope on the underlying photosphere. A lot of the radiation intercepted by the envelope is simply reflected back into the photosphere (*reflector effect*), and forces the photosphere, in order to transport the correct flux, to have a steeper temperature gradient, and a higher emitted flux in the visible parts of the spectrum. This in turn implies that the visible spectrum of the star will mimic that of a star of higher effective temperature. In fact, Abbott and Hummer showed that for O stars (and *a fortiori* WR stars) the effective temperature one deduces for a given visible spectrum depends significantly on the mass-loss rate in the star's wind. This fact should be borne in mind both in model construction and in spectral classification.

The Munich models give excellent agreement with the observed spectra of WR stars in both the visible and ultraviolet. Detailed features of the observed line profiles are reproduced accurately; in particular the terminal velocity of the computed flow produces the shape of the violet-shifted component of the P-Cygni profiles quite well. This global agreement means that these models do indeed yield accurate values of $T_{\rm eff}$ and $\dot{M}$. Ultimately theories of stellar evolution of the upper main sequence should take this information into account. By any measure, this modelling effort represents a real milestone, and sets a new standard for work on the subject!

2.5. WHERE ARE WE GOING ('90s)?

Finally, I would like to make some comments about some work in progress, and some directions to explore in the future. Like everyone else I know, I do not know how to predict the future, so I will be brief.

2.6. Atomic Data

One persistent problem in astrophysics is lack of atomic data. We always have needed comprehensive sets of bound-bound transition probabilities and photoionization cross-sections. And to do a nonLTE analysis we need collisional excitation and ionization rates. Only very recently have adequate bodies of data been generated. The radiative data mentioned above are required both for detailed simulation of an observed spectrum and for the calculation of mean opacities to be used in stellar structure computations. The most comprehensive set of radiative data available for spectrum synthesis has been assembled by Robert Kurucz over many years. With it he can achieve a remarkable fit to the observed solar spectrum, often down to very faint features in blends. By a combination of some careful detective work and a good deal of heavy computation he was able to show recently that even the ultraviolet spectrum can be fitted very accurately once one accounts for the presence of many thousands of lines of the first few ions of iron, thus obviating the need for hypothetical continuum opacities from exotic mechanisms. No other stellar spectrum has been modeled with such care, thoroughness, and success; Kurucz's work should stand as a challenge to all of us.

Optical data suitable for opacity calculations has recently been produced by two independent multiyear efforts: the OPAL data generated by F. Rogers and C. Iglesias at Livermore, and the OP (Opacity Project) data generated by a large international team lead by M. J. Seaton in London. The OPAL results were published last year in the *Astrophysical Journal*, and the OP results are currently in press in the *Monthly Notices of the Royal Astronomical Society*. Neither dataset is suitable for precise spectral synthesis like that done by Kurucz because both use theoretical energy levels to compute line positions, which are therefore uncertain by 1% or more. Errors of that size would spoil the fit between computed and observed line profiles of overlapping lines, but essentially do not affect the accuracy of an opacity calculation. In both cases atomic structure calculations were made for all ions of all astrophysically abundant elements up through iron. More recently most of the iron-group elements, Mn, Cr, Ni, ... have been added. In the coming years, the OP atomic structure calculations (carefully archived) will be used to calculate collisional rates.

The OPAL and OP methodologies differ in detail, so it is very reassuring that the two sets of results agree extremely well. I believe that we can now say with confidence that stellar opacities are finally known to $\pm 10\%$ or at worst $\pm 15\%$, i.e., the accuracy to which these two independent calculations agree. Similar accuracies have long been attributed to the Los Alamos opacities, which have been the only comprehensive source of astrophysical opacities for many years. The new work shows that those earlier estimates were seriously over-optimistic: in fact, both calculations show that in certain ranges of temperature and density the Los Alamos results are in error (typically too small) by factors of three to five! The deficiency of the earlier work is mainly that too few lines were included, basically because the atomic structure of the atoms/ions was not worked out in sufficient detail.

Large changes in the opacity must have important implications for stellar structure and evolution. Certainly one should recompute the evolution of upper main-sequence stars, whose envelopes are essentially completely radiative. For the later types the envelopes are convective, and changes in opacity should not be too important. Opacity changes might be important in the core (at those densities the methods used by OPAL still yield numbers, but those used by OP break down). It is also possible that the new opacities might have noticeable effects on the solar oscillation spectrum. By far the most important effects found thus far have been in stellar pulsation. As first suggested by Norman Simon several years ago, an increase in the opacity in the temperature-density domain responsible for driving the pulsations can reconcile the masses found for double-mode Cepheids from pulsation theory on one hand and stellar evolution theory on the other. Remarkably, the opacity increases predicted by OPAL and OP are of just the right size to remove the discrepancy, thus validating Simon's conjecture. Certainly a great deal of work is needed to explore all the possible effects of the opacity changes.

As far as stellar atmospheres work is concerned, both OPAL and OP data can be used to construct reliable opacity distribution functions (ODFs) even though individual line positions are somewhat incorrect, because the essence of the ODF method is that the actual position of a line in a specified spectral band is unimportant. These data will thus be very valuable in computing more accurate theoretical colors, both broad-and intermediate-band, and spectral distributions compatible with typical spectrum scanners, for a wide range of stellar types.

2.7. Accelerated Lambda Iteration

As mentioned above, one of the severe limitations on the nonLTE spectrum studies of the '70s was that the computing time for realistic systems became prohibitive. In retrospect one can recognize that the original methods were all based on *direct* solutions of the coupled statistical equilibrium and transfer equations. And, as has been found in many other applications, it is often possible to find much cheaper *iterative* methods to solve the same problems. Just in the last few years a wealth of new iterative methods for the solution of transfer equations have been developed by L. Auer, R. Buchler, P. Kunasz, G. Olson, K. Werner, and others. These turn out to be *very* fast, and completely change our notions about how to solve transfer equations. In practical terms, problems which were unthinkable two decades ago are now trivial, not because of improvements in computers, but because of improvements in the algorithm. One now finds in the literature statistical equilibrium calculations allowing *hundreds* of spectral lines in a transition array. These give such a complete description of the spectrum, and the radiative exchanges among levels of the atom studied, that one expects, and gets, an excellent fit to the observed spectrum. Indeed, we are back to the position that lack of atomic data (e.g. collision rates) sets the limits on the accuracy we can achieve. Some of these methods lend themselves to massively parallel computers; the combined speedup of the algorithm and parallel computation will be phenomenal.

2.8. Dynamics

The spectra of most stellar atmospheres usually show some kind of hydrodynamical effects, often small-scale velocity distributions referred to as "microturbulence." Far more dramatic effects occur in pulsating stars like Cepheids and RR Lyrae stars, where we see line splitting, appearance and disappearance of line components, and emission features. As is well known, the pulsations are driven by a "thermodynamic engine" located in the $He^+ \leftrightarrow He^{++}$ ionization zone. As a result of the temperature-density behavior of the opacity, radiation is trapped, heats the material, and creates a significant overpressure in that zone. The trapped radiation drives an ionization front outward in the envelope. This front is a D-front, preceded by a shock. It eventually burns its way to the surface, at which point the star reaches maximum luminosity. Near maximum light the trapped radiation escapes freely, and the star finds itself overexpanded, and lacking pressure support. The star then begins to contract. The back and forth motion of the ionization front becomes strictly periodic, and drives the stellar pulsation. The primary shock preceding the ionization front, quickly runs out through the atmosphere, lofting overlying material to greater heights; during these phases we see line emission and line splitting. Later, during the contraction phase, when the ionization front is being buried deeper and deeper in the envelope, the material lofted by the primary shock falls back in on the contracting envelope and produces a secondary accretion shock of moderate strength.

Many calculations of these pulsation phenomena have been made by several investigators. They are certainly adequate to find periods as a function of mass, luminosity, and composition. On the other hand, they yield extremely poor light curves, and at best marginal velocity curves. A good example of the problems is provided by the Goddard Model, where several investigators, using indpendent codes, all computed pulsation models for an assigned mass and luminosity. The computed light curves have "random" discrepancies and fluctuations of the order of ± 0.4mag or $\pm 40\%$, and the curves computed with different codes often don't agree even qualitatively. By contrast the observations are typically accurate to ± 0.01mag or $\pm 1\%$! A "fit" to the data of such a poor calculation is illusory, and almost insulting to the observers! What is the problem? Put simply, the ionization front is very thin, perhaps only 0.001 scale heights wide. It is therefore very hard to resolve numerically with "standard" (Lagrangean) methods. The erratic behavior of the computed light curves can largely be attributed to artifacts of radiative transfer through the thick Lagrangean zones. Happily, powerful adaptive grid methods which allow one to resolve the ionization front (and the shocks) have been developed in the past few years, and Michael Gehmeyr and I plan to use them to caluclate more accurate Cepheid light curves.

From adaptive-grid models we will have no difficulty computing colors; we hope that these results will facilitate many different astrophysical investigations ranging from analyzing the physical structure of the envelope to setting the cosmic distance scale. Copious data pertaining to line strengths, splitting, and emission have existed for over 30 years thanks to the work of George Preston, Bodhan Paczynski, Robert Kraft, Helmut Abt, Bev Oke, and others. These data simply cry out for interpretation, but that has never been done because it requires the solution of *two* tough problems: the dynamics of the atmosphere, and full scale nonLTE effects at the densities appropriate to supergiants. This will be a large order, but I believe that we are now in a position to do it.

Finally, I would like to mention a fruitful interaction between a well established result of spectral classification and calculations of pulsating atmospheres. Back in 1947 Art Code, working under Morgan's guidance, published a short paper showing that all Cepheids at maximum light have the same spectral type; at minimum light they have progressively later spectral types with increasing period. These remarkable correlations have not been interpreted for over 45 years! But this year, using a relatively primitive adaptive-grid code, the "T-grid code"(originally developed by John Castor), Norman Simon, Shashi Kanbur and I were able to match Code's empirical results. We found that maximum light always occurs just as the ionization zone burns out of the star, and the very precisely defined temperature and density of the front locks the spectrum and colors to a distinctive value, independent of pulsation period. The same calculations explain the observations at minimum light, but the details are a bit more complicated than can be addressed here. My guess is that in relatively short time we should be producing reliable light curves.

3. Conclusion & A Caveat

In this talk I have been able to touch upon only a few of the many threads composing a complex fabric. I hope that they are adequate to show that the theory of stellar atmospheres is making significant progress. Yet I want also to offer a caveat: virtually all of our modeling assumes that the atmosphere is *smooth*, i.e., is stratified into smooth planar or spherical layers. One argues (or believes) that even if there is structure, the derived models might reproduce reasonable average values of the fluctuating physical quantities in each layer. Yet when one looks at high-resolution photographs of the Sun, such as taken by R. B. Dunn at Sacramento Peak Observatory, and by Alan Title and his colleagues at La Palma and in space, one realizes that the belief mentioned above is probably only a pious hope. Such photographs show an *immense* amount of structure on all scales in the photosphere, chromosphere, and corona. And the radiative processes taking place (transfer, statistical equilibrium) are inherently *nonlinear*. A linear average is probably not possible, and even when possible, not useful. For example, the average longitudinal magnetic field of the Sun is only about a gauss, but that is the result of near cancellation of $\pm 1000G$ fields which are concentrated into slender flux tubes only 75 - 100 km in diameter, much below the usual limit of ground-based resolution. The average that we see is a thousand times smaller than the fluctuation around that average, and thus hardly tells the whole story!

Much of the structure and many of the phenomena we see on the Sun don't affect the photospheric spectrum much because they are optically thin (corona), occupy only a very small filling factor (flux tubes), or are most easily seen at wavelengths outside the visible range. But it is not hard to imagine that in other stars many of these structures could be optically thick, and/or have a much larger filling factor. After all, the solar wind transports only about 10^{-14} solar masses per year, and we know many stars that have mass fluxes that are 10^8 or 10^9 larger. The sun's wind is optically thin, and produces no appreciable effect on the observed spectrum. In contrast, the radiation from the more massive winds in early type stars *is* the spectrum. The point I am trying to make is that we must remain alert about oversimplifying our models, and that we must keep an open mind and remain sensitive to nagging "little" discrepancies between observed and computed spectra, which may be indicators of major structural differences between real stars and our models.

Finally, I would like to make a philosophical remark. I view intellectual activity as being organized into a kind of hierarchy. The metaphor I have in mind is that of climbing up the side of a big mountain. The broadest layer at the bottom I think of as "*information*"; it tells us "what, when, where,. . ." about the objects of our study. Here we are caught in a veritable rain of bits pouring out of our computers and CCDs. If we push upward we reach a second, less broad, layer I think of as "*knowledge*", where we begin to understand "how, why" our specimens work, and behave the way we observe. Next we reach a narrow layer called "wisdom", often draped in clouds, where we learn the "*but(s)*" that show us the limitations of our knowledge. And if we push on a little farther, through the clouds, we arrive at the peak, where in friendly, warm sunshine we finally achieve "insight", the "ahah!" experience, from which we not only gain deeper understanding about the subject of our study but also see how to generalize our ideas and how to apply them in other contexts. Climbing this mountain is hard work! But it is very important for the future of our fields that we constantly concentrate on making our way upward, and that we constantly remain aware of where we *are*, so that we don't confuse mere information with insight. One of the nice things about working in these fields is that the spectra themselves provide the ultimate standard of reality, honesty, and accuracy. So my advice is to stick as close to the spectra as possible, and let the theory tag along as best it can.

Acknowledgments. I have been taught a great deal about both stellar atmospheres and stellar classification by a large number of astronomers; too many to acknowledge here. But it would be unfair if I did not at least thank W.W. Morgan for helping me learn (what little I could!) about stellar classification, and Nolan Walborn for stimulating and fruitful discussions. On the theoretical side I'm happy to thank Larry Auer both for what he taught me, and for the good time we had doing some of the research described above. Some of the work described herein has been supported in part by unrestricted research funds made available by the University of Illinois, for which I am most grateful.

Discussion

Garrison: I have a comment about the Code diagram you showed (1947, cepheid classifications around the light cycle). This is one of those straight-forward pieces of work which contains the potential for great insight, but which is ignored for many years. In the MK Process volume (1984), there is a paper by Gauthier, based on his Ph.D. thesis at Toronto, which strengthens and reinforces the original diagram, but with more data.

Weaver: Our attempts to do detailed NLTE calculations a couple of years ago were thwarted by a severe lack of collisional data. Is there an effort comparable to the opacity project to determine collisional data?

Mihalas: Collisional data have indeed been badly lacking for a long time. Our collective progress has been seriously impeded by this lack. Happily help is on the way. A number of members of the original OP team have undertaken a huge new project to calculate collisional cross sections and rates using existing OP wave functions. Results will be put in a form convenient for users (functional fits to collision rates). The timescale for the project is between five and ten years.

The MK Process at 50 Years
ASP Conference Series, Vol. 60, 1994
C.J. Corbally, R.O. Gray, and R.F. Garrison (eds.)

The MK Process and OB$^+$ Stars

John S. Drilling

Department of Physics and Astronomy, Louisiana State University, Baton Rouge, LA 70803

Abstract. The MK process has been applied to stars classified as OB$^+$ in the Case-Hamburg-LSU objective-prism surveys. The result has been the discovery of more than 400 new O and B supergiants plus a number of astrophysically important peculiar stars. The latter include most of the known extreme helium stars and 16 of the hottest known sdO stars. The calibration of physical parameters for these stars is discussed.

1. Introduction

The OB$^+$ stars are stars which show continuous or nearly continuous spectra at a resolution of 10 Å. The most complete survey for OB$^+$ stars made to date has been that of the Case and Hamburg Observatories (see Stephenson and Sanduleak 1971). This survey covers the entire Milky Way and contains 998 OB$^+$ stars which did not show Balmer emission lines. Because those objects at the highest galactic latitudes tended to be the most interesting astrophysically, these surveys were extended by Drilling (1987) to b = ±30° for l= ±60° using the Curtis Schmidt telescope and 4° objective prism at Cerro Tololo. Together, these surveys cover 28% of the sky to photographic magnitude 12.

2. Normal OB$^+$ Stars and Galactic Structure

Most OB$^+$ stars are normal O-type stars or early B-type supergiants. These stars are of interest in the study of galactic structure because it is these stars, and their associated HII regions, which illuminate the spiral arms in spiral galaxies. This was, in fact, the reason that the Case-Hamburg surveys were carried out. When these surveys were completed, I began a project to obtain photoelectric *UBV* photometry and accurate MK spectral types for all non-emission OB$^+$ stars which lacked such data, so that accurate distances could be determined and used to study the space distribution of these stars, associated HII regions, and intervening interstellar dust. *UBV* photometry has now been published for 786 stars (Drilling 1991), and accurate MK spectral types have been determined for 446 stars from spectrograms obtained with the CTIO 0.9-m spectrograph (Drilling and Perry 1981), the CTIO 1.0-m image-tube spectrograph (Vijapurkar and Drilling 1993), and the KPNO 0.9-meter image-tube spectrograph (Drilling 1993).

3. Peculiar (Astrophysically Interesting) OB$^+$ Stars

The OB$^+$ stars are of astrophysical interest because they include the hottest subdwarf O stars, hot hydrogen-deficient stars, some dwarf novae at maximum light, VV Cephei stars with luminous O- and B-type components, FG Sagittae, SN 1987A, and nearly all of the optically identified X-ray sources which lie in the Milky Way. The rest of my talk will concern the calibration of physical parameters for extreme helium stars and very hot subdwarf O stars.

The extreme helium stars are early-type giants which have atmospheric H abundances of around one per 10,000 atoms, and are of interest because here the material of the nuclear-burning core is exposed to direct observation without contamination by the original envelope material. Four of the 21 known extreme helium stars are N-rich, and are well understood to be the result of type BB binary evolution (Plavec 1973; Schönberner and Drilling 1983). The other 17 are C-rich and show no evidence of binary nature (Drilling 1986; Jeffery, Drilling, and Heber 1987). Thirteen of the 21 known extreme helium stars were discovered as a result of the MK classification described in the last section.

The subdwarf O stars occupy a position in the HR diagram which is intermediate between normal main-sequence O stars and the DO white dwarfs, and are thought to be the immediate progenitors of the latter. The 22 OB$^+$ stars which have been classified as subdwarf O represent some of the hottest and/or most helium-rich subdwarf O stars known, and 16 of these were discovered as a result of the MK classification described in the last section.

3.1. Effective Temperatures of Extreme Helium Stars from Continuum Measurements

The effective temperatures of 14 extreme helium stars were determined by Drilling *et al.* (1984) from *UBVRIJHKL* photometry and low-resolution IUE spectrophotometry. The observed fluxes were corrected for interstellar reddening using the standard reddening law of Seaton (1979) and the strength of the 2200 Å feature. The effective temperatures were then determined using the method described by Underhill (1982) and Böhm-Vitense (1981), where H-poor, LTE model atmospheres were used to determine the angular diameters of the stars and the contribution to the flux from outside the observed wavelength range. The resulting effective temperatures ranged from 8,400 to 32,000 K.

3.2. Effective Temperatures of Very Hot sdO Stars from Continuum Measurements

The effective temperatures of 12 very hot subdwarf O stars were determined from *UBV* photometry and IUE spectrophotometry by Schönberner and Drilling (1984). The observed fluxes were corrected for interstellar reddening as described above, and also by assuming that $(B-V)_o$ = -0.35. The probable error in E_{B-V} using either method is estimated to be $\pm$ 0.03. The effective temperatures were then determined by fitting pure H/He LTE model atmospheres to the de-reddened fluxes. The resulting effective temperatures were found to be relatively insensitive to surface gravity, but tended to be about 25% higher for a given star if the composition was assumed to be helium-rich rather than normal or helium-poor, a choice which could be made using the classification spectra described above. The resulting effective temperatures ranged from 59,000 to 110,000 K. *UBV* photometry and low resolution IUE spectrophotometry have now been obtained for a total of 32 very hot subdwarf O stars, including 11 of the hottest objects from the Beers-Preston-Shectman high-latitude surveys (Beers *et al.* 1992). These surveys were also done with the Curtis Schmidt telescope and 4° objective prism at Cerro Tololo, but made use of an interference filter which isolates the region around the CaII H and K lines and allows OB^+ stars as faint as B=16 to be detected.

3.3. The Fine Analysis of High-Resolution Spectra of Very Hot sdO Stars and Extreme Helium Stars

High-resolution ESO-Caspec (3950 - 4950 Å; 0.25 Å resolution) and IUE (1150 - 2100 Å; 0.1 Å resolution) spectra of the Case-Hamburg-LSU sample of very hot sdO stars have been obtained in collaboration with various workers who at some point in their careers were associated with K. Hunger or U. Heber at the University of Kiel (with whom this collaboration was begun). The results of NLTE fine analyses of these spectra which have been completed to date are summarized in Table 1. In addition, the results of LTE fine analyses of six extreme helium stars are available in the literature (Schönberner and Wolf 1974; Schönberner 1978; Kaufmann and Schönberner 1977; Walker and Schönberner 1981; Hamann, Schönberner, and Heber 1982; Heber 1983), and the effective temperatures determined from the fine analysis of line spectra for all of these stars are plotted against the continuum temperatures described above in Figure 1. The agreement is quite good when one stops to consider that the strongest lines of carbon and nitrogen in the very hot subdwarf O stars arise from highly excited levels which are close to degeneracy, requiring very elaborate model atoms to be used in the NLTE calculations and rendering the treatment of line broadening difficult (Heber 1992). Also, elements heavier than H and He were neglected entirely in determining effective temperatures from the continua of the very hot sdO stars, and these results are effected by interstellar reddening and possible errors in the IUE calibration.

3.4. Birthrates for the Immediate Progenitors of White Dwarfs

Drilling (1992) has determined absolute magnitudes and distances for the stars in the Case-Hamburg-LSU sample of very hot sdO stars from either the known color excesses (using the run of color excess with distance for stars nearby in the sky) or from the surface gravities and bolometric corrections from the fine analyses (using masses from stellar evolution theory). These absolute magnitudes and effective temperatures derived as described above show that these stars occupy a region in the HR diagram bounded by the central stars of planetary nebulae, other sdO stars, and the hottest white dwarfs. Because these stars, when added to the the central stars of planetary nebulae detected in surveys for planetary nebulae, form a complete sample (with respect to apparent magnitude) of the hottest stars in the Galaxy, and because all evolutionary tracks leading from the main sequence to the white dwarfs funnel through the region of the HR diagram defined by these stars ($T_{\rm eff}$ > 50,000 K), one can use them to determine the relative birthrates of the various types of white dwarf progenitors. The results are given in Table 2. The first birthrate is that given by Phillips (1989) for

Table 1. Results of NLTE fine analyses of very hot subdwarf O stars.

Star	Ref.	T_{eff}	log(g)	n_H/n_{H+He}	Remarks
LSS 1362	1	100,000	5.30	0.90	C-poor
LSE 153	2	70,000	4.75	≤0.10	C-rich
LSE 125	3	78,000	5.00	0.95	
LSIV-12°1	4	60,000	4.50	0.90	C-poor
LSE 259	2	75,000	4.40	≤0.05	C-rich
LSE 263	2	70,000	4.90	≤0.10	C-poor
LSIV+10°9	5	44,000	5.60	≤0.10	C-rich

Heber, Werner, and Drilling 1988
Husfeld, Butler, Heber, and Drilling 1989
Mendez *et al.* 1988
Heber and Hunger 1987
Dreizler 1992

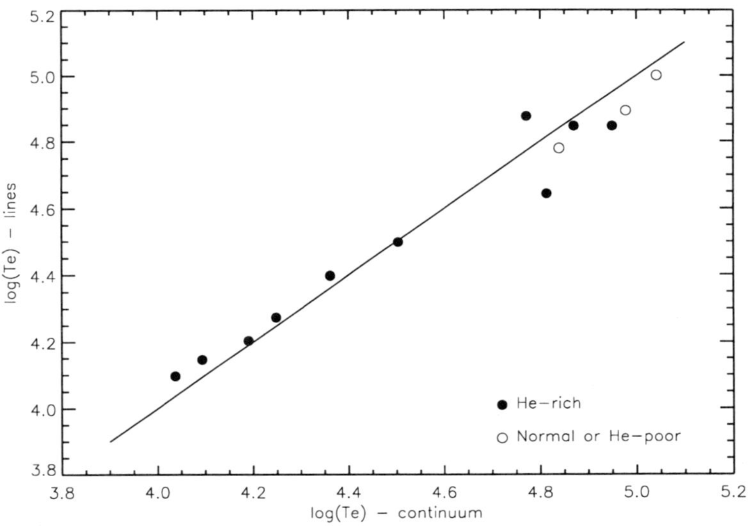

Figure 1. Effective temperatures determined from the fine analysis of high resolution line spectra versus those determined from continuum observations.

the central stars of planetary nebulae detected in surveys. The second is for the stars in our sample which lie on the post-asymptotic giant branch tracks of Schönberner (1981) and Wood and Faulkner (1986) in the HR diagram, but do not show planetary nebulae bright enough to be detected in surveys for planetary nebulae. The third is for stars which lie on the post-extended horizontal branch tracks of Paczynski (1971) and Caloi (1989). In all cases, the space densities were determined from the known distances of the stars, and the times that it takes the stars to cross the region of the HR diagram defined by them from the evolutionary calculations. We conclude that all but a negligible fraction of the white dwarfs are descended from post-AGB stars, but that 20% of these stars did not produce planetary nebulae bright enough to be detected in surveys for planetary nebulae.

Table 2. Birthrates ($pc^{-3}yr^{-1}$) for white dwarf progenitors.

Central Stars of Known PN	230×10^{-14}	Phillips 1989
Other Post-AGB	46×10^{-14}	11 stars
Post-EHB	1×10^{-14}	10 stars

Acknowledgments. This work was supported in part by grants from the National Aeronautics and Space Administration (NAG5-71) and the National Science Foundation (AST 8514574).

References

Beers, T.C., Preston, G.W., Shectman, S.A., Doinidis, S.P., and Griffin, K.E. 1992, AJ, **103**, 267

Böhm-Vitense, E. 1981, ARA&A, **19**, 295

Caloi, V. (1989), A&A, **221**, 27

Dreizler, S. 1992, in *The Atmospheres of Early-Type Stars*, U. Heber and C.S. Jeffery, Berlin: Springer-Verlag, 270

Drilling, J.S. 1986, in *Hydrogen-Deficient Stars and Related Objects*, K. Hunger, D. Schönberner, N.K. Rao, Dordrecht: Reidel, 9

Drilling, J.S. 1987, in *The Second Conference on Faint Blue Stars*, A.G Davis Philip, D.S. Hayes, and J.W. Liebert, Schenectady: L. Davis Press, 489

Drilling, J.S. 1991, ApJS, **76**, 1033

Drilling, J.S. 1992, in *The Atmospheres of Early-Type Stars*, U. Heber and C.S. Jeffery, Berlin: Springer-Verlag, 257

Drilling, J.S. 1993, in preparation

Drilling, J.S., and Perry C.L. 1981, A&A*S*, **45**, 439

Drilling, J.S., Schönberner, D., Heber, U., and Lynas-Gray, A.E. 1984, ApJ, **278**, 224

Hamann, W.-R., Schönberner, D., and Heber, U. 1982, A&A, **116**, 273

Heber, U. 1983, A&A, **118**, 39

Heber, U. 1992, in *The Atmospheres of Early-Type Stars*, U. Heber and C.S. Jeffery, Berlin: Springer-Verlag, 233

Heber, U., and Hunger, K. 1987, in *The Second Conference on Faint Blue Stars*, A.G Davis Philip, D.S. Hayes, and J.W. Liebert, Schenectady: L. Davis Press, 599

Heber, U., Werner, K., and Drilling, J.S. 1988, A&A, **194**, 223

Husfeld, D., Butler, K., Heber, U., and Drilling, J.S. 1989, A&A, **222**, 150

Jeffery, C.S., Drilling, J.S., and Heber, U. 1987, MNRAS, **226**, 317.

Kaufmann, J.P., and Schönberner, D., 1977, A&A, **57**, 169

Mendez, R.H., Kudritzki, R.-P., Herrero, A., Husfeld, D., and Groth, H.G. 1988, A&A, **190**, 113

Phillips, P. 1989, in *Planetary Nebulae*, S. Torres-Peimbert, Dordrecht: Kluwer, 425

Plavec, M.J. 1973, in *Extended Atmospheres*, A.H. Batten, Dordrecht: Reidel, 216

Paczynski, B. 1971, *Acta Astr.*, **21**, 1

Schönberner, D. 1978, *Mitt. A.G.* **43**, 266

Schönberner, D. 1981, A&A, **103**, 119

Schönberner, D., and Drilling, J.S. 1983, ApJ, **268**, 225

Schönberner, D., and Drilling, J.S. 1984, ApJ, **278**, 702

Schönberner, D., and Wolf, R.E.A. 1974, A&A, **37**, 87

Seaton, M. J. 1979, MNRAS, **187**, 73P

Stephenson, C.B., and Sanduleak, N. 1971, *Publ. Warner and Swasey Obs.* **1**, 1

Underhill, A.B. 1982, ApJ, **263**, 741

Vijapurkar, J., and Drilling, J.S. 1993, ApJS, **89**, 293

Walker, H.J., and Schönberner, D. 1981, A&A, **97**, 291

Wood, P.R., and Faulkner, D.J. 1986, ApJ, **307**, 659

Discussion

Liebert: There has been a need to improve upon the old, rudimentary definitions of hot subdwarfs used traditionally by Greenstein and others. Do you approve the system defined at Kiel about two years ago?

Drilling: These stars are the hottest of the sdO stars analyzed by the Kiel group. Due to the large differences in the effective temperatures, surface gravities and abundances of this small sample of stars, I think that it would be premature to try to further subdivide the sdO classification at the present time.

Walborn: A comment. The classification of hot subluminous stars is not two-dimensional, because discontinuous evolutionary stages are represented. However there are very well defined categories of peculiar spectra among them with substantial memberships in each.

Sion: Do I understand that you have found a larger percentage of white dwarfs that are descended from sdO stars than the few percent you estimated with Detlev Schönberner?

Drilling: The more recent paper is based on a sample of 21 stars for which IUE spectrophotometry has been obtained, as compared to twelve in the original work, plus the NLTE fine analyses completed so far. The conclusions are similar: although only a few percent of the white dwarfs have evolved directly from the extended horizontal branch, one in six have evolved from post-AGB stars which did not evolve fast enough to illuminate planetary nebulae which would have been detected in existing surveys for planetary nebulae.

The MK Process at 50 Years
ASP Conference Series, Vol. 60, 1994
C.J. Corbally, R.O. Gray, and R.F. Garrison (eds.)

The Spectroscopic Classification of White Dwarfs: Unique Requirements and Challenges

James Liebert

Steward Observatory, University of Arizona, Tucson, AZ 85721

Edward M. Sion

Department of Astronomy and Astrophysics, Villanova University, Villanova, PA 19085

Abstract. We present a historical review of the development and evolution of spectroscopic classification schemes for white dwarf stars and suggest changes to the presently used system, engendered by the need for refined temperature classification of the coolest and hottest degenerate stars, and by the need for anchoring the present classification to model-independent spectrophotometric standard stars.

1. Historical Introduction

The spectroscopic classification of white dwarf stars had its birth 83 years ago when 40 Eridani B was described as being "A-type" by Willamena Paton Fleming and Edward C. Pickering, while all other known low luminosity stars were class G or later! Professor Pickering, with singular prescience, reportedly mused "it is just such discrepancies which lead to the increase of our knowledge". Henceforth, spectroscopy of white dwarfs required an exceptional classification to accomodate spectroscopic "discrepancies" out of keeping with normal stars. White dwarfs were never incorporated into the MK system. At the time the system was first defined, relatively few white dwarfs were known and most were so faint that good quality spectra could only be obtained at lower than the MK resolution and/or on the very largest of telescopes. But most importantly, it became clear that their spectra are fundamentally unlike those of the normal MK types, because they lack a solar-like relative distribution of elemental abundances.

Kuiper (1941) appears to have been the first investigator systematically to attempt classifying white dwarf spectra. He introduced several principal spectral types. He used a lower case "w" preceding the spectral type to designate a white dwarf (just as the Mt. Wilson system uses "d" for dwarf, "g" for giant, etc.). The types then were (1) A, for those showing hydrogen lines, (2) F, when Ca II lines are the dominant or only feature, (3) G, when weaker ultraviolet metallic blends more resembling a G-type spectrum are also present with Ca II, and (4) "Con" for a continuous, featureless spectrum. No white dwarfs showing helium lines had been discovered at the time of Kuiper's work. Kuiper's only attempts at adding subtypes were to insert a "5" after A or F if the principal lines were particularly strong, or an "s" if the lines appeared sharp.

Recognizing that these objects are the ultimate in dwarf stars, Luyten (1952) substituted the capital "D" for Kuiper's "w". He added the "B" type for those showing only lines of He I. But his principal innovation was an attempt to add a temperature subtype based on a combination of the width of the lines and the approximate color. He recognized that the hottest DA stars showed broader and generally stronger hydrogen lines than the coolest ones. However, Luyten also used the qualifying letters "s" and "n" for sharp and nebulous lines, respectively. Any current investigator familiar with the approximate temperatures of these now well-studied stars will realize that Luyten's numerical subtypes do not correlate that well with the stellar temperature. For example, L870–2 (WD0135–053) is nearly the coolest of the DA stars in his sample, but he assigned the number "4", intermediate between the extremes of "0" and "7". Also, there was little attempt to use the subtypes for non–DA spectra.

Jesse Greenstein's avid interest in this subject area insured that the largest telescope in the world during the 1950s – the Palomar 200-inch – would be used extensively to sort out the intracacies of white dwarf spectra, certainly to find and study many more of these objects. In the "Stars and Stellar Systems" volume on stellar atmospheres, Greenstein (1960) gave an extensive review of the types of white dwarf spectra, including detailed line profile measurements. He dropped the Luyten temperature subclass, preferring to use a color index or direct photometric measurement to estimate this physical quantity. Otherwise, the system used by Greenstein was similar to that of Luyten. Additional spectral types were introduced for stars showing He II (type "DO"), and for stars showing apparent molecular bands such as the Swan system of C_2. This system remained in general use during the 1960s and 1970s, when Greenstein, Eggen, Green and many other investigators added hundreds of new spectroscopically-identified white dwarfs to the known sample. The enormous increase of the sample size led to the recognition of new spectroscopic subtypes, the enlargement of the sample size in already existing but sparsely populated subgroups, and broader ranges of effective temperature bounding each subgroup. These developments mandated a classsification system which could better embrace the increased spectroscopic complexity and diversity as explained below.

2. The Physical Basis of Modern Spectral Classification

The very nature and compactness of degenerate stars dictates spectral classification requirements and criteria outside the mainstream of MK classification philosophy and techniques traditionally applied to lower gravity, non-degenerate stars. There are two principal reasons for this: (1) their spectra, which are primarily mono-elemental, directly manifest the action of gravitational diffusive separation (the lightest element floating at the surface), convection, possibly accretion and mass loss. The presence of spectral features other than those of the dominant atmospheric element are the exception rather than the rule. When present however, these secondary features betray a process which opposes the otherwise inexorable operation of gravitational diffusion to create a pristine, mono-elemental spectrum. These same-envelope physical processes can radically alter their dominant surface compositions thus transforming their spectra along the cooling sequence so that criteria applied to normal stars such as absorption line ratios as indicators of temperature along a well-defined sequence, are useless; (2) Since white dwarfs undergo their slow loss of thermal energy (their cooling evolution) at essentially constant radius, the color of a white dwarf indicates both its temperature and its luminosity (and hence the age of the white dwarf since its formation). Hence temperature information gleaned from a color index and/or spectrum carries profound evolutionary significance.

During the decade of the 1970's and beyond, the advent of ever more advanced detector technology, the gradual disappearence of photographic spectra, greatly improved spectral resolution and higher quality spectra for increasingly fainter objects, led to the concomitant recognition of ever more varied hybrid-composition spectral groups. Two factors greatly hindered efforts to gain astrophysical understanding of the origin and evolution of the spectroscopic subgroups: (a) the complex interplay of accretion, convective dilution and mixing, mass loss, accretion, magnetism, and spectral evolution of one type to another; and (b) the lack of clear spectroscopic temperature distinctions between hydrogen-dominated stars across a broad temperature range (5,000 K to 100,000 K!) and among helium-dominated degenerates as well. This confusing myriad of spectroscopic subgroups, and the unique importance of temperature information for a white dwarf engendered among the community of researchers, a new mandate for the spectral classification of white dwarfs: a classification scheme which incorporates quantitative temperature information coupled with hybrid types designated by combinations of symbols for primary and secondary chemical constituents, the net result being a more physical spectroscopic classification system which can actually help to elucidate the very physical processes in white dwarfs which produced their spectra.

3. The Classification System for the 1990'S

At the beginning of the "CCD era," Sion *et al.* (1983) – representing a true consortium of the community doing research on white dwarf stars – redefined the classification system to use high quality digital spectrophotometry. The principal additional features are (1) widespread use of secondary spectral types for the trace elements showing up with increasing frequency in high S/N spectra, and (2) a revised temperature index. The latter was initially defined by a spectrophotometric slope, using approximate temperatures estimated from model atmospheres. The original theta in a model atmospheres analysis is defined as $5040/T_{\rm eff}$. Here we use ten times that number or "theta" $= 50400/T_{\rm eff}$. Thus a DA4 star has a temperature near 13,000 K. In order to have a set of principal spectral types all using single letters, the λ4670 stars showing carbon features (especially a band at 4670 Å) are now called DQ, the magnetic stars DP if polarized or DH if showing magnetic Zeeman features in the spectrum, or DXP if magnetic and detectably polarized, with unknown features. The system of primary and secondary spectroscopic symbols is summarized in Table 1.

All of the above primary composition symbols may be assigned to denote secondary or weaker features seen in any part of the electromagnetic spectrum,

Table 1. Primary Spectroscopic Symbols

DO – He II lines are the dominant features in the optical spectrum

DA – Hydrogen (Balmer) lines are the dominant features in the optical spectrum

DB – He I lines are the dominant features in the optical spectrum

DC – Continuous optical spectrum; no lines deeper than 5% of the continuum

DZ – Metal lines (here any element heavier than helium) are the strongest absorption features in the optical spectrum (the only exception is carbon which has its own symbol).
NOTE: This class implies no restriction to cool degenerates.

DQ – Carbon lines, either atomic or molecular, are the dominant features in the optical spectrum.

Other Symbols:

P – magnetic white dwarfs with detectable polarization

H – magnetic white dwarfs without detectable polarization

X – peculiar or unclassifiable spectrum

E – emission lines are present

? – uncertain assigned classification; a colon (:) may also be used

V – optional symbol to denote variability

as illustrated in Table 3 of Sion *et al.* (1983) and the atlas of Wesemael *et al.* (1993).

During the past decade it has become apparent that the system of white dwarf spectral classification could be further refined in step with the increasing accuracy of derived atmospheric parameters. The challenges and need for revisions to the Sion *et al.* (1983) classification scheme was the theme of a special session of IAU Commission No. 45 (Stellar Classification) on White Dwarf Spectral Classification chaired by one us (EMS) at the XXIst General Assembly of the IAU in Buenos Aires. However the unfortunate disruption of the scheduled sessions by a fire forced a deferral of any formal discussions. It is therefore particularly appropriate to discuss suggested revisions to the Sion *et al.* (1983) scheme at this Vatican Conference celebrating the 50th Anniversary of the MK System. These refinements will be implemented in the upcoming fifth edition of the Catalog of Spectroscopically Identified White Dwarfs. In the remaining sections, these refinements are discussed.

4. The Use of Spectrophotometric Standards

In Figure 1 we show sets of (left) DA spectra with H lines and (right) DB spectra with He I ordered by temperature class, taken from the new spectrophotometric atlas of Wesemael *et al.* (1993). The hottest DA star at the top left exceeds 50,000 K in its estimated $T_{\rm eff}$ and shows metallic features at ultraviolet wavelengths; hence the additional subtype of "Z" for metals. The temperature of such a white dwarf may be estimated from either the slope of the energy distribution or from line profiles. The set of stellar spectra shown here may permit these stars to be used as standards for simple visual comparison of fluxed spectrophotometry of any star one observes. The subtype thus does not depend on the accuracy of model atmospheres, or on results from any particular analysis. Thus, the DA4 star (second from bottom) is fixed in type; we do not change its type to DA5 if a better estimate of the temperature turns out to be 10,000 K.

5. An Expanded, Refined Temperature Index

The limitations of the temperature index in the Sion *et al.* (1983) system become obvious at the lowest ($T_{\rm eff}$ 5040 K) and highest surface temperatures ($T_{\rm eff}$ 50400 K) when it no longer provides the needed indicial temperature discrimination. Sion *et al.* (1983) suggested the extension to a double digit temperature index for the coolest degenerates and the double digit index was implemented by Greenstein (1986) and subsequently by others. For the hottest white dwarfs the assignment of "zero" was proposed by Sion *et al.* (1983) but never implemented principally because of the inevitable confusion with the letter O. This difficulty is exacerbated by the nearly imperceptible difference of these characters in widely used computer operating systems (a computer zero with a "slash" is no longer ubiquitous) but secondarily because a zero temperature index is simply undefined and unphysical. A viable solution is to assign the hottest white dwarfs (stars with $T_{\rm eff}$ 50,400 K) a temperature index of decimal 9 to decimal 1 (.9 to .1). Note that for $T_{\rm eff}$ = 200,000 K, a value higher than any presently known object at log g $\geq$ 7, the index would be .25. Note that a leading zero before the decimal point is specifically prohibited to avoid confusion with the hot primary spectral type of O.

With the increasing accuracy of derived white dwarf temperatures (and gravities) using ever more sophisticated LTE and NLTE synthetic spectra codes with improved atomic physics, a number of advantageous refinements to the temperature index suggest themselves. The use of a decimal half-integer temperature index allows for refined temperature classification. Thus, for example the sequence of DA stars would extend from DA.5, DA1, DA1.5, DA2, DA2.5.... DA13 star. A DA2 star would have a temperature in the range 22,400 K to 28,800 K while a A2.5 would have $T_{\rm eff}$ in the range 18,327 K to 22,400 K. The clear advantage of a decimal half-integer over the use of the plus symbol ("+") is that further subdivision is possible as temperatures become even more accurate. The same decimal subdivion is also possible for the hottest white dwarfs; for example an extremely hot star with $T_{\rm eff}$ = 175,000 K and showing lines of oxygen (the strongest features), carbon and HE II becomes DZQO.29.

6. An Optional Gravity Indictor

Due to the increasing accuracy of gravity determinations by fitting the all of the Balmer lines using the techniques of Bergeron, Saffer and Liebert (1992), it is possible to reliably identify the relatively rare white dwarfs with masses above 1.0 $M_\odot$ up to the Chandrasekhar Limit and below the minimum mass for

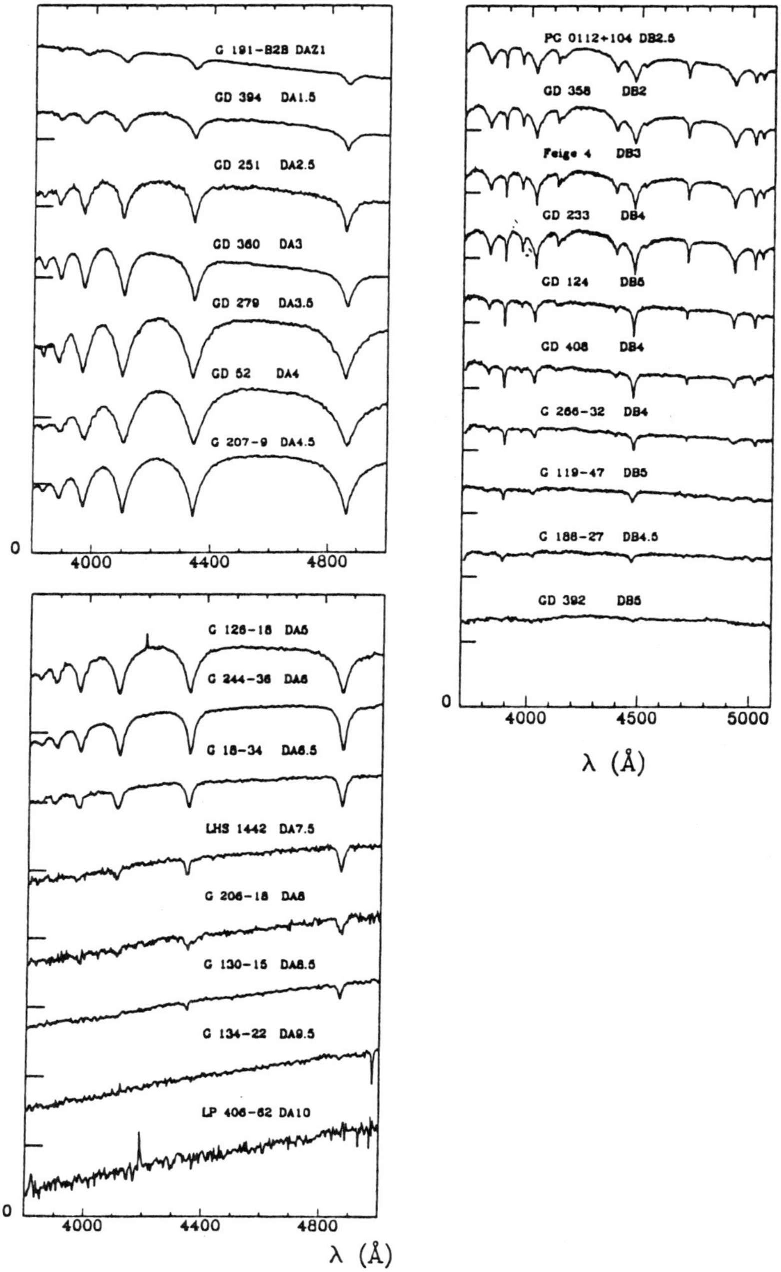

Figure 1. DA and DB $T_{\rm eff}$ sequences as described in the text.

helium burning. These may correspond to gravities (log g) above 9 or below 7, comparable to the differences between giants and supergiants, etc. Since the spectral features at these two extremes appear very diffuse or broad at high gravity and very narrow or sharp at low gravity, the symbols "d" and "n" (always taken as lower case), originally used to designate the appearence of lines on photographic spectra, could be assigned to objects of exceptional gravity. For example, the massive DA white dwarf GD50 (Bergeron, Saffer and Liebert 1992) could be classified DA1d.

7. Conclusions

The spectral classification of white dwarfs has traditionally been based upon what the optical spectrum actually shows. However the chemical complexity and the multifarious physical processes which oppose the simple monoelemental spectrum resulting from gravitational diffusive separation and the evolutionary significance of white dwarf temperature information, speak to the need for a more sophisticated, quantitative classification. We have introduced suggested modifications and refinements to the white dwarf spectral classification scheme presented by Sion *et al.* (1983). These suggestions include adoption of a decimal half-integer temperature index from 1 to 13 for white dwarfs with $T_{\rm eff} \leq 50,400$ K and decimal 9 to decimal 1 for stars hotter than 50,400 K (defined by $10\times\theta_{\rm eff} = 50{,}400/T_{\rm eff}$), the rejection of zero as a temperature index, a gravity indicator for white dwarfs with exceptionally low and exceptionally high gravity, and the use of white dwarf spectrophotometric standards which would be model-independent.

Acknowledgments. This paper is dedicated to the Memory of the late Rev. Edward Felix Jenkins, O.S.A., Ph.D., Founder and First Chairman of the Department of Astronomy and Astrophysics at Villanova University.

We are grateful to Harry Shipman, Francois Wesemael, Pierre Bergeron and Gordon Hammond for useful discussions. We thank numerous colleagues who participated in discussions of classification issues at the 182nd meeting of the American Astronomical Society in Tucson. This work was supported by NSF grant AST90-16283 (EMS) and AST92-17961 (JL).

References

Bergeron, P., Saffer, R., and Liebert, J. 1992, ApJ, **394**, 228

Greenstein, J. L. 1960, in *Stars and Stellar Systems, Stellar Atmospheres*, ed. J. L. Greenstein (Chicago: University of Chicago Press), Vol. **6**, p. 676

Greenstein, J. L. 1986, ApJ, **304**, 334

Kuiper, G. P. 1941, PASP, **53**, 248

Liebert, J. 1980, *Ann.Rev.Astr.Ap.*, **18**, 363

Luyten, W. J. 1952, ApJ, **116**, 283

Sion, E. M., Greenstein, J. L., Landstreet, J., Liebert, J., Shipman, H. L., and Wegner, G. A. 1983, ApJ, **269**, 253

Wesemael, F., Greenstein, J. L., Liebert, J., LaMontagne, R., Fontaine, G., Bergeron, P., and Glaspey, J. W. 1993, PASP, **105**, 761

Discussion

Garrison: There is an enormous advantage to keeping the classification of the appearance of the spectrum autonomous and independent of the calibration; i.e, there is important information in the interface, which you throw away if you use the interpretation to influence the classification.

If, as you say, you can *see* the differences in the spectra, then I would urge you to rank the spectra accordingly and *then* calibrate. The purity of the MK Process will be rewarding by providing you with new insight into the astrophysical processes.

The spectral features are *real*; the interpretation is transitory. Keep them separate.

Sion: We have tried as much as possible to base the system on what is already seen in the spectrum, but the enormous range in $T_{\rm eff}$, for example, among the DA stars makes a quantitative temperature indicator highly desirable. The use of spectroscopic standards will help to minimize the effect of model-dependent interpretation on the classification purity of the system.

Abt: A strength of the MK system is that it is independent of other physical parameters. Morgan has warned that it is a mistake to modify MK classifications because of knowledge of other data, such as colors, parallaxes, etc.

Liebert: The DA stars have the same H lines in the spectrum from perhaps 100,000 K to about 5,000 K. But you want to specify in some way whether the star is hot or cool. Some kind of color, energy distribution or line strengths can be used to break this up into a reasonable number of bins akin to the MK types and subtypes. The problem is that the color information is quite heterogeneous. When the McCook and Sion catalogue of white dwarfs (1987, ApJS, **65**, 603, and 1994 in preparation) was initially published, we took various color – temperature relations and translated to a rough temperature (the theta index) as a way of getting a uniform scale.

Roman: I also believe that classification must be an empirical description of the appearance of the spectrum independent of the calibration. Which of your descriptions is truly classification and which is calibration?

Liebert: The primary and secondary composition symbols refer to what is actually seen in the spectrum and thus are purely classification. The use of color-color and color-temperature correlations utilizing model atmospheres (Shipman 1972, ApJ, **177**, 723) to generate $\theta_{\rm eff}$ for each degenerate is a departure from "pure" classification we readily admit.

Rountree: If you are seeing stars with truly continuous spectra, how do you know they are white dwarfs and not some other type of object?

Sion: Model atmosphere continuum fits, proper motion, and color indices together can confirm a DC object as a true degenerate star. A number of objects classified as featureless DC spectra among color-select survey objects have turned out to be BL Lac objects (Fleming, T. A. *et al.* 1993, AJ, in press). Therefore the DC spectral type must be applied cautiously. It would not be impossible that some DC objects could be featureless because of rapid rotation (near breakup). However, Milton (1974, ApJ, **189**, 543) showed this to be unlikely.

PART III.

PRECISION SPECTRAL CLASSIFICATION

The MK Process at 50 Years
ASP Conference Series, Vol. 60, 1994
C.J. Corbally, R.O. Gray, and R.F. Garrison (eds.)

Precision Classification

R. O. Gray

Department of Physics and Astronomy, Appalachian State University, Boone, NC

Abstract. The potential of precison classification as a powerful tool for astrophysical insight is examined. The development of the MK spectral classification system through the use of precision classification to discover and define localized third dimensions is discussed and examples are given. Practical problems associated with the definition of such third dimensions and their notation are discussed and a number of guidelines are suggested. Finally a number of current and future projects in precision MK classification which have the potential for yielding astrophysical insight are mentioned.

1. Introduction

MK spectral classification has been one of the most powerful tools of modern astronomy during the past 50 years. The MK spectral class, when properly obtained by careful reference to the standards, represents a fundamental datum of astronomy. Its usefulness lies in its ability to place a star within the context of a broad population of stars and to isolate interesting stars that deviate in some way from normality. MK spectral classification has been one of the primary driving forces behind stellar astronomy and astrophysics for the past five decades. It is not an exaggeration to say that many of the important discoveries in stellar spectroscopy were first made with, or at least intimated by MK spectral classification. It is my conviction that MK classification can continue to play this important role for the <u>next</u> 50 years, but certain changes of emphasis, especially along the lines of increased precision will be required to keep MK classification at the leading edge.

Let me take just a few moments to review some of the momentous discoveries of MK classification during the past 5 decades. Morgan (1933), even before the formal development of MK classification was the first to use what we would recognize today as the MK Process to study systematically the Ap stars, although isolated members of the group were known earlier. Morgan distinguished six types of Ap stars, the Mn, 4200, Si, Cr, Eu and Sr stars. The Am stars in the Hyades were discovered by Titus and Morgan (1940). These two major discoveries, by one of the founders of the MK system, have provided and still do provide thousands of hours of pleasure and frustration to many astronomers. Discovery or definition of the following groups of stars are all due to spectral classification studies: the weak-line stars (Roman 1950), the barium stars (Bidelman and Keenan 1951), λ Bootis stars (Morgan, Keenan and Kellman 1943), the helium-strong stars (first defined by Bidelman 1952), the helium-weak stars (first defined by Garrison 1967) and many others which I do not have the time to enumerate. These discoveries or definitions of new classes were not made by astronomers who considered the process of classification routine and boring; rather they were made by astronomers who were keenly interested in what they were doing, and were alert to any peculiarities, however small, they saw in their spectra. This, I think, is the true basis of precision classification.

2. The Meaning of Precision Classification

MK classification has three roles in the future, as far as I can see: its traditional role of placing a star within the context of a broad population of stars will continue, but be extended to fainter and fainter stars. This is best performed by projects like the monumental work of Nancy Houk in the reclassification of the HD catalog. Large numbers of fainter stars will be placed on the MK system using automated techniques. The second role is the use of MK classification in more precise modes to discover and define new groups of peculiar stars and to bring into its sphere wider populations of stars by the definition of third (or more) local dimensions. These two directions are not mutually exclusive. As the MK system reaches to fainter and fainter stars new groups and dimensions will be discovered of which we have no inkling now. The third direction for the MK Process in the future will be its extension to other wavelength regions. Good progress has already occurred along these lines in the ultraviolet, with the work of Walborn *et al.* (1985) in the O stars and Rountree and Sonneborn (1991) in the B-type stars and in the near-IR with the work of Torres-Dodgen and Weaver (1993) and Kirkpatrick *et al.* (1991). Chris Corbally will talk specifically about the second and third roles on Thursday; my comments will be mostly restricted to the second role - the further development of precision MK spectral classification.

What exactly do we mean by precision classification? What is normally meant by "precision" classification is the division of already existing MK classification boxes into finer boxes or improving the discrimination of the system by a tighter mesh of standards. This is along the lines of what I did in my PhD thesis with Bob Garrison (Gray and Garrison 1987, 1989a&b, Garrison and Gray 1994) in the A-type stars where I increased the precision of the luminosity and temperature classifications by the introduction of a tighter mesh of standards, and the definition of a system of rapidly rotating standards. One of the reasons why this type of precision classification is important is because it allows a detailed confrontation of the MK system with other types of stellar classification, such as Strömgren photometry. If one is strict in the classification process and leaves the comparison with photometry to the last, this confrontation can uncover all sorts of interesting effects, as I demonstrated in my papers with Bob Garrison with respect to rotation in the A-type stars.

But the term "precision" classification is not limited to dividing the MK boxes into ever finer boxes. The emphasis that I want to make in this talk is that high precision classification can lead to the definition of additional localized dimensions. These in turn can lead to further astrophysical insight and discoveries. Let me give you a few examples which I hope will illustrate a number of points I consider important in this regard.

3. The Definition of Localized Third Dimensions

Walborn's work on the O stars is perhaps one of the best examples of how precise spectral classification can lead not only to a much better understanding of the range of phenomena present in a group of stars, but also can lead to new astrophysical insight. We have already heard in Dimitri Mihalas's splendid talk on the physical basis of classification how Walborn's work on the Of stars and his discovery of the O((f)), O(f), Of sequence (if you will, the definition of a localized third dimension) enabled Mihalas (1971) to give a theoretical explanation of the emission-line phenomenon in these stars in terms of autoionization and dielectronic recombination. This interaction between a classifier and a theorist sets, I think, an example for all of us to follow. The ultimate purpose of our work, as classifiers of stars, should be greater insight, whether into the astrophysics of stars or the structure and evolution of our galaxy and other galaxies. While some insight can come about through our development and elaboration of the MK system, the real rewards will come from our interaction from the theorists, and the deep understanding of the physical basis of classification such as we glimpsed in Dimitri Mihalas's talk. Incidently, I believe that the very future of our field, and our ability to attract young astronomers to spectral classification, depends upon this interaction, and the excitment that it brings to our work.

Getting back to our subject, there are more lessons that we can draw from Walborn's work on the O stars. More recently, Walborn has extended further the notation for the Of stars by introducing definitions for the f^* and the f^+ stars (Walborn 1977), which denote different combinations of emission lines present in the spectra of O stars. Walborn has come under criticism from some other classifiers for this notation (c.f. Jaschek and Jaschek 1987); some say that this notation is too complex or that it borders on a "description" of a spectrum instead of a classification, others that many stars so classified may be spectrum variables, and so the possibility exists that, say, a star may be an f^* star at one time and not at another and so on.

I think that we should listen to such criticisms carefully, but they should not prevent careful classifiers, such as Walborn, from taking spectral classification to the furthest practical limits in the pursuit of greater precision in spectral classification. One of the end results of this greater precision can be the development of additional local dimensions, such as Walborn's Of sequence. The development of these additional dimensions to the traditional 2-dimensional MK system will necessarily mean the creation of new types of notation. In order for these dimensions and the accompanying notation to be meaningful, I feel that they should conform to the following criteria.

First, they must be defined in terms of stable standard stars, and second, these dimensions must reflect what appears to be a real spectral group or sequence present in a fairly sizeable population of stars, not just in two or three specimens. These two criteria are necessary if our ultimate goal – astrophysical insight – is to be obtained. They are also necessary if these extra dimensions are to gain broad acceptance in the astronomical community. Third, the notation must be simple enough to be within what we call a "classification", and not a "description", but we should not hesitate to give enough information in the classification so that our colleagues know what is going on. For instance, the classification "Ap" is almost useless; it is much better to write A2 V SrCr(Eu), or even *kB9 hB8 HeB9 III (MnHgSi)* (a notation that Bob Garrison and I adopt in a new paper on the classification of the late B-type stars: Garrison and Gray 1994), and be accused of prolixity.

Walborn's work has demonstrated that additional dimensions can be defined even within the population I stars. However, opportunities for the definition of localized third and multiple dimensions also lie with extending the MK spectral classification system to the metal-weak and metal-strong populations. Chris Corbally will talk about this in much more detail on Thursday, but let me make just these few brief comments. The pioneer in this field, of course, is Keenan who, in the past decade or so has defined and/or elaborated numerous local dimensions in the late- type stars, such as the metal-weak stars, CN strong, CN weak stars, the barium stars and so on (c.f. Keenan and McNeil 1976, 1989). We should note that his dimensions are defined, as I stipulated above, in terms of fairly stable standards, and do actually reflect sizeable populations of stars. His notation is relatively simple, but conveys important information to the astrophysicist in an unambiguous way. I might also mention my own extension of the MK spectral classification system to the metal-weak F-type stars (Gray 1989), in which I set up a grid of metal-weak standards, instead of relying on the old method of classifying these stars by comparing them to a number of solar-composition MK standards.

4. The Future Development of the MK Classification System

What opportunities exist for the future development of the MK spectral classification system along these lines? Opportunities for real advances through precision classification in the early-type stars exist in abundance. This includes not only the familiar Ap and Am stars, but early-type stars in the halo and external galaxies which may have quite different abundance patterns from those in the solar neighborhood. As an example, I can cite the work of Roberta Humphreys. She and her colleagues have investigated a group of peculiar A supergiant stars in the Magellanic Clouds which have hydrogen lines which are much too broad for their luminosities. In a recent paper, Humphreys, Kudritzki and Groth (1991) have concluded that these stars are enhanced-helium or hydrogen-weak stars, and probably represent helium-burning post-red supergiants, similar to the predecessor of SN 1987A. If this is the case, such stars may exist in our own galaxy, but have been passed over because of their relatively broad hydrogen lines. To locate and classify the galactic counterparts of these stars and to learn, through precision classification, how to distinguish these stars from "normal" A-type giants and bright giants would make a superb PhD thesis and a worthy field for the elaboration of the MK system. This will not be easy; with the aid of spectral synthesis and a hydrogen-weak model from Robert Kurucz, I have found that such luminous hydrogen-weak metal-poor stars will be difficult to distinguish from normal, solar-composition A-type giants. Chris Corbally and I have already experienced this challenge in our attempts to classify about 60 A-type stars in the galactic halo. These stars are from Dave Philip's list of FHB candidates. Remarkably, most of these stars appear to be normal dwarfs or subgiants, and yet have balmer jumps (i.e. c_1 indices) more characteristic of giants or even bright giants. We wonder if our stars may be lower-luminosity counterparts of the anomalous A-type supergiants of Humphreys and be understood from the standpoint of enhanced helium abundances.

Let me mention at this point that Chris Corbally and I have devised a technique, termed by us "e-mail classification", which we have found to be extremely useful in obtaining the full classification precision possible with a set of stellar spectra. Chris normally obtains the observations at Kitt Peak, and then e-mails or ftps the processed spectra in the form of ASCII files to me in Boone NC. We both then classify the spectra independently, and, when finished with the first cut, e-mail each other our classifications. We are usually in quite good agreement, but surprisingly, each notices individual features in the spectra missed by the other. A series of e-mail messages ensue in which we exchange ideas about the classification of certain recalcitrant stars. Finally, however, we come to agreement, sometimes only after additional spectra are obtained, and then we have a set of classifications that we know are as good, and as precise as the data set allows. We have now used this method in three different classification projects together, and can recommend it to everyone. It is not only effective, but fun!

In the few minutes remaining to me, let me give you two examples of instances in which precision classification is yielding or has the potential to yield astrophysical insight. My examples will come from the A-type stars with which I am most familiar; I hope during our discussion periods that the rest of you will share examples from other parts of the HR diagram.

One example is from my own work on the λ Bootis stars. In 1988 I published a paper on the λ Bootis stars (Gray 1988) in which, among other things, I demonstrated that the λ Bootis stars fall into two classes – ones with normal hydrogen-line profiles, and those in which the hydrogen-line profiles are peculiar, with broad, shallow wings and rather weak, narrow cores. I designated these two types NHL and PHL for normal and peculiar hydrogen lines (see Figure 1). This classification – the definition, if you will, of a very localized third dimension – did not get much attention until just last year when two astronomers, Dr. Ilian Iliev (Iliev and Barzova 1993) of the National Astronomical Observatory of Bulgaria and Dr. Sturenburg of Kiel Observatory demonstrated, quite independently, the reality of these two classes, by using high dispersion spectroscopy and theoretical Balmer-line profiles. Sturenburg (1993) has gone on to show that the PHL λ Boo stars are associated with gaseous circumstellar material, whereas the NHL stars tend to show only weak indications of a gas shell.

Another project awaiting precise, systematic treatment from a careful classifier is the Am stars. The classification of the Am stars has never been completely systematized; a symptom of this is that Am stars are still classified with respect to the normal abundance MK standards; this just doesn't work, for much the same reasons that Garrison gave for the metal-weak stars. A phenomenon associated with the Am stars, and which needs to be investigated thoroughly from a classification standpoint is the so-called anomalous luminosity effect, discovered and defined by Abt and Morgan (1976). This anomalous luminosity effect refers to the observation that luminosity sensitive lines in the near UV part of

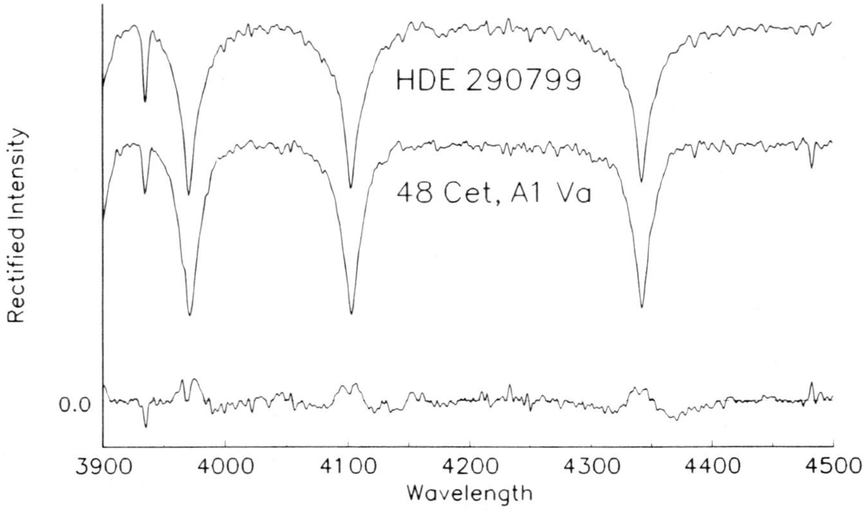

Figure 1. The spectrum of HDE 290799, a λ Bootis star compared with the A1 Va standard. The slightly peculiar nature of the hydrogen-line profiles of HDE 290799 (broad wings, weak cores) can be seen clearly from the difference spectrum at the bottom of the graph.

the spectrum of many Am stars yield a much higher luminosity than lines in the "blue" region; the most extreme example is HD 103877 where the contrast between the two regions is Ib to IV. It is my opinion that we cannot claim to have understood the Am phenomenon until this anomalous luminosity effect has been thoroughly explored and explained theoretically; the first step is a careful, systematic, precision classification program.

These are just two out of the many examples I could have mentioned this morning, but time does not permit. I hope that during our discussions – both during and outside of our formal sessions – we can discuss more ways and examples of using MK spectral classification in a precision mode.

References

Abt, H.A. and Morgan, W.W. 1976, ApJ, **205**, 446

Bidelman, W.P. 1952, AJ, **116**, 227

Bidelman, W.P. and Keenan, P.C. 1951, ApJ, **114**, 473

Garrison, R.F. 1967, ApJ, **147**, 1003

Garrison, R.F. and Gray, R.O. 1994, in press

Gray, R.O. 1988, AJ, **95**, 220

Gray, R.O. 1989, AJ, **98**, 1049

Gray, R.O. and Garrison, R.F. 1987, ApJS, **65**, 581

Gray, R.O. and Garrison, R.F. 1989a, ApJS, **69**, 301
Gray, R.O. and Garrison, R.F. 1989b, ApJS, **70**, 623
Humphreys, R.M., Kudritzki, R.P. and Groth, H.G. 1991, AS **245**, 593
Iliev, I. and Barzova, I.S. 1993, IAU Colloquium #138, *Peculiar Versus Normal Phenomena in A-type and Related Stars*, eds. M.M Dworetsky, F. Castelli & R. Faraggiana, ASP Conf. Ser., **44**, 423
Keenan, P.C. and McNeil, R.C. 1976, *An Atlas of Stellar Spectra of the Cooler Stars: Types G, K, M, S and C* (Columbus: Ohio State University Press)
Keenan, P.C. and McNeil, R.C. 1989, ApJS, **71**, 245
Kirkpatrick, J.D., Henry, T.J. and McCarthy, D.W. 1991, ApJS, **77**, 417
Jaschek, C. and Jaschek, M. 1987, *The Classification of Stars*, (Cambridge: Cambridge University Press)
Mihalas, D. 1971, ApJ, **170**, 541
Morgan, W.W. 1933, ApJ, **77**, 330
Morgan, W.W., Keenan, P.C. and Kellman, E. 1943, *An atlas of stellar spectra*, Chicago Univ. Press
Roman, N. 1950, ApJ, **112**, 554
Rountree, J. and Sonneborn, G. 1991, ApJ, **369**, 515
Sturenburg, S. 1993, A&A, **277**, 139
Titus, J. and Morgan, W.W. 1940, ApJ, **92**, 257
Torres-Dodgen, A.V. and Weaver, B. 1993, PASP, **105**, 693
Walborn, N.R. 1977, ApJ, **215**, 53
Walborn, N.R., Nichols-Bohlin, J. and Panek, J. 1985, *IUE Atlas of O-type Spectra from 1200 to 1900 Å* (NASA Ref. Pub. 1155)

Discussion

Walborn: I have three comments which are to be expanded in subsequent sessions:
1) Spectral classification must maintain principles and precision, but it must also evolve in response to technical and sociological changes. Insistence on traditional approaches will limit its future usefulness.
2) While external information must be excluded in the application of a classification system, it is permissible and indeed essential to the development of such systems. Ample historical examples exist. All criteria must be contained in the classification spectrograms, but other information may guide their initial definition. Interactions with astrophysicists should be mutual.
3) For the OB stars, very important wavelength extensions to the 900–1200 Å and 2 μm regions in progress are not on the program of the meeting and will be mentioned later.

Unfortunately, the Humphreys *et al.* anomalous A-stars in the Magellanic Clouds may not be real. E. Fitzpatrick has found that in some cases apparent hydrogen-line strength excesses may be due to small spectral-type errors from old image-tube data. In other cases the data are consistent with a normal hydrogen-luminosity relation at class Ib. Even if such peculiar stars do exist in the Magellanic Clouds, it does not follow that there should be galactic counterparts (except possibly in the outer galaxy), since they could result from evolution at lower metallicities.

Gray: External information has had an important role in the development of the MK spectral classification system. To deny that is to deny obvious facts. However, it is extremely important to be very certain that the external information which influences, for instance, the development of additional dimensions to the MK system is not model dependent. If for instance, one blindly follows a model which says that a certain spectral feature is temperature dependent, and then next year, a better model is produced which tells us that the behavior of that spectral feature is actually due to NLTE effects, then the classification system is in trouble. So, yes, we must not follow tradition blindly, but sometimes tradition encapsulates wisdom.

As for the Humphreys *et al.* anomalous A-type supergiants in the Magellanic Clouds, if you read the Humphreys *et al.* (1991) paper, you will see that their conclusions are based almost exclusively on new CCD data taken with CASPEC. Some minor changes to the image-tube spectral types were made with this new data, but the basic results remain – that there is a population of A-type supergiants in the Magellanic Clouds which have anomalously strong hydrogen lines. I have not heard Fitzpatrick's arguments in detail, but I certainly found the Humphreys *et al.* paper convincing. I also would not want to rule out the possibility of the existence of galactic counterparts to these stars. With the existence of a metallicity gradient in the disk, we certainly might expect to find such stars in the outer disk. Nor would I want to rule out the existence of such stars in the halo. We are certainly finding out that the stars in the halo are not exclusively "classical" population II stars.

Drilling: I would like to say that I had no difficulty in using the Walborn standards to obtain luminosity classes for the O-stars from the Case-Hamburg OB surveys. I just treated them the same as the MK standards, and had no more or less difficulty with them than with the MK standards.

Gray: Yes, if standards related to third dimensions are properly defined, then there should be no difficulty in their use.

Drilling: Did the similarity of your synthetic helium-rich A giant spectrum to normal A giants spectra extend to all observable wavelengths, or just to the visible spectrum?

Gray: The synthetic helium-rich spectrum showed a number of interesting details. First, the model had log $g = 2.5$, but the width of the hydrogen-line wings were comparable to solar composition models of the same temperature, but log $g = 3.5$. The model was also metal-weak ($[M/H] = -1.0$), but the metallic lines were of almost normal strength. What is happening here, of course, is that when we remove the hydrogen and replace it with helium, the continuous opacity is significantly reduced. Since line strengths go roughly as the line opacity over the continuous opacity, the metallic lines were strengthened, and thus the metal weak spectrum showed almost normal line strengths. The Balmer jump of the helium rich model was only slightly greater than the solar composition model with the same effective temperature and gravity. The synthetic spectrum covered the region from 3200–5000 Å only, so I can't say anything about the rest of the spectrum.

Warren: Almost all Am classifications lack luminosity types, certainly because luminosity criteria have not been defined for Am stars. Yet, we do see certain stars classified as, for example, A2 Vm, A3 IIIm, etc. How is it that some classifiers seem to be able to assign luminosity types to Am stars while the majority of classifiers cannot?

Gray: Perhaps most classifiers have not tried because of the uncertainty introduced by the anomalous luminosity effect. When I specify a luminosity type, it is based on the 4172-8 Å luminosity sensitive blend, but I place it in parentheses so that people know that it probably has no relation to the true luminosity of the star.

Abt: To my knowledge no one has ever constructed a model atmosphere that fits even the main features of metallic-line (Am) stars. Perhaps the best way to proceed with better classifications of such stars is to work together (perhaps by e-mail!) with a person who constructs model atmospheres so that you can mutually decide what features need to be studied and explained.

Gray: Agreed.

Levato: Let me make a final comment to agree with you about the Am stars. We do not know the very nature of these stars. We do not know how to explain the luminosity effect that you mentioned and also we have been observing a list of Am stars for the Hipparcus and we have experimented with the red region where it is supposed that the spectrum of the companions should be, but we could not find any of them. Something is wrong.

Gray: I agree. The Am stars are among those classes of stars that we believe we understand but really know very little about.

The MK Process at 50 Years
ASP Conference Series, Vol. 60, 1994
C.J. Corbally, R.O. Gray, and R.F. Garrison (eds.)

The O3 Stars in 1993

Nolan R. Walborn
Space Telescope Science Institute, 3700 San Martin Drive, Baltimore, Maryland 21218, USA

Abstract. The spectroscopic criteria which distinguish the O3 class in both the optical and ultraviolet wavelength domains are reviewed. An updated census of the class now comprises 27 definite and 43 possible members, for a total of 70, in the Galaxy and the Magellanic Clouds; 55 of them are in the LMC, and only 2 are in the northern hemisphere. Several current and near-future programs applying state-of-the-art ground-based and space-based instruments to these objects will provide improved determinations of the physical parameters for the most massive hot stars. Some recent spectrograms from the Hubble Space Telescope are illustrated.

1. Introduction

How massive are the most massive stars and what are their relative numbers? These questions are relevant to many areas of current astrophysical interest, including star formation and the initial mass function (IMF), the ionization and dynamics of giant H II regions and starbursts, stellar evolution, supernovae, and chemical enrichment. The answers must be established observationally, but the optical and near-UV colors of the O stars are essentially degenerate, while the far UV is highly sensitive to extinction. Hence, spectroscopy is required, with high resolution and signal-to-noise to support quantitative analysis, but as always the efficiency of those techniques will benefit from the prior identification and categorization of optimum candidates by spectral classification.

2. Classification Criteria

The first O3 stars, located in the ionizing clusters of the Carina Nebula, were recognized on blue-violet photographic spectrograms of dispersion 60 Å/mm and widening 1.2 mm, through detailed comparison with equatorial standards of types O4–O5 (Walborn 1971). The fundamental criterion was the non-detection of He I λ4471 in the O3 spectra with those observational parameters, whereas the line is definitely present, albeit weak, in types O4–O5. For the original O3 V spectra, that was the only criterion; that is, the category is bounded on the low-temperature side only, since the horizontal classification discriminant of He I/He II has disappeared. At types O3 III and O3 I, however, strong absorption lines of N V $\lambda\lambda$4604, 4620 and emission of N IV λ4058 appear which qualitatively distinguish them from the O4 counterparts (Walborn 1982); an

N IV emission stronger than the classical Of N III features at $\lambda\lambda 4634, 4640, 4642$ is denoted by Of*, and is a characteristic of O3 spectra. Digital observations of O3 spectra in the blue-violet are shown by Walborn & Fitzpatrick (1990) and Walborn & Parker (1992).

The International Ultraviolet Explorer and the Hubble Space Telescope have provided high-quality observations of many early-type spectra in the space ultraviolet down to about 1200 Å, including a number of type O3. The distinctions between O3 V and O4–O5 V spectra in the UV are again rather subtle, but the higher-luminosity O3 spectra also present a unique signature in this wavelength range, namely a strong O V $\lambda 1371$ stellar-wind feature. Some examples from the IUE archive are shown by Walborn, Nichols-Bohlin, & Panek (1985), and some recently obtained by HST later in this paper.

Optical observations of O3 spectra at high resolution and S/N detect He I $\lambda 4471$ in some of them, leading to a range in the derived effective temperatures within the class (Kudritzki *et al.* 1992a, 1993), as expected from the classification criterion discussed above. It may be noted that some of the O3 V spectra display weak N III emission features, while others such as HD 64568 (Walborn 1982) and LH10:3058 (Walborn & Parker 1992) have weak N IV and N V features instead. None of the latter objects has been analyzed quantitatively nor observed in the UV as yet. Depending on the results of further studies, it may become necessary and desirable to introduce an O2 spectral class in the future, if for instance the He I/He II ratios at high resolution establish the weak N IV, N V features in some O3 V spectra as indicators of a significantly higher temperature.

3. Current Census

A review of the O3 class by Walborn (1982) listed 10 known members at that time. The numerous subsequent discoveries scattered throughout the literature are compiled here in Table I, which is followed by discovery references and notes on individual stars.

Most early O stars are naturally located in bright emission nebulosities, which introduce uncertainty in the horizontal spectral types due to contamination of the weak stellar He I absorption lines by the corresponding nebular emission lines at low spectral resolution. As a consequence, only spectral-type ranges can be given in some cases, the extent of the uncertainty being judged from the overall strength of the nebular emission features. When no He I absorption can be detected underlying the emission, the possible range includes type O3. The most extreme cases are potentially very young O stars embedded in dense nebular knots, classified as O3–6 V (Walborn & Blades 1987; Walborn & Parker 1992). In other cases, the range of uncertainty may be only O3–O4. Most of the "possible" members of the O3 class distinguished in Table I arise from this cause, although a few are due to a need for confirmation with better observational material.

Table I lists 27 definite and 43 possible members of the O3 class, for a total of 70. The membership is dominated by the Large Magellanic Cloud, and particularly the 30 Doradus region with over half the total, although all of the latter cases are in the "possible" category due to the bright H II, except for the

TABLE I

1993 Census of the O3 Stars

	O3 V		O3 III		O3 I	/WN-A	Totals		
	Def.	Poss.	Def.	Poss.			Def.	Poss.	Grand
GALAXY									
Carina Nebula	5				1		6		6
Cygnus OB2				1	1		1	1	2
Singles	2		1				3		3
Totals							10	1	11
LMC									
30 Doradus		26		3		5	5	29	34
30 Doradus Periphery		3		2		1	1	5	6
N11	1	2	2				3	2	5
N180			1	1			1	1	2
Singles	1	3	3			1	5	3	8
Totals							15	40	55
SMC									
Singles		2	1			1	2	2	4
Grand Totals							27	43	70

References and Notes to Table I

Carina Nebula: Walborn (1971, 1973a, 1982)
Cygnus OB2: Walborn (1973b), Massey & Thompson (1991)
Galaxy Singles:
HD 64568/Puppis OB2: Walborn (1982)
BD−16°4826: Garmany & Vacca (1991)
LSS 4142: Vijapurkar & Drilling (1993)
30 Doradus: Melnick (1985), Walborn (1986, 1991), Walborn & Blades (1987, 1994), Parker (1993)
30 Doradus Periphery:
30 Dor B, C: Testor & Schild (1990), Schild & Testor (1992)
DD13/N159A: Conti & Fitzpatrick (1991)
N11A, B: Parker *et al.* (1992), Walborn & Parker (1992)
Wo599/N11C: Heydari-Malayeri *et al.* (1987)
N180: Conti *et al.* (1986), Garmany & Walborn (1987), Massey *et al.* (1989)
LMC Singles:
HDE 269810 = Sk−67°211: Walborn (1977, 1982)
Sk−67°22: Walborn (1982, 1986)
Sk−68°137: Conti *et al.* (1986), Garmany & Walborn (1987)
Sk−67°274: Conti *et al.* (1986)
LH58:10A/N144: Conti *et al.* (1986)
N44C: Stasińska *et al.* (1986)
Sk−71°51/N214: Garmany & Walborn (1987)—initially classified O3 III(f*), but subsequent digital data from CTIO 1 m indicate O3 V((f*)) preferable, with weak N IV, N V features as in HD 64568 and LH10:3058/N11B.
Sk−66°172: Kudritzki *et al.* (1993)
SMC Singles:
AB2: Azzopardi & Breysacher (1979), Walborn (1986)
NGC 346 No. 3: Walborn & Blades (1986), Niemela *et al.* (1986)
AV14 = Sk9: Garmany & Conti (1987)
N81: Heydari-Malayeri *et al.* (1988)

intermediate O3 If*/WN-A objects. Except for the one definite member and one possible member in Cygnus OB2, all of the known O3 stars lie in the southern hemisphere.

4. Current and Future Programs

In order to investigate the questions noted in the Introduction, one must determine the effective temperatures, luminosities, metallicities, mass-loss rates, and masses of the O3 stars. Such studies, of course, require high-resolution and high S/N spectroscopy in both the optical and ultraviolet domains. A further observational problem is presented by the strong tendency of massive stars to form in compact groups; in deriving physical parameters it is essential to know whether one is dealing with a single or multiple object, which may be especially difficult to determine at the distance of the Magellanic Clouds. Several current and near-future programs addressing these problems for the O3 stars will be briefly described.

A comprehensive, comparative investigation of the physical parameters of massive stars in the Galaxy and the Magellanic Clouds, based upon optical data from the ESO 3.6 m echelle and UV data from the HST Faint Object Spectrograph, is being led by the Munich group (Kudritzki *et al.* 1992b, 1993). The HST/FOS data for three stars of type O3 III(f*) are shown in Figure I. Note the marked O V λ1371 wind profiles in all of them; Sk$-66°$172 was first identified as a member of the O3 class from this feature, and subsequently confirmed by the optical data. The most recent analyses yield effective temperatures of 50,000–60,000 K for these stars, and a mass of 190 $M_\odot$ for Sk$-67°$211 = HDE 269810. The masses of Sk$-68°$137 and $-66°$172 are determined as 110 and 100 $M_\odot$, respectively (Kudritzki *et al.* 1993).

The multiplicity problem can be addressed by direct imaging and, at smaller physical scales, by radial-velocity studies. State-of-the-art ground-based imaging can now achieve a resolution of $0''.3$; a nice example in the LMC is presented by Heydari-Malayeri *et al.* (1993). With HST, however, $0''.1$ is routine. Considerable attention has been paid to the 30 Doradus central cluster (Weigelt *et al.* 1991; Walborn *et al.* 1992; Campbell *et al.* 1992; De Marchi *et al.* 1993). This author and collaborators are obtaining UBV frames of other Magellanic Cloud fields containing O3 stars to investigate the upper IMF; some of the brighter objects are resolved into multiple components while others remain unresolved, either result being of significance! At the short end of the range of separations, radial-velocity monitoring of the brighter objects in these same fields has been undertaken by A. Moffat and collaborators using the Argus multiple-fiber instrument at the CTIO 4 m, to detect spectroscopic binaries. These programs will provide valuable new information about the upper IMF. However, even $0''.1$ corresponds to 5000 AU at the LMC, whereas large-amplitude early-type spectroscopic binaries have typical separations of order 0.1 AU. Hence, future interferometric observations will be required to achieve closure of this problem in the Magellanic Clouds.

Finally, the vital 900–1200 Å range, which contains numerous high-ionization resonance lines including the unique O VI $\lambda\lambda$1032, 1038 feature, has remained unexplored in O3 spectra. The first examples have been observed by R. Kudritzki with the ORFEUS instrument during this meeting. Also, this author and collaborators have been allocated time with the Hopkins Ultraviolet Telescope during the Astro-2 mission scheduled for late 1994, to observe several of the same Galactic and Magellanic Cloud O3 stars contained in the current HST and ground-based programs. These data should offer useful new constraints for physical models of the atmospheres and winds of the most massive hot stars.

References

Azzopardi, M., & Breysacher, J. 1979, A&A, **75**, 120

Campbell, B. *et al.* 1992, AJ, **104**, 1721

Conti, P. S., & Fitzpatrick, E. L. 1991, ApJ, **373**, 100

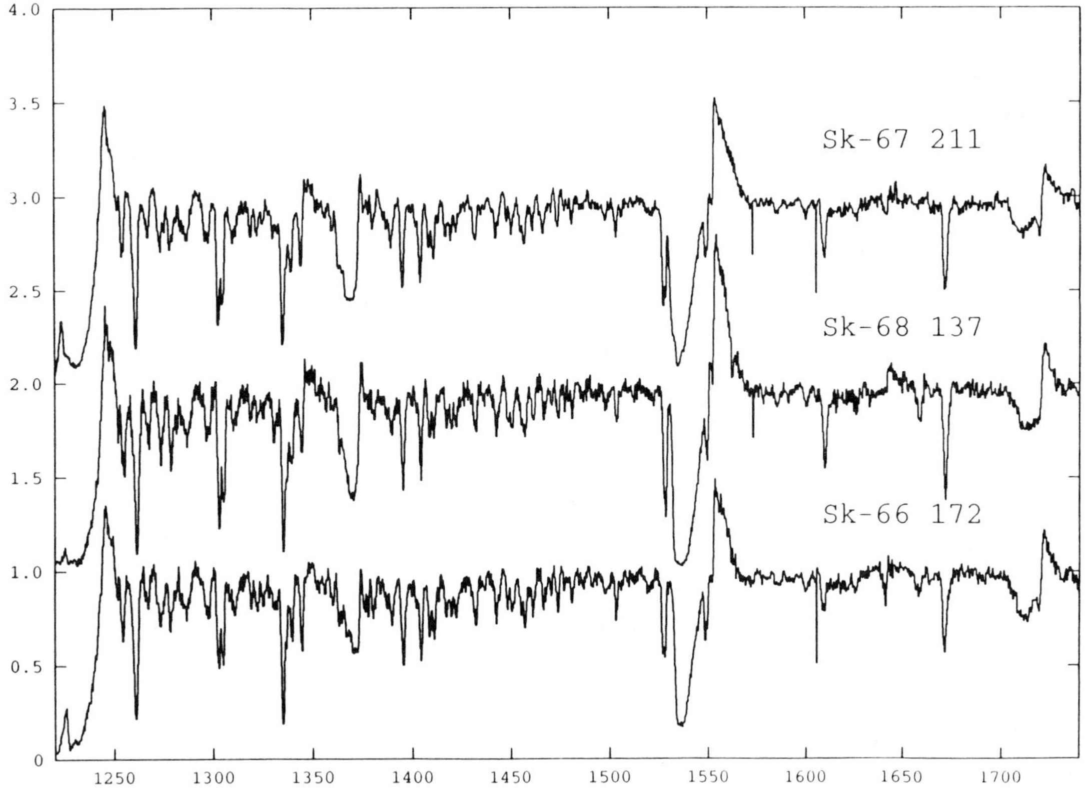

FIGURE I Rectified HST/FOS spectrograms of three O3 III(f*) stars in the LMC, from the program of R. Kudritzki *et al.* Sk−67°211 = HDE 269810. The prominent stellar-wind features are N V λλ1239, 1243; O V λ1371; C IV λλ1548, 1551; and N IV λ1718. The plot was kindly prepared by D. Lennon.

Conti, P. S., Garmany, C. D., & Massey, P. 1986, AJ, **92**, 48

De Marchi, G., Nota, A., Leitherer, C., Ragazzoni, R., & Barbieri, C. 1993, ApJ, **419**, 658

Garmany, C. D., & Conti, P. S. 1987, AJ, **93**, 1070

Garmany, C. D., & Vacca, W. D. 1991, PASP, **103**, 347

Garmany, C. D., & Walborn, N. R. 1987, PASP, **99**, 240

Heydari-Malayeri, M., Grebel, E. K., Melnick, J., & Jorda, L. 1993, A&A, **278**, 11

Heydari-Malayeri, M., Le Bertre, T., & Magain, P. 1988, A&A, **195**, 230

Heydari-Malayeri, M., Niemela, V. S., & Testor, G. 1987, A&A, **184**, 300

Kudritzki, R.-P., Hummer, D. G., Pauldrach, A. W. A., Puls, J., Najarro, F., & Imhoff, J. 1992a, A&A, **257**, 655

Kudritzki, R.-P. *et al.* 1992b, in *ST-ECF/STScI Workshop, Science with the Hubble Space Telescope*, ed. P. Benvenuti & E. Schreier (ESO Proceedings No. 44), 279

Kudritzki, R.-P. *et al.* 1993, in *Brussels Workshop, Evolution of Massive Stars: A Confrontation between Theory and Observations*, in press

Massey, P., Garmany, C. D., Silkey, M., & DeGioia-Eastwood, K. 1989, AJ, **97**, 107

Massey, P., & Thompson, A. B. 1991, AJ, **101**, 1408

Melnick, J. 1985, A&A, **153**, 235

Niemela, V. S., Marraco, H. G., & Cabanne, M. L. 1986, PASP, **98**, 1133

Parker, J. Wm. 1993, AJ, **106**, 560

Parker, J. Wm., Garmany, C. D., Massey, P., & Walborn, N. R. 1992, AJ, **103**, 1205

Schild, H., & Testor, G. 1992, A&AS, **92**, 729

Stasińska, G., Testor, G., & Heydari-Malayeri, M. 1986, A&A, **170**, L4

Testor, G., & Schild, H. 1990, A&A, **240**, 299

Vijapurkar, J., & Drilling, J. S. 1993, ApJS, **89**, 293

Walborn, N. R. 1971, ApJ, **167**, L31

———. 1973a, ApJ, **179**, 517

———. 1973b, ApJ, **180**, L35

———. 1977, ApJ, **215**, 53

———. 1982, ApJ, **254**, L15

———. 1986, in *IAU Symp. 116, Luminous Stars and Associations in Galaxies*, ed. C. W. H. de Loore, A. J. Willis, & P. Laskarides (Dordrecht: Reidel), 185

———. 1991, in *IAU Symp. 148, The Magellanic Clouds*, ed. R. Haynes & D. Milne (Dordrecht: Kluwer), 145

Walborn, N. R., & Blades, J. C. 1986, ApJ, **304**, L17

———. 1987, ApJ, **323**, L65

———. 1994, in preparation

Walborn, N. R, Ebbets, D. C., Parker, J. Wm., Nichols-Bohlin, J., & White, R. L. 1992, ApJ, **393**, L13

Walborn, N. R., & Fitzpatrick, E. L. 1990, PASP, **102**, 379

Walborn, N. R., Nichols-Bohlin, J., & Panek, R. J. 1985, *International Ultraviolet Explorer Atlas of O-Type Spectra from 1200 to 1900 Å* (NASA RP 1155)

Walborn, N. R., & Parker, J. Wm. 1992, ApJ, **399**, L87

Weigelt, G. *et al.* 1991, ApJ, **378**, L21

Discussion

Houk: For these newer types of O stars do you need standard spectra both in the visual (blue) and in the UV, and are these still MK types as defined in the visual? Do you think different spectral regions in general should be utilized, as in the case of the cool red stars, to get MK types?

Walborn: No, all of these categories are defined in the visual blue-violet region. However, the UV features I showed are almost uniquely correlated with the optical types, and could be used to identify some of the categories if only UV data were available. In general, classification criteria in different wavelength regions must be investigated, starting with the optical standards; if good relationships are found, the new material might be used to infer MK types. However, there is no guarantee that correlated criteria will be found, for either observational or physical reasons. The approach will depend on the available data and scientific objectives. For instance, Conti is now investigating the 2 μm spectra of OB stars with a view toward classifying very young embedded objects which are not observable at shorter wavelengths.

Rountree: For the B stars, it is possible to classify directly in the ultraviolet by comparing the relevant features in the ultraviolet spectrum of the program stars with the corresponding features in the MK standards. But there is one caveat – you have to be sure to use photospheric lines as classification criterion, because in the B stars the wind lines are not always consistent within a spectral class.

Walborn: The stellar-wind features in the B stars, especially at the lower luminosities, present a different situation from those in the O stars. In the latter, the radiation field dominates, which is probably the reason for the detailed correlations observed between wind features and optical spectral types: both are determined ultimately by the fundamental stellar parameters. In low-luminosity B stars that is no longer true, and other physical effects become competitive. For instance, in Be, He-rich, and Algol stars the wind features may be strongly enhanced for the spectral types, and highly variable to the point of complete disappearance.

Garrison: It seems that the discrimination of the types O4-O3-(O2) is more dependent on resolution than is the case for most types, mainly because it relies on the absolute strength rather than ratios. My question is, from your experience, what is the required resolution (and signal-to-noise) for discriminating types O4-O3- (O2)?

On another topic, I am quite concerned about your comment that theory should dictate the direction and structure of your classification schemes. Supposing a theoretician now tells you that the criterion is a pure temperature effect, then next year, another theoretician tells you it is really a luminosity effect and in another year a third theoretician tells you it is an autoionization effect or a magnetic field effect, etc. Will you change your scheme to satisfy each of these models?

Walborn: The question reduces to the minimum requirements for detection of He I λ4471 in O4 spectra, which would require detailed investigation as a function of resolution and S/N for a precise answer, but probably lie in the ranges 1-3 Å and 50-100, respectively, for digital data. The same question could be asked about the distinction of type O4 from O5, the latter being the earliest in the original MKK atlas. A hypothetical O2 type would depend on detection of weak N IV emission and N V absorption features in some spectra currently included in the O3 V type; the observational requirements are probably the same as above. As has been the case in the history of the MK system, external information may guide the selection of criteria visible in the spectra or may be opposed to the subsequent application of the system. Validation of the above criteria for an O2 type will depend on detection of He I or not in O3 spectra at high resolution and S/N. Any empirical classification system is hypothetical until its utility is confirmed by calibration in terms of physical parameters.

Crawford: It was nice to see and learn of the composite spectrum. I have always "worried" about the impact of such stars (double and multiple) on MK classification and on photometry. I don't think we consider it enough. Can we discuss this, in general, on Friday? (Clearly, these very early type stars are prime candidates.)

Walborn: It is a serious problem which affects the questions of the most massive stars and upper IMF systematically and increasingly with distance. Even at the LMC, $0\farcs1$ is 5000 AU! The highest possible spatial resolutions and astute spectroscopic interpretations must be applied. The problem is worst for massive stars which favor compact groupings, but other types can be affected, especially in dense clusters. An example is the late-type "super luminous giants" in LMC clusters which have recently been shown to have been composite images. But we must not go overboard and assume that every very massive candidate will turn out to be composite, as some seem to be doing. A 12-14th mag object in the LMC which has a pure O3 spectrum and remains unresolved at $0\farcs1$ is very likely to be a very massive star.

The MK Process at 50 Years
ASP Conference Series, Vol. 60, 1994
C.J. Corbally, R.O. Gray, and R.F. Garrison (eds.)

MK Spectral Classification in the Orion OB1 Association

Hugo Levato[1] and Stella Malaroda[2]
Complejo Astronómico El Leoncito, San Juan, Argentina

Monica Grosso
Complejo Astronómico El Leoncito

Nidia Morrel[1]
Facultad de Ciencias Astronómicas y Geofísicas de La Plata

Abstract. We have obtained spectroscopic observations of 350 stars in the Orion OB1 association. We derived MK spectral types and $v \sin i$ values for them. The results show (1) three Be and Shell stars, (2) two λ Bootis stars, (3) one 'sn', (4) two stars with C II strong, (5) four Hg-Mn stars, (6) fifteen metallic line stars, (7) five Si-Sr-Cr stars, and (8) five silicon stars. The $v \sin i$ values indicated that the stars of the Association rotate on the average similarly to the field stars of the same temperatures for the older subgroup (a) and 30% higher for the younger subgroup (d).

1. Introduction

The Orion OB1 association in our galaxy has been studied by several researchers. The investigation made by Warren and Hesser (1977a,1977b, 1978) which hereinafter we will call WH1, WH2, and WH3 respectively, is one of the most comprenhensive studies of the stellar population of the association. They have also discussed properties of the interstellar material. Extensive references on papers prior to 1977 may be found in the WH series of papers. From the data compiled by WH1 the lack of homogeneous MK types for many of the 526 stars included was evident. By the time of WH publications, Levato and Abt (1976) and Abt and Levato (1977) published spectral classification for 26 stars in the Orion Nebula Cluster, and 150 stars in the Northwest region (a), the Belt region (b) and the Upper Sword region (c). Later Guetter (1981) published MK types for stars in the Belt region, and Morrell and Levato (1991) published a radial velocity study on the 100 brightest members of the WH list. In the 80's we started to obtain the observations for the remaining 350 stars which had

[1]Member of Consejo Nacional de Investigaciones Cientificas y Tecnicas de la Republica Argentina. Visiting Astronomer, Cerro Tololo Inter-American Observatory. CTIO is operated by AURA, Inc. under contract to the National Science Foundation

[2]Member of the Comision de Investigaciones Cientificas de la Provincia de Buenos Aires

not been observed previously by Abt and Levato. The complete results will be published elsewhere but in the present paper we will show some of the findings on the spectroscopically anomalous objects.

2. Observations

We have obtained the observations with the CTIO 0.9m and 1.5m telescopes and their spectrographs. We have recorded two sets of spectra on KODAK IIaO photographic plates: (1) at 128 Å/mm reciprocal dispersion and (2) at 39 Å/mm reciprocal dispersion. All the spectra were widened between 0.9 and 1.2 mm. A set of MK standards from the MK78 Atlas (Morgan, Abt and Tapscott, 1978) was also obtained using the same instrumentation and treatment of the photographic material. In addition we wanted to use the medium dispersion spectrograms to derive the projected axial rotational velocity for each star observed in the sample. With this purpose we also obtained spectrograms of the standard stars which define the rotational velocity system (Slettebak *et al.* 1975).

3. Results and Discussion

The classification was performed using an spectral comparator. SM and MG classified the stars independently, and HL gave the final spectral classification. The same visual procedure was used to derive the $v \sin i$ values for the different objects. The comparison of the line widths of the program stars with those of the proper net of Slettebak's standards permits the derivation of $v \sin i$ values with an estimated error of 10%. A comparison with the spectral types and luminosity classes derived by Guetter (1981) for 50 stars in common shows that our types differ from Guetter's types by 0.08 spectral classes and in luminosity by 0.2 luminosity classes. During the process of spectral classification we have discovered a number of spectroscopically anomalous objects, namely: 3 Be-shell stars, 2 λ Bootis, 1 sn, 15 Am, 4 Hg-Mn, 2 CII strong, 5 Si-Sr-Cr, and 5 Si stars. Tables 1, 2, and 3 provide a listing of the Am stars, the Bp-Ap stars, and the λ Bootis stars, respectively, found among the 526 stars included by WH1. The columns of the Tables are self-explanatory and the last column includes a remark if evidence of non-membership was found in WH or was derived from the radial velocities of Morrell and Levato (1991) or from the present spectral classification. An asterisk by the WH numbers means that the star's nature was discovered, as far as we know, in the present paper. The group assignment indicated in the Tables was taken from WH1 and the $v \sin i$ values are those derived in this investigation. Let us comment briefly about some of the groups of anomalous objects.

3.1. Metallic line Stars

Among the 30 Am stars in Table 1 there are 17 probable members of the Ori OB1 association, while 4 are probable non-members, and 9 are considered definitely non-members. The probable members are located in the visual absolute magnitude range between 0.1 and 2.3 magnitudes. We have used the photometric data compiled by WH1, the proper correction for reddening and the distances to the different subgroups derived by WH2 and WH3. The frequency of Am stars among the stars in the same absolute magnitude range, found for the different subgroups of the association, are indicated in Table 4 where we have also listed the average $v \sin i$ value of the stars in each subgroup with respect to the average for fields stars of the same temperature. The average for field stars was taken from Abt (1970).

Table 1. Am Stars in Orion OB1.

WH	HD	DM	Spectral Type K/H/M	$v \sin i$	GR	Membership
10*	33900	+4 868	B9.5/?/A1	80	a	
15	34317	+1 957	B9/A0V/A1p		a	NM
81*	36017	-4 1141	A6/A9/F2	80	c	NM?
82*	36016	-4 1142	A4/F1/F2	40	c	
128	36366	-6 1204	A1 V+ B9/A/A2	80 + <40	c	
135	36412	-5 1281	A7/F1V/F3		c	
154*			A8/?/F0			
157	36559	-4 1163	A0/A1V/A4	122	c	
165	36606	-5 1285	A4/A2V/A9		c	NM?
180	36671	-4 1165	A3/F0IV/F2		c	NM
182	36670	-4 1166	A3/A5V/F1		c	
187	36697	-7 1115	A1/A1V/A5	3	c	
188*	36694	-0 991	A1/A5/A9	<40	b3	NM?
194	36726	-0 993	A1/A1V/A5		b3	
205*		-5 1296	A3/?/?/		c	
208	36811	-2 1294	A1/A2/A5	<40	b2	NM
223*	36863	-1 952	A5/A7/F0	<40	b2	NM?
225	36843	-4 1167	A4/A8V/F1	<40	c2	NM
228	294265	-5 1302	A2/A8V/A9	180:	c3	
230*			A3/A/A7		c	
263*			A1 Vn+B9/?/A2	210,45	d	
323	37111	-0 1016	A1/A1V/A4		b3	
400	37410	-4 1198	A4/A2V/A7	80	c	NM
410	37444	-5 1347	A4/A8V/F3	<40	c	NM
464*	37789	-1 1006	A1/?/A3	<40	b1	
497*		-5 1377	A5/?/A7	50:	c	
499*	38311	-0 1073	A0/?/A2	<40	a	
505*	38662	+1 1131	A0-1/?/A3	<40	a	
514*	39033	+0 1187	A2/?/A7	<40	a	NM
524*	39572	+1 1155	A2-3/?/A8-9	<40	a	NM?

3.2. The Bp-Ap Stars in Orion

Table 2 presents the 37 Bp-Ap stars found among the 526 stars from WH list. Five of them are non-members or probable non-members of the association. Among the remaining 32 stars there are 11 Si stars, 6 Sr-Cr-Si stars, 5 He-weak and 5 Hg-Mn stars.

Table 2. Ap-Bp Stars in Orion OB1

WH	HD	DM	Spectral Type	$v \sin i$	GR	Membership
2*	32867	+2 860	B9 III (Si,Mn,Hg)	<40	a	
22	35008	-1 872	B9 Vp (Si)	180	a	NM
33	35298	+1 996	B4 Vp (He wk)	<40	a	
40*	35456	-2 1237	B9 III p (C II)	<40	a	
42	35502	-2 1241	B5 IVsnp (He wk)	200:	a	
43	35548	-0 945	B9.5 p (Hg,Mn,Fe)	<40	a	
70*	35901	-7 1083	A1 p (Sr, Si, Cr:)	60	c	NM
117*	36313	-0 977	B8 IIIp (Si)	140:	b3	
136	36429	+2 986	B4 IVsnp (He wk)	105:	a	
144*		-0 984	A0 p (Hg-Mn)	<40	b3	
146*	36513	-7 1105	A0 Vb pec	<40	c	
151	36540	-4 1162	B7 IVp (He wk)		c	
156	36549	+1 1053	A0 p (Mn,Sr,Si)	<40	a	
242		-5 1306	A2p V (K/A0.5)	240:	d1	
247	36955	-1 955	Ap (Si,Cr,Sr)		b2	
248	36916	-4 1173	B9 IVp (Si)	90:	c1	NM?
257*	290665	-0 1008	Ap (Sr,Cr,Si)	<40	b2	
285	37017	-4 1183	B1.5 Vp (He st)	133	c1	
326*	37114	-5 1331	B9.5p V (K/B8)	222	d1	
332	37140	-0 1018	B7 IVp (He wk)	<40	b3	
340	37151	-7 1131	B8 Vp (Si)	<40	c	
346	37210	-6 1254	A0 Vp (Si)	<40	c	
376*	37333	-2 1319	A1p (Cr,Sr)	80	b1	
413*	37467	+2 1028	B9 p (Hg,Mn)	<40	a	
414	37479	-2 1327	B2 IV-Vp (He st)		b1	
423*	37492	-8 1183	B8 p (Si-4200)	<40	c	
435*	37592	+0 1146	A0p (Sr)	<40	a	
438	37633	-2 1332	A0p (Si)		b1	
440	37635	-9 1197	B7 Vp (Mg II wk)		c	
443	37642	-3 1167	B9 p (Si)		c	
454*	37687	-3 1168	B7 IIIp (C II)	<40	c	
509*	38824	-8 1219	B8.5p (Si)		c	
512	38912	+1 1143	A0p		a	
515*	39082	+4 1054	Ap (Sr,Cr,Eu)	<40	a	NM?
518*	39254	-4 1256	Ap (Si?)		c	
521*	39419	-6 1337	B9.5p (Si?)	<40	c	
525*	39773	+5 1044	B9.5/A0p (Hg,Mn,Si)		a	NM

The frequency of Bp-Ap stars among the normal main sequence stars in the same absolute magnitude range for each subgroup of the association is shown also in Table 4.

3.3. The λ Bootis Stars

We found two new λ Bootis stars. They are listed in Table 3. Jointly with HD 290799 discovered by Gray and Corbally (1993) there are three stars of this class in Ori OB1. All of them are probable members of the association. The average visual absolute magnitude for the three λ Bootis stars is $< M_v > = 1.34 \pm 0.88$ (σ). The range is between 0.46 to 2.23 magnitudes. The three objects fall on the ZAMS (WH452) or very near it (WH352, WH452). It is very important to confirm the membership of these λ Bootis stars in the Ori OB1 association in order to know more about the time scale involved in the development and evolution of the spectral anomalies of this kind of peculiar star.

As a preliminary conclusion of this project we have shown that the Ori OB1 association contains among its stellar population a fairly large number of Ap, Am and at least three λ Bootis stars. In the complete paper we will provide more details about the statistics and a description of some of the most interesting spectra.

Table 3. λ Bootis Stars in Orion OB1

WH	HD	DM	Spectral Type	$v \sin i$	GR	Membership
352*	294253	-3 1154	A0 Vp λ Boo		c	
452	290799		A2 V λ Boo		b2	
472*	37886	-3 1172	A0p λ Boo		b1	

Table 4. Frequency of Ap and Am Stars in Ori OB1

Group	$< v \sin i >/< (v \sin i)_{FS} >$	Ap%	N	Am%	N
a	1.02 ± 0.04	17	10	21	3
b	1.16 ± 0.05	9	7	15	4
c	1.08 ± 0.05	12	10	18	11
d	1.29 ± 0.08	22	2	25?	1

Acknowledgments. We appreciate very much the support of the Vatican which permitted HL to attend this meeting. We also thank Helmut Abt for his comments. Support from the CONICET is also appreciated.

References

Abt, H. A. 1970, in *IAU Colloquium 4, Stellar Rotation*, ed A. Slettebak. Dordrecht:Reidel, 193

Abt, H. A., and Levato, H. 1977, PASP, **89**, 795

Gray, R.O., and Corbally, C.J. 1993, in *Peculiar vs Normal Phenomena in A-Type Stars*, ASP Conference Series **44**, ed M. Dworetsky, F.Castelli, R. Faraggiana, 418

Guetter, H. H. 1981, AJ, **86**, 1057

Levato, H., and Abt, H. A. 1976, PASP, **88**, 712

Morgan, W. W., Abt, H. A., and Tapscott, J. W. 1978, *Revised MK Spectral Atlas for Stars Earlier than the Sun*, University of Chicago and Kitt Peak National Observatory

Morrell, N., and Levato, H. 1991, ApJS, **75**, 965

Slettebak, A., Collins, G. W., II, Boyce, P. B., White, N. M., and Perkinson, T. D. 1975, ApJS, **29**, 137

Warren, W. H. and Hesser, J. E. 1977a, ApJS, **34**, 115 (WH1)

—— 1977b, ApJS, ApJS, **34**, 207 (WH2)

—— 1978, ApJS, ApJS, **36**, 497 (WH3)

Discussion

Corbally: This is excellent work, and especially exciting to me are the two new λ Bootis stars.

Gray and I while, looking at about 50 A-type stars in the Orion and Lacertae OB1 Associations, found 9 stars whose metal lines were slightly weak; and not only were their λ4481 lines weak, but other metal features as well. Did you find such spectra among your survey? If so, where did their photometry put them with respect to the ZAMS?

Levato: Yes, we have found some stars, e.g., WH 356, with a metallic spectrum earlier than the Hydrogen type. We have also found discrepancies in the intensity of the G band. As far as I remember, some of these stars are located between 0.75 and 1.5 magnitudes above the ZAMS.

Drilling: Are the 'sn' stars shell stars or binaries?

Levato: After the discovery of the 'sn' stars made with Helmut Abt in the papers on the Ori OB1 Association published in the late seventies, we believed, and still do, that the 'sn' stars may represent a certain kind of shell star.

Abt: Regarding the 'sn' stars, there is a recent model by Steve Shore involving strong magnetic fields and a complex structure.

Parthasarathy: Are the three λ Bootis stars that you have mentioned definite members of the Orion Association?

Levato: According with the available information there are no reasons to think that they are non-members.

The MK Process at 50 Years
ASP Conference Series, Vol. 60, 1994
C.J. Corbally, R.O. Gray, and R.F. Garrison (eds.)

MK Spectra of the Bright A-Type Stars

Helmut A. Abt
Kitt Peak National Observatory, Box 26732, Tucson, AZ 85726-6732, USA

Abstract. We obtained new rotational velocities and MK classifications for about two thousand A-type stars. Frequencies of various peculiarities are given. The bimodal distributions in rotational velocities published by van den Heuvel disappear when more detailed MK classifications are used. The deconvolutions of rotational velocities, assuming random orientations of axes, show that all slow rotators have abnormal spectra, all rapid rotators have normal or λ Boo spectra, but that there is a overlap of about 10% that is not explainable by diffusion. However it is possible that some hitherto undetected abnormalities occur, such as those of many of the A1-3 IV stars, and that that might explain the overlap.

1. Introduction

This project has two goals. One is to confirm and explain the evidence by van den Heuvel (1967) that among many of the B- and A-type stars there are bimodal distributions of rotational velocities. The other goal is to provide a set of consistent rotational velocity measures in the new Slettebak *et al.* (1975) system for most of the A-type stars in the Bright Star Catalogue. A preliminary report on half of the observational material has been published by Abt and Morrell (1992).

2. Data

Nidia Morrell and I obtained CCD spectra of all the 1740 A0-F0 stars in the Bright Star Catalogue between declinations $-30°$ and $+70°$. Those spectra were obtained with the Kitt Peak Coudé Feed telescope at a dispersion of 15 Å mm^{-1}. We measured the half-widths of 4481 Mg II and (in most spectra) 4471 Fe I; those were calibrated against the rotational velocities by Slettebak *et al.* (1975) to give rotational velocities accurate to about 13 km s^{-1}. We also obtained the equivalent widths of 4481 Mg II. In addition I obtained a separate set of wide photographic spectra at 39 Å mm^{-1} of 2010 A0-F0 stars between $-30°$ and the north pole, using the Kitt Peak 2.1 meter Cassegrain spectrograph. Those were classified against MK standards. The agreement with Gray and Garrison (1987, 1989a. 1989b) is excellent for normal stars.

3. Stellar Statistics

The frequencies of various kinds of stars among the 2010 stars are shown in Table 1 for three intervals of spectral types.

Table 1. Frequencies of the stars

Spectral range	A0-A1	A2-A4	A5-F0
Normal V, IV	59%	75%	53%
Normal III-I	15%	7%	12%
Ap	10%	7%	3%
Am	0%	2%	28%
λ Boo, 4481 weak	15%	9%	4%
Shell	1%	1%	0%

Some comments about these are appropriate. The Ap stars are mostly of the "Si", "Sr,Cr,Eu", or "Hg,Mn" varieties. Nearly all have $v \sin i < 120$ km s^{-1} except for a few in very short-period binaries. Although there are a few early A-type Am stars known, such as Sirius, most Am stars have hydrogen types later than A4 and all have $v \sin i < 120$ km s^{-1}. The "HR 4049" stars are rare; only that one example is known among these 2010 stars.

The true λ Boo stars have weak metallic lines for their hydrogen types and they usually have very broad lines. Because the hydrogen lines pass through a broad maximum in strength in the early As, hydrogen types are difficult to assign accurately. Therefore the classical λ Boo stars are only those few stars with extremely weak metallic lines. However a way to discover less extreme λ Boo stars may be to note the strengths of λ4481 Mg II. This line also passes through a broad maximum in the A-type stars, which means that, if this line is weak, the weakness cannot be attributed to an error in spectral type. Fortunately we obtained quantitative measures of the λ4481 equivalent widths from the CCD spectra that confirm our visual classifications, so we were able to discover many less-extreme cases of λ Boo stars. We also checked the Ca II K-line photoelectric photometry by Henry and Hesser (1971) and learned that for those stars with weak Mg II λ4481, the Ca II K-lines are also weak by an average of one spectral subtype. The current best explanation for the λ Boo phenomena is not that of diffusion when the rotational velocities are low, as seems to be valid for the Ap and Am stars, but rather the accretion of interstellar gas that has been depleted of certain elements, such as Ca and Mg, that went into the formation of grains (Charbonneau 1991).

I do not know how to distinguish λ Boo stars from Population II metal-poor stars (if such stars exist), except that the latter seem to have low rotational velocities.

Extreme shell stars such as HR 1968 = 12 Lep are rare. A way of identifying much less extreme shell stars is to look for sharp Ti II absorption lines at $\lambda\lambda$3685, 3759, and 3761 in broad-lined spectra (Abt and Moyd 1973); those often provide the first evidence for the existence of an inner high-excitation shell or disk.

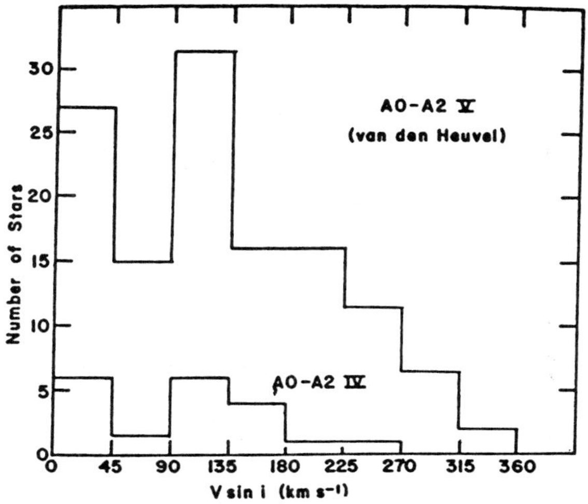

Figure 1. An example of van den Heuvel's (1968) data showing that at some spectral types the rotational velocities show a bimodal distribution of slow and intermediate rotational velocities.

4. Rotational Velocity Results

Figure 1 below shows an example of van den Heuvel's data for A0-A2 stars of luminosity classes V and IV. Note the peaks for slow ($v \sin i < 45$ km s^{-1}) and intermediate ($90 < v \sin i < 135$ km s^{-1}) stars. Why would normal stars show such a bimodal distribution? However, when the new MK classifications are used, such peaks disappear within our statistical errors. It seems to be true that when more of the spectral peculiarities are identified, the rotational velocity distributions for normal or for abnormal stars are smooth distributions with a single broad maximum. In particular, many sharp-lined Ap stars were called "A0" because they showed no helium lines.

Figure 2 shows the result of a deconvolution of the rotational velocities of the A5-F0 stars. Note that the Ap and Am stars are slow rotators with < 150 km s^{-1} in nearly all cases and that all the rapid rotators are normal or λ Boo stars. But there is an overlap of 13%. e.g. a star with v = 100 km s^{-1} can be either an abnormal or normal star. We would not expect an overlap if diffusion alone determines whether a star will be abnormal or normal.

5. Discussion

We can think of two possible reasons for an overlap in the rotational velocity distributions of abnormal and normal stars.

First, perhaps many of the slow rotators among the normal stars are members of short-period binaries in which their rotational velocities, which were originally large, are being reduced by tidal interactions but that diffusion has

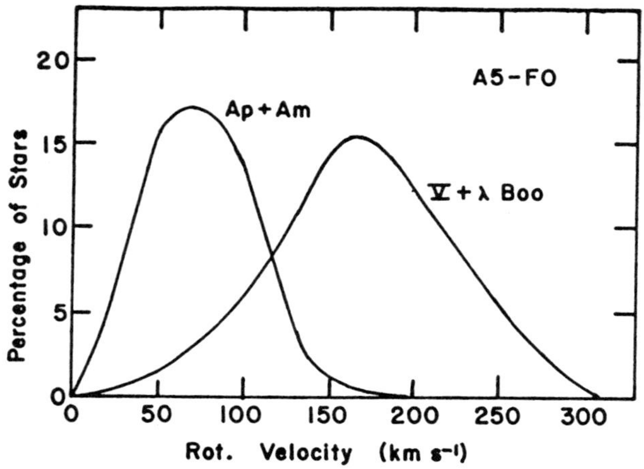

Figure 2. Deconvolution, assuming a random orientation of rotational axes, for the A5-F0 stars along the main sequence.

not yet shown the effects of element separation. We are checking that possibility by studying the binary characteristics of a sample of normal sharp-lined stars. This explanation is unlikely to be valid for the following reason. The time scale for the substantial reduction of rotational velocities by tidal interactions is of the order of 10^9 yr (Goldman and Mazeh 1991) whereas the time scale for element separation by diffusion must be of the order of 10^6 yr because Am stars occur in the Orion Nebula cluster (Levato and Abt 1976), Orion OB1 association (Abt and Levato 1977), and other groups of such an age. Thus stars would become Am stars almost as quickly as their rotational velocities drop below about 120 km s^{-1}.

The second explanation for an overlap arises when one considers the stars classified approximately A2 IV. Most such stars have sharp lines. We collected observational data regarding A2 IV stars compared with those for A2 V, A2 III, and A2p stars. They are summarized in Table 2:

Table 2. Observational data for A2 IV stars compared with other A2 stars.

	A2 V	A2 IV	A2 III	A2p
$< v \sin i >$ (km s^-1)	131	61	70	46
$< \lambda 4481$ E.W.$>$ (Å)	0.500	0.452	0.478	0.466
$< M_v >$ (mag.)	1.55 ± 0.21	1.43 ± 0.18		

We would expect the A2 IV stars to have a mean rotational velocity intermediate between that (131 km s^{-1}) of class V and that (70 km s^{-1}) of class III stars; however it is lower and approximately like that (46 km s^{-1}) of A2p stars. The equivalent width of λ4481 Mg II should also be intermediate between that (0.500 Å) of class V and that (0.478 Å) of class III stars, but it is much lower and like that of the A2p stars. Finally, parallax measures of the nearer A2 IV stars (for which the parallaxes are significantly larger than their errors) yield absolute magnitudes that are not significantly brighter than for class V stars.

Therefore many of the A2 IV stars seem to be a species of slow-rotating metal-weak class V stars that had not been previously recognized as being peculiar. Thus we suspect that some of the slow-rotating stars called "normal" are really abnormal and that there might not be any overlap in the rotational velocity distributions if all the abnormal stars were so identified.

References

Abt, H. A., and Levato, H. 1977, PASP, **89**, 797

Abt, H. A., and Morrell, N. 1992, in IAU Coll. 138, *Peculiar Versus Normal Phenomena in A-Type and Related Stars*, ed. M. M. Dworetsky, F. Castelli, and R. Faraggiana, 384

Abt, H. A., and Moyd, K. I. 1973, ApJ, **182**, 809

Charbonneau, P. 1991, ApJ, **372**, L33

Goldman, I., and Mazeh, T. 1991. ApJ, **376**, 260

Gray, R.O ., and Garrison, R. F. 1987, ApJS, **65**, 581

———. 1989a, ApJS, **69**, 301

———. 1989b, ApJS, **70**, 623

Henry, R. C., and Hesser, J. E. 1971, ApJS, **23**, 421

Levato, H., and Abt, H. A. 1976, PASP, **88**, 712

Slettebak, A., Collins, G. W., II, Boyce, P. B., White, N. M., and Parkinson, T. D. 1975, ApJS, **29**, 137

van den Heuvel, E. P. J. 1968, BAN, **19**, 309

Discussion

Roman: In grouping stars, on what characteristic were the Am stars classified?

Abt: By their hydrogen types because they are the closest to representing Am colors, temperatures, and luminosities.

Roman: In your number vs. velocity plots, were projection effects taken into account?

Abt: Yes, the deconvolutions from $v \sin i$ to v assume a random orientation of axes. But even in a distribution by $v \sin i$, all Ap stars show low velocities and all stars having high $v \sin i$ are normal or λ Bootis.

Gray: In reference to your conclusion that the A2 IV stars may be a previously undetected class of peculiar stars, I wonder how many A2 III stars you have in your sample. In my sample of A-type stars, I found only a handful (3 or 4) out of 1,100+ stars, and so I wonder if your conclusions about the peculiarities of the A2 IV stars (gained from comparisons with the adjacent classes A2 V and A2 III) may be due to the statistics of small numbers?

Abt: Out of 2,000 stars, about 10% are above the main sequence so roughly 100 are of Class III. But you and I agree very well regarding both types and luminosity classes.

Hensberge: In the range of early A-type stars, magnetic chemically peculiar stars are recognized quite early through their 520 nm depression feature, either by photometry or spectroscopy. Did you check the photometric literature in that respect concerning your A2 IV stars, or did you look spectroscopically at that wavelength region?

Abt: No, I did not look yet. However, there is a paper to come by Dr. Rao in which he has listed these A2 IV stars. If the "peculiarity" of A2 IV is not very obvious, so the 5200 Å feature may be very subtle.

Osterbrock: Continuing from Nancy Roman's question about the difference between rotational $v \sin i$ and v, it seems to me that you have shown that if you assume the normal A stars and the Ap and Am stars have their axes of rotation oriented in random directions, then there is an overlap in rotational velocity between the two groups. But I think the possibility remains that the Ap and Am stars might be preferentially pole on toward us, either because these peculiarities are more concentrated at the poles of the stars, or because you cannot detect the peculiarities so well if the lines are broadened by rotation.

Abt: Good point. The partial separation of abnormal and normal stars occurs in both the $v \sin i$ and v distributions. However, Deutsch showed that the magnetic Ap stars are not spherically symmetric. So there may be some evidence that some Ap stars do not show a random orientation of rotational axes. The problem is not with the Ap and Am stars, which all have $v \sin i < 200$ km s^{-1}, but with there being too many sharp-lined normal stars.

The MK Process at 50 Years
ASP Conference Series, Vol. 60, 1994
C.J. Corbally, R.O. Gray, and R.F. Garrison (eds.)

The A2 IV Stars

S. Sreedhar Rao
Center of Advanced Study in Astronomy, Osmania University, Hyderabad 500 007, India

M. Parthasarathy
Indian Institute of Astrophysics, Bangalore 560 034, India

C. Raghavender Rao
Center of Advanced Study in Astronomy, Osmania University, Hyderabad 500 007, India

Abstract. A preliminary analysis of the high resolution spectra of eight A2 IV stars are presented. As reported by Abt and Morrell (1993) the slowly rotating A2 IV stars have relatively weak Mg II 4481 lines compared to the rapidly rotating A2 IV stars. However, the slowly rotating A2 IV standard star HR 4343 shows a very strong Mg II 4481 line.

We have also examined the IRAS fluxes of several A2 IV stars in order to find any link between the weakness of Mg II 4481 line in slowly rotating A2 IV stars to the presence of warm circumstellar dust.

1. Introduction

Abt and Morrell (1993) state that there is something strange about most of the stars classified at or near A2 IV. They have found that most of the stars classified as A2 IV have sharp lines and in some cases the 4481 Mg II, 4077 and 4215 Sr II and some other lines are too weak. They have found that the strength of the 4481 Mg II line is less in A2 IV stars (0.452 ± 0.015 Å) than in A2 V stars (0.500 ± 0.008 Å). Abt and Morrell (1993) mention that could it be that the A2 IV stars represent an undiscovered class of peculiar stars that may not even be class IV stars, despite their narrower hydrogen lines.

2. Observations

In order to further understand the nature of A2 IV stars we have started a high resolution spectroscopic study of these stars. We have selected the A2 IV stars from the list of refined MK classification of the early A type stars by Gray and Garrison (1987). Their rotational velocities are taken from the Bright Star Catalogue (Hoffleit and Jaschek 1982). We have obtained high resolution CCD spectra of eight A2 IV stars with the coudé echelle spectrograph on the 1 meter telescope of the Vainu Bappu Observatory, Kavalur. The spectral resolution is

0.45 Å. In this paper we report the results of preliminary analysis of the spectra of these eight A2 IV stars. We also report an analysis of the IRAS fluxes of stars at or near A2 IV.

3. Analysis

The list of eight A2 IV stars for which we have obtained high-resolution spectra are given in Table 1, along with their rotational velocities taken from the Bright Star Catalogue. We have derived the Mg II 4481 Å equivalent widths which are also given in Table I. In the slowly rotating A2 IV stars HR 4689 and HR 7194 the equivalent widths of Mg II 4481 lines are similar to those found by Abt and Morrell (1993). However, in HR 4343 which is also a slowly rotating A2 IV star we find the Mg II 4481 equivalent width to be 0.828 Å. Corbally and Gray (1993) consider HR 4343 as a low $v \sin i$ standard A2 IV star. In the Bright Star Catalogue its spectral type is given as A2 III. The lines of Ti II, Cr II, Fe I, Ba II etc., in the spectrum of HR 4343 are relatively stronger compared to the strength of these lines in the spectra of HR 4689 and HR 7194 (Table 1). The b-y and β indices (Gray and Garrison 1987) and the Hβ line strength of HR 4343 suggest that it is an A2 IV star.

Table 1. Mg II 4481 Equivalent Widths of A2 IV Stars

Star	V	Sp. Type	$v \sin i$	b-y	m_1	c_1	β	W (mÅ)
HR 4689	3.89	A1 IV	34	0.017	0.162	1.132	2.86	464
HR 4343	4.48	A2 IV	58	0.010	0.161	1.201	2.87	828
HR 7194	2.6	A2 IV-V	72	0.03	0.2	1.03	2.91	416
HR 6446	4.33	A1.5 IV	115	0.02	0.161	1.11	2.89	535
HR 5867	3.67	A2 IV	170	0.046	0.160	1.132	2.866	535
HR 5107	3.37	A2 IV	173	0.067	0.180	1.010	2.875	563
HR 4875	5.89	A3V/A2IV[a]	183					376
HR 7254	4.11	A2 IV-V	201	0.018	0.185	1.062	2.886	650

[a]Renson *et al.* 1990

The 12 μm IRAS flux of HR 4343 indicates the presence of slight IR excess and hence the presence of warm dust. Detailed analysis of the spectrum of HR 4343 may enable us to understand this star. The strengths of Mg II 4481 line in the spectra of rapidly rotating A2 IV stars HR 6446, HR 5867, HR 5107 and HR 7254 are found to be similar to those in A2 V stars except in the star HR 4875. HR 4875 has been listed by Renson, Faraggiana and Boehm (1990) as an A2 IV star and a suspected λ Boo type which is, however, classified as an A3 V star in the Bright Star Cataloge. We have measured the equivalent width of its Mg II 4481 line to be 376 mÅ which is much weaker compared to those in the other stars. Based on the weakness of Mg II 4481 we conclude that HR 4875 is a new λ Boo star. Detailed abundance analysis of this star is needed.

4. IRAS Fluxes

The IRAS fluxes of the stars at or near the MK class A2 IV are listed in Table II. The IRAS fluxes of several A2 IV stars are found to show the presence of warm circumstellar dust. Detailed analyses of high resolution spectra of these stars are required to study the abundance pattern of C, N, O, S, Zn and Fe, Ca, Mg, Si, Ti, etc. to find any relation between these stars and λ Boo stars.

Table 2. IRAS Data of A2 IV and Related Stars

Star	V	Sp. Type[a]	IRAS flux (Jy)			IR
			12 μm	25 μm	60 μm	Excess
HR 63	4.61	A2 IV	0.68	---	---	---
HR 383	4.76	A2 IV	.63	---	---	---
HR 887	4.63	A2 IV	0.64	---	---	---
HR 1324	4.61	A2 IV	0.8	---	---	Yes
HR 1389	4.29	A2 IV	0.94	---	---	---
HR 1483	5.01	A2 IV	0.6	0.1	0.12	Yes
HR 2946	4.99	A3 IV	0.55	---	---	---
HR 3131	4.61	A2 IV	0.79	---	---	Yes
HR 3311	6.3	A2 IV	0.49	---	---	Yes
HR 3615	4.0	A2 IV	1.33	0.34	0.15	Yes
HR 3690	3.82	A2 IV	1.61	---	---	Yes
HR 3863	5.32	A3 IV	0.5	0.19	0.35	Yes
HR 3894	4.59	A2 IV	0.74	---	---	---
HR 4033	3.45	A2 IV	1.81	0.52	---	---
HR 4192	5.08	A2 IV	0.45	---	---	---
HR 4343	4.48	A2 IV	0.8	---	---	---
HR 4689	3.89	A1 IV	1.36	---	---	Yes
HR 4828	4.88	A0 V	0.62	---	---	Yes
HR 4875	5.89	---	0.37	---	---	Yes
HR 5107	3.37	A2 IV	2.76	0.70	---	Yes
HR 5367	4.05	A0 IV	1.04	0.47	---	Yes
HR 5867	3.67	A2 IV	1.94	0.47	---	Yes
HR 5972	4.83	A3 V	0.61	---	---	Yes
HR 6332	5.25	A3 V	0.78	---	---	Yes
HR 6549	5.25	A5 IV	0.48	0.26	0.22	Yes
HR 7194	2.6	A2 IV-V	4.3	0.98	---	---
HR 7254	4.11	A2 IV-V	1.15	0.5	---	Yes
HR 7779	5.59	A2 V	0.25	0.14	0.15	---

[a] Gray & Garrison 1987

5. Conclusions

We confirm Abt and Morrell (1993) findings that the slowly rotating A2 IV stars are peculiar with respect to their slightly weaker Mg II 4481 lines compared to A2 V stars while the rapidly rotating A2 IV stars show similar strengths as in A2 V stars. The weakness of Mg II 4481 in HR 4875, a fast rotator, suggests that it is a new λ Boo star. Analysis of all the 30 odd stars classified as A2 IV in the Bright Star Catalogue should be carried out to determine whether they form a new peculiar class among the early A type stars.

Acknowledgments. We gratefully acknowledge the travel support extended to SSR by the Vatican.

References

Abt, H. A. and Morrell, N. I. 1993, in *Peculiar versus Normal Phenomena in A-type and Related Stars*, eds. M. M. Dworetsky, F. Castelli, and R. Faraggiana, ASP Conf.Ser., **44**, 384

Corbally, C. J. and Gray, R. O. 1993, ibid, 432

Gray, R. O. and Garrison, R. F. 1987, ApJ, **65**, 581

Hoffleit, D. and Jaschek, C. 1982, *The Bright Star Catalogue*, (New Haven: Yale Univ. Obs)

Renson, P., Faraggiana, R. and Boehm, C. 1990, Bull.Inform.CDS, No. 38

Discussion

Roman: In view of your and Abt's comments on the A2 IV stars, do you feel that these are really between III and V or might the appearance indicate a form of peculiarity?

Rao: The A2 IV class could be distinguished from III and V from the wings of their Balmer lines. However, the effect is subtle and often we tend to assign a III-IV or IV-V class. These stars may show peculiar abundances.

Corbally: Gray has found that near A0 there is quite a variability in the strength of the Mg II 4481 line in *normal* stars, and I would invite him to comment whether this continues for the A2 stars.

Gray: Yes, indeed, the Mg II 4481 line varies significantly in strength from star to star in the early A-type stars (including A2) and so I am not surprised that you found such a variation in the A2 IV class. One thing that I am surprised about, however, is that you found that rotation seemed to strengthen Mg II 4481. When I did my classifications of 1100+ A-stars, I had the distinct impression that rotation tended to *weaken* Mg II 4481, and slightly strengthen other blends, such as λ4417.

I might also add that a weak Mg II 4481 line is *not* enough to classify a star as a λ Boo star – it must also show significant weakening in the general metallic-line spectrum. Thus, you may be premature in announcing HR 4875 as a λ Boo star. In addition, you attribute the types in Table 2 to Gray and Garrison, but in a number of instances they differ significantly from our published types: one of the stars in that table (HR 4828) is a λ Boo star, and another one (HR 3311), we mentioned in the notes to our paper as having mild λ Boo characteristics including a weak 4481 line.

Rao: The fast rotators in our sample, with the exception of HR 4875, show normal strength for their Mg II 4481. HR 4875 has a weak Mg II 4481 and it should be studied in detail to find out if it exhibits λ Boo characteristics.

Abt: With regard to what Richard Gray just said about rotation effects, of the λ Bootis and 4481-weak stars that we find, almost all have broad lines.

Rao: The rapidly rotating A2 IV stars in our sample show broad lines but their 4481 Mg II equivalent widths are similar to those in A2 V except for HR 4875.

The MK Process at 50 Years
ASP Conference Series, Vol. 60, 1994
C.J. Corbally, R.O. Gray, and R.F. Garrison (eds.)

Spectral Classification and Incidence of the Upper Main Sequence Chemically Peculiar Stars

Francesco A. Catalano

Istituto di Astronomia, Città Universitaria, I–95125 Catania, Italy

Francesco Leone

Osservatorio Astrofisico di Catania, Città Universitaria, I–95125 Catania, Italy

Abstract. The spectral type incidence of the stars currently known or suspected to be chemically peculiar (CP) is compared with the spectral type incidence of all the stars up to 8.5 visual magnitude. The 8.5 magnitude limit has been assumed because it is just above the completeness limit of early type stars in the HIPPARCOS Input Catalogue. The CP star sample has been extracted from the *General Catalogue of Ap and Am Stars* (Renson et al. 1991). The same general increase in the number of stars per unit magnitude interval occurs for CP and non-CP stars up to the 6.5th magnitude. However, in the range from 6.5 to 8.5 magnitude the number of CP stars suddenly decreases, as if an observational selection was present. The incidence of the various subgroups of CP stars is also evaluated and the problem of their spectral classification is discussed.

1. Introduction

Since the recognition of the existence of peculiarities in the spectra of a number of A stars in the HD Catalogue, the first systematic study of the spectral peculiarities of the so called Ap stars was done by Morgan (1933), who demonstrated the *existence of a peculiar branch of the spectral sequence in the interval B8–F0* in which such elements as Si, Mn, Cr, Sr, Eu show abnormally strong lines. It is a lucky coincidence that while celebrating with the present workshop the fifty years anniversary of the MK process, at the same time we can also celebrate the sixty years anniversary of the first recognition of the Ap and Bp stars as a class by Prof. Morgan.

After Babcock (1947, 1958) discovered large magnetic fields to be present in a number of Ap and Bp stars, great interest has arisen and many studies have been devoted to understanding the properties of such a class of stars. Among the several review papers and conference proceedings available in the literature, one of the most comprehensive discussions of the properties of the peculiar A and B stars is given in Wolff's (1983) monograph.

Recently a *General Catalogue of Ap and Am Stars* has been published (Renson et al. 1991), in which almost all Bp, Ap, and Am stars known up to the end of 1991 are listed. In this catalogue the Ap and Am terms are used because they are largely known and representative: there the term Ap stands for Bp, Ap, and Fp. Related stars, as for example λ Boo, are not included. Moreover in the group of Am stars the Fm and δ Del stars are also included.

For the sake of brevity here we will use Preston's (1974) nomenclature, according to which the abbreviation CP stands for *upper main sequence chemically peculiar stars*, and the various subgroups are indicated by a number (1 through 4) appended to the term CP.

In this work we will briefly illustrate some relevant characteristics of CP star subgroups with the aim of investigating their incidence among all stars in the same spectral range.

2. The CP Star Subgroups

At present 6684 upper main sequence stars are known (or suspected) to be chemically peculiar (Renson et al. 1991). Among them, 3492 stars (52.2%) are classified as Am, or CP1 stars, according to Preston's (1974) classification, 2922 (43.7%) stars are classified as Ap and Bp stars, i.e. CP2, CP3, and CP4 stars according to Preston's classification, while the remaining 270 stars (4.1%) are suspected to be peculiar, but no peculiarity subgroup is assigned. As far as the reliability of the peculiarity classification is concerned, 42 stars are quite surely erroneously classified and should be rejected (we will do it in the following), 1726 stars have a somewhat dubious classification which needs to be confirmed, while 4916 stars can be considered almost reliably classified. In the following we will consider the CP sample as composed of the 6642 more or less well classified stars. The subdivision of CP star subgroups we shall refer to here is given in Table 1.

Table 1. CP star subgroups and their designations.

Subgroup	peculiarity type	designation	N. of stars
CP1	Am, Fm, δ Del	CP1 or Am	3481
CP2	Cr, Eu, Sr and combinations	SrCrEu	965
	Si + Cr, Eu, Fe, Mg, Sr, ...	Si+	277
	Siλ4200, Si	Si	1356
CP3	Mn, Hg	HgMn	124
CP4	He weak, He strong, He var	He abnormal	185
—	Pecul. class. missing	n.c.	254

CP2 and CP4 stars are also commonly referred to as magnetic CP stars, since in many of them large scale magnetic fields stronger than a few hundred gauss have been measured. However, although stars in the CP2 and CP4 subgroups are observed to possess detectable magnetic fields, it is by no means clear that all the stars belonging to these subgroups do possess large-scale organized magnetic fields. For example, He weak (CP4) stars showing strongly enhanced lines of Ga and P are quite certainly non magnetic (Borra et al. 1983).

It is important to note that the classification into subgroups given in Table 1, as well as any other, is only schematic and for practical use. In fact, no two stars are found which are perfectly alike: large individual differences exist from star to star and cases intermediate between two subgroups are found, so that the boundaries of the subgoups in no case should be considered as rigid. However the necessity arises to select and order the occurence of the peculiarities: this may provide useful hints in picking-up common properties in each subgroup which can be useful for successive interpretation. In this respect the extension of Preston's classification proposed by Maitzen (1984) looks useful. The basis is the odd-even effect with respect to the presence of magnetic fields (CP1 and CP3 non-magnetic, CP2 magnetic), and it consists in introducing the subgroups CP4 and CP5 for the He-weak and the subgroups CP6 and CP7 for the He-strong stars. A recent thorough discussion of CP star detection and classification has been given by Faraggiana (1987) and will not be repeated here.

3. The Incidence of CP Stars

CP stars fall in the spectral range B0–F7; hence to determine their incidence we have selected all the stars contained in such a spectral range in the *HIPPARCOS Input Catalogue* (Turon et al. 1992, henceforth referred to as HIC) up to the 8.5 visual magnitude. It is important to note that the $m_V \approx 8$ is the limit up to which HIC should be complete, as far as the early type stars are concerned. In fact, according to the authors, all stars earlier than the spectral type G5 are listed up to $m_V = 7.9 + 1.1 \sin|b|$.

Among the stars in the B0–F7 spectral type range we have selected the stars of the luminosity classes V, IV, and III. Class III has been included since very frequently such a luminosity class is assigned, or at least it has frequently been, to CP stars. To avoid the above mentioned problem of assigning each CP star to a well defined subgroup, in this preliminary study we have divided CP stars in only two groups: CP1 and CP2/3/4.

Table 2. Spectral type distribution of CP and non-CP stars up to $m_V = 7.5$.

Spectral	CP1		CP2/3/4		All CP		Non–CP		All
range	N.	%	N.	%	N.	%	N.	%	N.
B0–B1	—	—	5	0.70	5	0.36	285	2.69	290
B2–B3	—	—	22	3.08	22	1.56	710	6.70	732
B4–B5	—	—	17	2.38	17	1.21	445	4.20	462
B6–B7	—	—	33	4.62	33	2.35	285	2.69	318
B8–B9	7	1.01	327	45.80	334	23.76	1582	14.92	1916
A0–A1	104	15.03	191	26.75	295	20.98	1719	16.21	2014
A2–A3	308	44.51	58	8.12	366	26.03	1362	12.84	1728
A4–A5	149	21.53	6	0.84	155	11.02	495	4.67	650
A6–A7	58	8.38	16	2.24	74	5.26	229	2.16	303
A8–A9	22	3.18	14	1.96	36	2.56	225	2.12	261
F0–F1	29	4.19	13	1.82	42	2.99	823	7.76	865
F2–F3	12	1.73	8	1.12	20	1.42	955	9.01	975
F4–F5	3	0.43	3	0.42	6	0.43	941	8.87	947
F6–F7	—	—	1	0.14	1	0.07	548	5.17	549
total	692	100	714	100	1406	100	10604	100	12010

In the tables and figures the spectral types were binned into two-spectral-subtype-wide bins to avoid the odd-even effect. V magnitudes were binned into 1–magnitude wide bins centered on V = 0.0, 1.0, 2.0, $\cdots$.

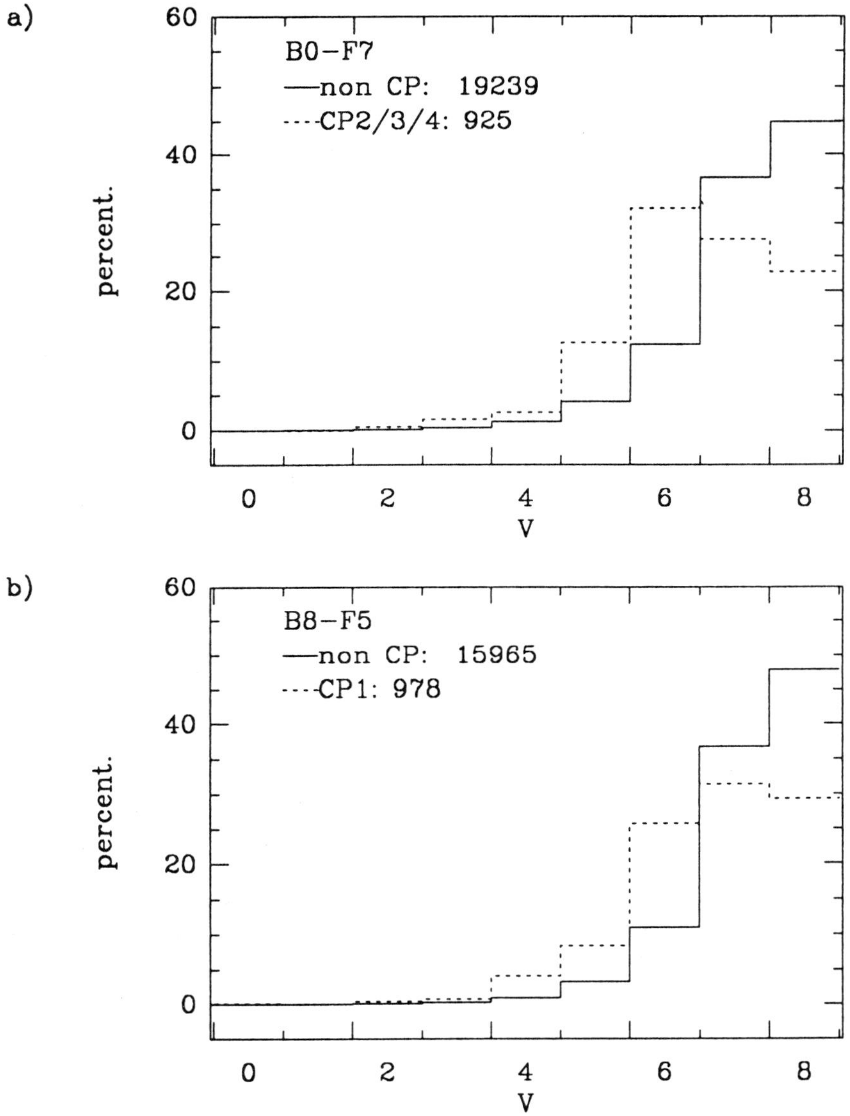

Figure 1. The frequency of CP and non-CP stars is shown as a function of the V magnitude up to V=8.5; normalization is with respect to the number of stars in each group within the appropriate spectral range:

a) CP2/3/4 and non-CP stars in the spectral range B0-F7
b) CP1 and non-CP stars in the spectral range B8-F5

In Figure 1 the incidence of CP and non-CP stars per unitary magnitude interval is shown up to the visual magnitude 8.5 in the appropriate spectral type ranges: a) CP2/3/4 and non-CP stars in the spectral range B0-F7, b) CP1 and non-CP stars in the spectral range B8-F5. From this figure we see that the completeness limit of knowledge of CP stars is at V = 6.5. Beyond this magnitude value the number of known CP stars rapidly decreases.

In Table 2 the spectral type distribution of CP and non-CP stars is shown up to V = 7.5: 81% of CP1 stars are found in the spectral type range A0–A5, while 72% of CP2/3/4 are found in the spectral type range B8–A1.

Table 3. Incidence of CP and non-CP stars up to $m_V = 6.5$.

Sp.	CP1		CP2/3/4		All CP		Non–CP		All
range	N.	%	N.	%	N.	%	N.	%	N.
B0–B1	—	—	3	2.05	3	2.05	143	97.95	146
B2–B3	—	—	15	3.61	15	3.61	400	96.39	415
B4–B5	—	—	13	6.10	13	6.10	200	93.90	213
B6–B7	—	—	28	16.28	28	16.28	144	83.72	172
B8–B9	4	0.59	202	29.62	206	30.21	476	69.79	682
A0–A1	46	11.95	119	18.51	165	25.66	478	74.34	643
A2–A3	179	28.78	40	6.43	219	35.21	403	64.79	622
A4–A5	85	33.20	3	1.17	88	34.38	168	65.63	256
A6–A7	37	23.27	10	6.29	47	29.56	112	70.44	159
A8–A9	14	11.97	12	7.55	26	16.35	91	57.23	117
F0–F1	13	5.46	7	2.94	20	8.40	218	91.60	238
F2–F3	6	2.00	4	1.33	10	3.33	290	96.67	300
F4–F5	1	0.43	2	0.85	3	1.28	231	98.72	234
F6–F7	—	—	1	0.44	1	0.44	224	99.56	225
B0–F7	385	8.71	459	10.38	844	19.09	3578	80.91	4422
B8–F5	385	11.84	399	12.27	784	24.12	2467	75.88	3251

In Table 3 the incidence of CP and non-CP stars is shown up to the visual magnitude 6.5, which is the completeness limit in magnitude, as evident from Fig. 1a and Fig. 1b. In the last two rows we compare the incidence of CP1, CP2/3/4, and non-CP stars in the spectral ranges B0-F7 and B8-F5, where all CP2/3/4 and CP1 respectively are found. If we consider the larger spectral range and include CP stars without respect of their subgroup membership, we find an incidence of 19%, whereas in the smaller spectral range, where all CP1 are found, a global incidence around 24% is found.

4. Conclusion

In the past the incidence of CP stars has been studied by several authors on the basis of selected samples or restricted spectral ranges, but the results are far from conclusive (Schneider 1993a, b). From the present study we can draw the following conclusions:
i) CP2/3/4 are mainly concentrated in the spectral type range B8-A2, with maximum frequency of ≈46% at B8-B9;
ii) CP1 are mainly concentrated in the spectral type range A0-A5, with maximum frequency of ≈45% at A2-A3;

iii) The CP star sample can be considered complete up to V = 6.5: fainter stars are underrated, mainly because of observational biases;
iv) Up to V = 6.5 the frequence of CP stars is:
- 11.84% for CP1 stars in the spectral type range B8-F5 (8.7% in the spectral range B0-F7),
- 10.4% for CP2/3/4 stars in the spectral type range B0-F7 (12.27% in the spectral type range B8-F5),
- 19.1% for all CP stars in the spectral type range B0-F7,
- 24.1% for all CP stars in the spectral type range B8-F5.

References

Babcock H.W., 1947, ApJ 105, 105

Babcock H.W., 1958, ApJS 3, 141

Borra E.F., Landstreet J.D., Thompson I., 1983, ApJS 53, 151

Catalano F.A., Renson P., 1984, A&AS 55, 371

Catalano F.A., Renson P., 1988, A&AS 72, 1

Catalano F.A., Renson P., Leone F., 1991b, A&AS 87, 59

Catalano F.A., Renson P., Leone F., 1993, A&AS 98, 269

Faraggiana R., 1987, Ap&SS 134, 381

Maitzen H.M., 1984, A&A 138, 493

Morgan W.W., 1933, ApJ 77, 330

Preston G.W., 1974, ARA&A 12, 257

Renson P., Gerbaldi M., Catalano F.A., 1991, A&AS 89, 429

Schneider H., 1993a, CP stars at High Galactic Latitudes: Photometry near the Galactic Poles. In Proc. IAU Coll. 138, Peculiar versus Normal Phenomena in A–Type and Related Stars, eds. Dworetsky M.M., Castelli F., Faraggiana R., ASP Conference Series 44, p. 592

Schneider H., 1993b, Statistics of CP stars in a magnitude–limited ample: The Bright Star Catalogue. In Proc. IAU Coll. 138, Peculiar versus Normal Phenomena in A–Type and Related Stars, eds. Dworetsky M.M., Castelli F., Faraggiana R., ASP Conference Series 44, p. 629

Turon C., Crézé M., Egret D., et al., 1992, The HIPPARCOS Input Catalogue. ESA SP–1136, ESA Publication Division, c/o ESTEC, Noordwijk, The Netherlands (HIC)

Wolff S.C., 1983, The A-type Stars: Problems and Perspectives. Monograph Series on Nonthermal Phenomena in Stellar Atmospheres, NASA SP-463

Discussion

Garrison: I must comment that I am not at all happy with the use of the term “chemically peculiar” for classifying the peculiar A-type stars. It implies a restrictive theoretical interpretation, which is already questionable. It does not allow for the evolution of the theory to include structural peculiarities and magnetic anomalies. I doubt that the Ap stars are really "chemically peculiar" overall.

PART IV.

COMPLEMENTARITY BETWEEN SPECTROSCOPY AND PHOTOMETRY

The MK Process at 50 Years
ASP Conference Series, Vol. 60, 1994
C.J. Corbally, R.O. Gray, and R.F. Garrison (eds.)

On the Complementarity between MK Classification and Strömgren Photometry

Erik H. Olsen
Copenhagen University Observatory, Brorfeldevej 23, DK-4340 Tølløse, Denmark

Abstract. The mutual interaction between the two different approaches to classification is described, beginning with Strömgren's early papers and Crawford's calibration work. The continuing symbiosis between the two classification systems to the present day is outlined with a heavy bias towards my own involvement on the photometry side.

The term 'complementarity' is used in a wider sense than usual, since I try to describe the early contacts between Strömgren and the Yerkes community. Also, the interdependence between spectroscopic and photometric surveys, in the sense that one type of survey is used to create an observing list for the other type, is regarded as 'complementarity'.

1. Introduction

I will restrict myself to the four-color and Hβ photometry developed by Strömgren and his colleagues. I will not comment on relations between MK classification and other photometric systems, like Geneva, Vilnius, DDO, or Johnson *UBV*, although much could undoubtedly be said about that. I will leave that to colleagues more well versed in those systems than I am.

Bengt Strömgren and William Morgan were both at Yerkes, and it is only natural that the development of Strömgren's quantitative classification work benefited greatly from a continuous interaction with Morgan and Keenan's classification of stellar spectra. I have chosen to look into this early interaction, instead of boring you with a lot of details about photometry and detailed comparisons between the two systems.

(I have noted that Dr. Keenan (1984) would like me not to use the word 'quantitative' as opposed to MK classification, since the latter is of course also a quantitative classification. I agree with Dr. Keenan and *Webster's Dictionary* that the words 'objective' and 'visual' would be better terms, but the use of the word 'quantitative' is so ingrained that we will probably have to live with it.)

I will try to describe the mutual interaction between the two different approaches to classification and show how both have benefited, the photometry far more so than the MK classification. The MK classification is, of course, the fundamental independent framework, especially when more dimensions – like abundances and rotation – are grafted onto it. Stars selected for quantitative classification, i.e. for photometric observations, have in many cases been screened by visual spectral classification to limit the spectral range and eliminate peculiar stars in advance.

Photometry is limited by using only a few selected features in the spectrum, making it more difficult to detect and specify peculiarities. Also, most photometric systems are only useful in part of the HR diagram. The power of photometry is speed, fainter limiting magnitude, and a higher internal precision, i.e. a higher resolving power. By paying sufficient attention to standard stars, this resolving power can be maintained within a standard system used by many observers, i.e. the high *internal precision* can be converted to a high *external accuracy*. This is not a "deification of quantization", as Garrison (1985) has warned against, but a necessary basis for accurate calibrations to more fundamental parameters.

Since photometry and MK classification are measuring spectra in two fundamentally different ways, comparisons between the results are very fruitful, especially when they do *not* agree. In those cases, we learn more about stellar atmospheres and also more about the classification systems. The combination is more powerful than the sum of the systems.

In describing the early history of complementarity and interaction between MK classification and Strömgren photometry, I can only rely on the published literature, since my own personal involvement on the photometry side, and my aquaintance with Bengt Strömgren, only date to the late 1960's, when Strömgren returned to Copenhagen. It may seem very presumptuous of me to try to do this with Dr. Keenan in the audience. I can only apologize in advance for any misinterpretations that may have crept in. I hope very much that Dr. Keenan will point out to me my errors, as I go along, and I regret very much that Dr. Morgan is not here today to do the same. Of course, Dave Crawford, who was also at Yerkes and played a very important rôle in these early developments, is also expected to correct me.

2. Early development, 1936-1959

2.1. First personal contacts, 1936-1948

Since both the photometry and the spectral classification are identified so thoroughly with the names of their originators, it seems natural to try to trace Strömgren's early association with Morgan and Keenan. In 1936-38, Strömgren spent a "particularly happy" 18 months at the University of Chicago and Yerkes Observatory as an assistant professor (Strömgren 1983). Among his most valuable experiences, he mentions "discussions with W.W. Morgan concerning the new system of MK classification (then almost fully developed)". Shortly after his appointment as extraordinary professor of astronomy at Copenhagen University in 1938, he wrote an account of the Yerkes Observatory (Strömgren 1938). It was written for a general public, which was sufficiently interested in astronomy to have become members of *Astronomisk Selskab*, which is a Nordic Astronomical Society open to anyone. He describes buildings, telescopes, and research fields of the scientists starting with George E. Hale. About Morgan he writes (translated from Danish, as are all other subsequent quotes from *Nordisk Astronomisk Tidsskrift*): "*Morgan* has been *Struve's* collaborator on certain of his spectroscopic investigations. For the moment, [*Morgan*] is primarily engaged in a systematic extension of the methods of very accurate spectral classification and determination of absolute luminosities by spectroscopic means using spectra of intermediate dispersion. An important application of these methods is the accurate prediction of the color of the light *emitted* by a star solely based on the nature of the absorption-line spectrum. A comparison with the color of the *observed* light gives a determination of color changes by the absorption in the interstellar space."

Here, Strömgren refers to the investigation by Morgan (1938), which was published with the title "On the Determination of Color Indices of Stars From a Classification of Their Spectra". This paper is an early example showing how new knowledge is created by comparing photometry and spectral classification. In view of the composition of Strömgren's circle of readers, it is interesting to note that, as always, Strömgren's choice of words is very precise. There is little doubt that this study by Morgan had an influence on Strömgren's later development of his photometric classification methods.

In the same article, Strömgren (1938) also mentions an astronomer at the University of Chicago, Dr. Keenan, "whose main scientific interest is solar research." Rather appropriate, one might say, considering that the first paper published jointly by Morgan and Keenan (1939) was "On the Spectral Type of the Sun", and that the solar spectrum has been close to the meeting/division point between Morgan and Keenan in later years.

After the war, in 1947-48, Strömgren was again visiting colleagues in the U.S. In four articles in *Nordisk Astronomisk Tidsskrift*, he has described his impressions from a number of American observatories, one of them the Perkins Observatory (Strömgren 1948). After a short historical introduction, he describes the main instrument with its two spectrographs. He mentions that Keenan and Hynek have "developed methods for spectral classification of red stars utilizing lines and molecular bands in the wavelength range 6900 - 8700 Ångström". He then singles out one of the research projects at the Perkins Observatory, "which concerns an especially significant subject." In an introductory digression of four paragraphs, Strömgren then presents his general readership with a concise survey of Baade's work on the two stellar populations.

> "Some years ago, *W. Baade*, at the Mt. Wilson Observatory, showed that the spiral galaxies on the one hand and the elliptical galaxies on the other hand show large differences with respect to the properties of their stars. The most conspicuous difference is that the most luminous stars in the elliptical galaxies are red giants, while for the spiral galaxies it is white giants. These studies led Baade to the assumption that two essentially different types of stars exist – *Population I* and *Population II* – which occur in strongly variable proportions in different parts of the metagalactic system. The stars in our neighbourhood of the Milky Way system belong mainly to Population I. In the globular clusters, one finds mainly (or exclusively) stars of Population II, among these red giant stars.
>
> Certain elliptical galaxies seem to consist exclusively of stars of Population II. Spiral galaxies contain both populations, albeit such that Population I predominates in the spiral arms, Population II near the centre of the system.
>
> It is probable that the core of our Milky Way system (around the Sagittarius centre) mainly contains stars of Population II. This is suggested by, among other things, the concentration of globular

clusters towards the centre of the Milky Way system. In our neighbourhood of the Milky Way system, Population I presumably predominates, but Baade supposes that a large fraction of the stars with large space velocities belongs to Population II, because these stars move in orbits, which reach relatively near the centre.

A quite clear and adequate definition of the two populations has not yet been put forward. There seems, it is true, not to be any doubt that, e.g., the luminous red giants and the Cepheids of short period belong to Population II, or that the O and B stars belong to Population I, but the dwarf stars comprise, in an unexplained way, both stars of Population I and of Population II."

After this digression, Strömgren (1948) continues: "Now, at the Perkins Observatory, Keenan, in collaboration with *W. W. Morgan* and *G. Münch*, Yerkes Observatory, has started investigating, if it is possible by spectroscopic means to distinguish between stars of Population I and Population II. For this purpose, they have studied stars in our vicinity, partly normal stars, partly stars with large space velocities relative to the Sun. In certain cases, peculiarities in the spectra of the stars with high velocities were actually found; peculiarities, which do not appear in the spectra of normal stars of the same effective temperature and absolute luminosity. Further investigations will be required, before these matters can be said to be elucidated, but it is an exceedingly interesting result that has been attained [cf. Keenan et al. (1948)]."

In the same article, Strömgren gives a short overview of the history of photoelectric photometry from 1907 in connection with his description of Stebbins and Whitford's work at the Washburn Observatory. He ends by mentioning recent work by Hiltner, Oort, and himself reaching 18th – 19th magnitude and concludes: "Since the [photoelectric] methods can now also be used for very faint objects, their importance will surely in the future be even greater than up to now."

It is interesting to note that in their Danish textbook on astronomy (Strömgren and Strömgren 1945), Elis and Bengt Strömgren do *not* mention the new two-dimensional MK classification. (The previous two-dimensional scheme, with prefixes *c*, *g*, and *d*, is described, of course.) The preface to the book is dated April 1945, and it seems reasonable to suppose that the diffusion of astronomical literature to Denmark was fragmentary in the war years. Strömgren (1983) himself has mentioned his "relative isolation and ... quite limited resources" during those five years. However, after Strömgren's trip to the U.S. in 1947-48, he gives a detailed account of the new MK system, describing both the MKK atlas (Morgan et al. 1943) and also results concerning metallic-line A stars (Strömgren 1949). A few years later, in a semi-popular book covering all fields of astronomy, Strömgren (1955a) describes both the MK classification and the two stellar populations.

2.2. Development of the c,l photometry, 1948-1958

The problem of deriving the *three* fundamental atmospheric parameters from photometry was discussed by Strömgren at least as early as 1937, viz., in his 241-page article "Aufgaben und Probleme der Astrophotometrie", appearing in *Handbuch der Experimentalphysik* (Strömgren 1937, Chapter 6, §2), which went to press while Strömgren was at Yerkes in May 1937. This article was reviewed by Struve (1938), who described it as "probably the only complete, modern exposition of practical photometry."

It seems clear that, in the late 1940's, Strömgren considered the time ripe for large-scale investigations of galactic structure, since the main problems of stellar atmospheres, interior structure, and evolution had been solved in the preceding two decades (cf. Strömgren 1952a). At that time, Morgan (Nassau and Morgan 1949, 1950) had also started investigations in galactic research leading to the recognition of spiral arms in 1951 (Morgan et al. 1952, 1953).

Thus, in 1948, as a visiting professor at Yerkes, Strömgren (1950, 1983) began working on spectral classification based on photoelectric photometry through interference filters, mostly done at the McDonald Observatory. After January 1951, when he had been appointed professor at the University of Chicago and director of Yerkes and McDonald Observatories, Strömgren began working "systematically to develop two-dimensional classification through photoelectric photometry ..., measuring [the] Balmer discontinuity and Balmer line strengths. ... Through my work with model stellar atmospheres, I had become convinced that the hydrogen criteria were reliable indicators of effective temperature and atmospheric gravity", and it is clear that the photometry was being tailor-made to derive these two main parameters of stellar atmospheres. This is in contrast to the MK process and explains why the photometrist with two *photometrically* indistinguishable stars must always go to the spectroscopist to learn if the stars *really* are twins.

The gradual evolution of the photometry was reported in several AAS abstracts, observatory reports, and a few papers during the 1950's (Strömgren 1951a,b, 1952b, 1953, 1954, 1955b, 1956a,b,c, 1957, 1958a; Strömgren and Gyldenkerne 1955). In the first exploration of the possibilities, a set of 26 filters, covering the wavelength range 3550-5500 Å, was used to measure 110 stars of all spectral types and luminosity classes. Of course, MK classifications by Morgan and Keenan were the foundation for the interpretation. The first mention of a metallicity index is in 1957, while age discrimination is reported the year before.

The first presentation of the photometric classification methods to a wider readership is given in *Vistas in Astronomy* in 1956. In his Fig. 1, Strömgren (1956a) shows the "very close correspondence" between his two photoelectric indices, l and c, and the MK classifications of F stars by Morgan. (The two indices, defined some years earlier [Strömgren 1952b], are the forerunners of the present-day β and c_1 indices.) Strömgren comments on the slight discrepancies for two stars (HR 3579 and 5185), nicely illustrating that the complementarity between the two approaches provokes the use of other types of data to explain these cases. Strömgren (1956a) goes on to derive numerical expressions giving Johnson's B-V and the absolute magnitude M_v as functions of l and c. Morgan's spectral classifications are used again here, now to derive absolute magnitudes. Very appropriately, Strömgren ends the paper by saying: "I am particularly indebted to ... W.W. Morgan for communicating results of his spectral classification work ... , and for many helpful discussions throughout this investigation."

An extension of this quantitative classification method to A stars was reported elsewhere (Strömgren 1956b), also in 1956.

One of the first thorough presentations of Strömgren's narrow-band photometry to a larger audience was given at the 1957 Vatican meeting on stellar populations in two papers, where Strömgren (1958b,c) summarizes work on B, A, and F stars. He discusses not only two-dimensional classification based on the *l* and *c* indices, but also a discrimination between populations, using his new *m* index. He presents results in two *c-l* diagrams, where he correlates the location of stars with their MK classifications and finds a "very satisfactory correspondence". This correspondence is, of course, utilized to gain insight into the capabilities of the photometry.

In Strömgren's (1958c) Fig. 1, we may note three features.

1. A band of stars of luminosity classes V, IV, and III. The width is about 0.080 mag at A2, but diminishes to nearly no width at F7.

2. A clear separation of classes II and I from classes III-V.

3. Am stars are in, or below, the lower envelope of the band of class III-V stars.

Guided by this comparison to MK classifications, Strömgren states that "it is clear that supergiants and metallic-line stars can be segregated from their location in the *c-l* diagram", and he immediately goes on to a quantitative calibration of the band of class III-V stars in terms of Johnson's *B-V* and M_v. He finds very little cosmical scatter in the relations.

In contrast to this, calibration of *c* and *l* in terms of Johnson's *U-B* indicates "that three-dimensional classification is necessary for satisfactory representation". He then demonstrates that, for F stars, the relation between Johnson's *U-B* and *B-V* depends on both M_v and metallicity (of course with reference to the earlier work by Johnson and Morgan (1953), Johnson and Knuckles (1955) and Roman (1954)). Therefore, "one cannot from [Johnson's] *B-V* and *U-B* draw accurate conclusions regarding chemical composition without additional knowledge concerning the absolute magnitude".

In discussing quantitative measures of composition differences, Strömgren (1958b) describes the solid foundation laid down by spectroscopists (Morgan, Keenan, Roman, Aller, Greenstein, and others) and also results of theoretical models of zero-age, homogeneous stars with different chemical composition, as well as evolved models in the giant region. He then defines the color index difference *m* and points out that since the wavelength baselines of the two color indices are similar, the index *m* should be "insensitive to changes in the energy distribution in the continuous spectrum due to changes in effective temperature and gravity. Also ... little affected by interstellar absorption".

Strömgren notes HD 113468 as having a large *m* index and a large deficiency in Johnson's *U-B*, i.e. metal richness is implied; but according to the discussion after Strömgren's presentation, Morgan was not been able to see any difference between the spectrum of this star and a normal star. This is an example of complementarity at work.

Now, let us return to Strömgren's (1958c) discussion of Population I stars. He interprets the lower boundary in the *c-l* diagram "as the locus of stars of age zero". Evolution takes a star "upwards" in the *c-l* diagram. He mentions that in the future the ZAMS in the *c-l* diagram can possibly be determined as a function of the index m. This has been accomplished recently by Nissen and Schuster (1991) for the successor system, *uvby-β*.

Strömgren then turns to the A stars and mentions that rough spectral types are necessary to separate B stars from A stars, since they occupy the same part of the *c-l* diagram. He identifies the group of A0-A2 stars, later christened the 'intermediate group' (cf. Strömgren 1966a), where fairly accurate MK classification is necessary to "resolve the ambiguity", i.e. to determine if a star is before or after the Balmer discontinuity maximum.

Let us now turn to Strömgren's (1958c) Fig. 2 with the B-type stars. Five features are conspicuous.

1. The line is drawn through association stars, i.e. it defines the ZAMS.

2. Class V stars are above that line.

3. Class III stars are above class V in the range B5-B9.

4. Class V, IV and III run together among the early B stars.

5. Class I stars are far above the line, but connect with the ZAMS at O9.

Thus, we see that separation of the luminosity classes improves with falling temperature. Strömgren calibrates classes V, IV, and III in terms of Johnson's $(U\text{-}B)_0$, which in turn allows a prediction of Johnson's $(B\text{-}V)_0$ and the spectral type on the MK system. The calibration in terms of intrinsic color does not involve the index l, and Strömgren points out that a combination of Johnson *UBV* photometry and the index l will be suitable for studies of B-type stars. This suggestion had already been taken up by Dave Crawford (1958), as I will get back to in a moment.

Next, Strömgren constructs a calibration of his Fig. 2 in terms of M_v, which is a difficult problem. Several associations are used to establish the calibration of the ZAMS from B9 to O9, while Pleiades stars are used to extend it somewhat above the ZAMS near B7-B8.

Finally, the age dimension is introduced. Strömgren notes that it is "undoubtedly an age effect" that, in the range B7-B9, the class V stars "generally lie considerably above" the ZAMS, and that the c,l photometry is better to discriminate in age than MK classification: "The spectroscopic age effect is here so relatively small that the stars are generally classed in luminosity class V". Next, the problem of age calibration of the *c-l* diagram is discussed in general terms. An empirical calibration may be made, as soon as a sufficient number of clusters and associations with B-type stars have been observed, and independent evaluations of their ages are available. The available observations of the Scorpio-Centaurus association and the Pleiades cluster allow Strömgren to estimate that for B8 V stars the p.e. of an age determination from the *c-l* diagram will be about ±30 million years, and better for stars further from the ZAMS. Such an empirical age calibration was made by Crawford (1958) utilizing Johnson's $(U\text{-}B)_0$ and the β index, which he introduced in that investigation. Crawford selected his cluster and association stars from among those with MK classifications and had the benefit of Morgan's continuing interest and advice during the study.

Strömgren (1958c), in his Vatican paper, also pointed to the future possibility of computing theoretical isochrones in the *c-l* diagram. One year later, in his

Halley Lecture (Strömgren 1958d), he presents the results of such a calculation for the B-type stars, emphasizing – as also Crawford (1958) did in his simultaneous paper – that the procedure is applicable to any star; there is no limitation to cluster and association stars. From his calibrated *c-l* diagram, Strömgren derives an age of about 70 million years for the Pleiades cluster, with a p.e. of 15-20 million years. A more detailed, numerical age calibration (of the *uvby* system), based on indices computed from theoretical model atmospheres and model interiors, was given a few years later (Strömgren 1963a, 1964, 1966b), and, for B stars, birth places in the local spiral arm and in the Perseus arm were pinpointed using the newly developed density-wave theory (Strömgren 1967).

Strömgren (1958c) ends his presentation at the Vatican meeting by noting that the limiting magnitude for Hβ photometry with the newly designed two-channel photometer "is about 5 mag fainter than that obtained with a conventional slit spectrograph for MK classification ... for the same telescope and exposure time". Recent CCD techniques have, of course, narrowed this gap considerably.

3. Four-color *uvby* and Hβ photometry, 1959-1982

The lessons and results of a decade of developing the narrow-band *c,l* photometry were distilled in the creation of the intermediate-band *uvby* system. It was worked out and tested in 1959 at Lick Observatory, and the first use of the system took place at Palomar in 1960, resulting in the still unpublished Strömgren - Perry (1962) catalogue. At the same time, Crawford developed the Hβ index at Kitt Peak (cf. Strömgren 1983).

This development had been followed closely by Morgan, who already in 1954 had commented favourably on the future of narrow-band photometry (Gyldenkerne 1993). When Morgan, in his dedication address delivered at Kitt Peak in March 1959, chose two "vistas in astronomical research", he described the vast possibilities in galactic research, which were opening up by Strömgren and Crawford's development of this type of photometry (Morgan 1960).

The next 15 years or so were characterized by the solid groundwork done by Crawford and his collaborators in establishing a net of standard stars and calibrations in terms of intrinsic color, absolute magnitude, and metallicity (cf. Crawford 1960, 1975, 1978, 1979; Crawford and Mander 1966; Crawford and Barnes 1970). I'll not go into details, since Dave Crawford can do that much better, and he gave most of you an excellent review of this work ten years ago in Toronto (Crawford 1984). Suffice it to say that this fundamental work was, of course, guided throughout by reference to the framework of MK classifications. As Strömgren (1963b) said it in his encyclopedic chapter in *Basic Astronomical Data*: "In the actual development of the quantitative methods, use was made of samples of stars already classified by the inspection methods ... In this way the obvious dangers connected with classification based on a very few criteria were very much reduced ... Usually the selection of criteria is guided by results already obtained ... in the process of development of the schemes of classification by inspection." Strömgren was referring to *all* quantitative methods, not just his own photometry.

In parallel with Crawford's calibration work, numerous problems in galactic structure and evolution were attacked with the new tool of *uvby*-β photometry by Strömgren, Crawford, Perry, Barnes, Bond, Knude, Twarog, and many others. Problems studied were past and present distributions of dust and gas; properties of open clusters; correlations between age, metallicity, and kinematics; the helium content of Population I stars; age determinations for B, A, and F stars; and birth places of B stars. Sadly enough, in the 1990's this type of galactic research, envisaged by Strömgren, is becoming more and more difficult, as small telescopes are rapidly becoming an endangered species.

4. Specific examples, 1958-1994

Let me finish by describing the interaction between spectral classifications and photometry in the case of a few large observing programs that I have been involved with. The goals of these programs are studies of the history of our Galaxy, based on large, complete, all-sky samples of stars.

In all three examples, a spectroscopic catalogue provides a basis for a photometric survey, which then provokes further spectroscopic classification work. The serendipity of large survey programs (cf. Garrison 1984) ensures that there is a spin-off consisting of many intriguing objects that may not have much to do with the purpose of the survey, but investigations of these objects are prime examples of complementarity.

4.1. The F-star catalogue

The first hint of an observing program to study the age-metallicity relation for F stars in the solar neighbourhood is found in one of Strömgren's (1958b) Vatican papers. After a discussion of the metallicity index m and prompted by a question from Schwarzschild, Strömgren asks "whether or not the present material is compatible with the assumption of a unique relation between initial chemical composition and epoch of formation of a star." In retrospect, this question is rather prescient, considering that such a unique relation has, until recently, often been taken for granted by people modeling the evolution of our Galaxy. Strömgren answers the question one page later: "We are led to the conclusion that there is in the galactic neighbourhood a detectable scatter in initial chemical composition among stars of the same age." He stresses the tentative nature of his conclusion and adds "... there is hope that further investigations along these lines may lead to more information regarding the correlation between epoch of formation and chemical composition at formation for stars of our galaxy."

At the conference on interstellar matter held at Princeton in 1961, Strömgren (1962) had the first results from the new Strömgren-Perry (1962) *uvby* catalogue and concluded that "stars of nearly the same chemical composition as the Hyades stars have been formed out of the interstellar medium through a period ... presumably longer than 4 billion years". The evidence for metallicity scatter among stars of the same age was still only tentative. Strömgren (1962) also explored the correlation between age, metallicity, *and* kinematics, and he concludes his presentation by mentioning a planned *uvby* observing program of F8 and G0 stars down to 8th magnitude, "so that [a] rather ... detailed analysis of the correlations between age, chemical composition, and space velocity would become possible." The same program is discussed in his *George Darwin Lecture* (Strömgren 1963a). The observations were initiated by Charles Perry (1969), who published *uvby* data for more than 800 F stars.

After Strömgren's return to Denmark in 1967, a four-channel *uvby* spectrometer and a two-channel Hβ photometer were designed, built, and used on the Danish 50 cm reflector (now renamed the *Strömgren Automatic Telescope*) on Cerro La Silla, ESO, Chile. In January 1974, in Brorfelde, Strömgren gave a talk, where he described the potential of *uvby*-β photometry in galactic research. I was fascinated by the prospect of learning details about the early evolution of our Galaxy, and when one of my colleagues, right after Strömgren's talk, told me about the new, very efficient cross-correlation methods for determination of radial velocities, I soon after went to Strömgren and told him that I would like to observe the 10,000 brightest F stars with the specific purpose of finding a large, kinematically unbiased sample of intermediate Population II stars within 100 parsecs.

To select the stars, we, of course, turned to spectral classifications. To get a sample as complete as possible, we used Miss Cannon's one-dimensional classifications in the all-sky HD catalogue (Cannon and Pickering 1918-1924) and selected all stars brighter than $m_v = 8.3$ and with spectral types A, A5, F, F0, F2, F5, F8, and G0. The early limit was chosen to ensure that even very metal-poor F stars were included. Also, the stars earlier than F5 could be used for investigations of the metallicity range among the younger Population I stars and studies of the distribution of interstellar matter near the Sun. This is an example of the indispensable groundwork done by spectroscopists, before quantitative classification can be profitably initiated.

The observations were made between 1976 and 1981 from La Silla and Kitt Peak. In that period, I prepared some lists of stars with interesting combinations of indices and tried to predict spectral classifications (Olsen 1979, 1980). Helmut Abt made MK classifications of many of these stars (cf. Abt et al. 1979, Abt 1984). His conclusions were that the photometry is a successful first step in discovering peculiar stars, but the detailed match between predictions and classifications is not always good. Again, we see the need for interaction between the two approaches. One of the more interesting groups of stars is the so-called 'Population II field blue stragglers' (cf. also Olsen 1978). Photometrically, they look like early-F main-sequence stars, but weak lines are indicated by a low m_1 index. Spectroscopically, they were first identified by Sandage (1964) and Bond (1970). It was very encouraging to this photometrist that Abt (1984) was able to confirm that 74% of these stars (20 out of 27) are indeed weak-lined main-sequence stars between A7V and F5V, i.e. they are earlier than the turn-off point for Population II field stars. Some of them may be metal-poor field analogues of the blue stragglers found in some old clusters. Hence the name 'Population II field blue stragglers'. Gray (1989) has later found that a small percentage of them are rapid rotators, *not* weak-lined stars.

On the basis of the F-star catalogue (Olsen 1983), Abt (1986) has also studied spectra of the metal-weak main-sequence stars in the 'normal' range of late-F stars and early-G stars. When Helmut Abt asked me for a list of these photometrically identified stars, I simply grabbed a list of about 800 stars that I had just sent to the Geneva people doing radial velocities. It contained intermediate Population II stars, but also a number of other stars with intriguing indices. When Abt then sent me a manuscript, in which he found that some of the stars were *not* metal-weak F and G stars, I had to backtrack and specify my predictions. Well, the final result was that Abt could confirm 97% of the proposed weak-lined dwarfs. Some of the contaminants are composites, and spectroscopic classifications are indispensable in finding these.

Nielsen (1984) made a comparison between the F-star catalogue and the *Michigan Spectral Catalogue* (Houk and Cowley 1975; Houk 1978). One important result was that, among the late-F stars, *all* of Nancy Houk's *w*-types have $[Fe/H] < -0.6$.

As you all know, Richard Gray (1989) has utilized the F- star catalogue to define an extension of the MK classification system to intermediate Population II stars and, in a preliminary way, to extreme Population II stars. Using the MK process, he has set up a grid of metal-weak and metal-poor classification standards. This is in contrast to earlier classification schemes, which relied on comparisons with the MK standards (cf. Morgan et al. 1943; Roman 1950, 1954; Bond 1970; Bidelmann and MacConnell 1973; Houk and Cowley 1975; Abt 1986; Corbally 1987; Jaschek et al. 1989). Gray finds a beautiful correlation between his new metallicity class and the metallicity index $\delta m_1(b\text{-}y)$ with a scatter of only 0.014 mag. He finds, like Abt (1986), that about 95% of the photometrically selected stars are indeed metal-weak and concludes "..., Strömgren photometry is a very effective tool for selecting intermediate Population II F type stars from the field, but spectral classification performs a useful complementary role."

For a carefully selected subset of the F-star catalogue, high-resolution, high S/N spectra have been analysed with model atmospheres by Edvardsson et al. (1993). To me, their investigation of the chemical evolution of the galactic disk is a quantum jump in galactic research and will probably be the most lasting result to come out of the F-star catalogue. It is fulfilling all the goals set by Strömgren around 1960 in his planning of this program.

4.2. The G-star catalogue

In 1978, during one of those long observing runs that promote quiet thinking and reflection, I became interested in the so-called 'G-dwarf problem' (cf. Schmidt 1963). Based on the correlations with MK classifications, which were evident among the primary *uvby* standards, I soon convinced myself that *uvby* photometry would be an excellent tool, both to separate late-type dwarfs from giants and to derive the metallicity of the dwarfs. In my observations for the F-star catalogue, I included 148 nearby G, K, and M dwarfs, and the discussion of the data (Olsen 1984) showed that, indeed, a single *uvby* measurement would provide all the necessary information. It therefore seemed entirely possible to isolate *all* G dwarfs down to some limiting magnitude. Since the *Michigan Spectral Catalogue* was still only in its early phase, it was again necessary to rely on the HD catalogue to select the observing list. All G5-type stars down to $m_v = 8.6$ were selected, while numerous smaller catalogues of stars with analysed spectra and subdwarfs provided calibration objects.

The observations were made from 1984 to 1987, and the catalogue is in print by now (Olsen 1993). An extension of the range of *uvby* photometry to stars that are cooler, more luminous, and more metal-poor than the original Crawford - Barnes (1970) *uvby* standards has been necessary.

One spin-off is a list of about 200 stars with unusual photometry, which Garrison and Gray are looking at. Garrison finds that HD 204848 (K2 III: Fe-3 CH+1) is a very weak-lined K star with no CN band, but with a CH band stronger than in a normal star, implying lots of carbon, but very little nitrogen (Garrison and Olsen 1992). Two stars that I independently judged to be hypergiants (HD 56126 and HD 179821) have recently been discussed in the literature as post-AGB objects in the pre-planetary nebula phase. HD 28004 looks like a extremely weak-lined K0 dwarf, but a classification by Chris Corbally shows that it is a reddened B star. Nice example of complementarity.

Based on the catalogue, it will be possible to give a list of perhaps 1500 'photometric dwarfs' in the range G2-G8 with photometrically determined metallicities. This list will be contaminated by some class IV stars; by composite stars; and also by giants with enhanced barium, CN, or CH features. Therefore, we need to exploit the complementarity with spectroscopy, and the next step will be careful MK classifications from slit spectra, especially for the stars in the metal-poor tail of the metallicity distribution. I hope that Bob Garrison will find the time to do this. In the meantime, some of my colleagues are getting at least two radial velocities of each 'photometric dwarf' (cf. Nordström and Andersen 1989). A thorough discussion of the sample will require information on the kinematics of the stars, and also detection of binaries.

4.3. A volume-limited sample of long-lived dwarfs

The magnitude-limited G-star catalogue, just described, is not expected to be complete with respect to spectral types, say, in the interval from G2V to G8V. This is due to the fuzziness and large discontinuities of the HD types, and as long as we only had that catalogue, it was difficult to define a handy observing list, which would result in a well-defined dwarf sample to, say, K2. Fortunately, we now have at least part of the *Michigan Spectral Catalogue*. This fine catalogue has made it possible to select photometric observing lists, which are much more well defined than lists selected from the HD catalogue. Using the first three volumes (Houk and Cowley 1975; Houk 1978, 1982), I have recently observed all potential G0 to K2 dwarfs within 40 pc, i.e. with a magnitude limit dependent on temperature type. Among the G0 to K2 stars, I selected not only all dwarfs, but also all stars with an uncertain luminosity class. Of 3900 stars observed in *uvby*, about 1300 seem to be 'photometric dwarfs' (Olsen 1994).

After definition of a volume-limited sample, it will be highly desirable to make MK classifications *based on slit spectra* of the stars, especially in the metal-poor tail of the distribution. The reason is that many of these 'photometric dwarfs' have a luminosity class III - although uncertain - by Nancy Houk. They could very well be giants masquerading as dwarfs due to enhanced barium or carbon features.

5. Concluding remarks

I will end this, already too long talk by apologizing to all the people, who have exploited the complementarity between Strömgren photometry and MK classification, but whose research I have not discussed. I could have talked at length about Howard Bond's (1970, 1980) early work on metal-deficient stars (he was, perhaps unknowingly, following up on a suggestion by Morgan (1958)); Chris Corbally's (1987) extension of the MK system; Richard Gray and Bob Garrison's thorough investigations of the influence of rotation on both classification systems (Gray and Garrison 1987, 1989a,b, Garrison and Gray, 1994); Richard Gray's (1991, 1992) recent calibration work on supergiants; his re-examination of the λ Bootis stars (Gray 1988); and the Jascheks survey of F stars (Jaschek *et al.* 1989).

And, I have probably overlooked many equally important studies.

References

Abt H.A., 1984, ApJ, **285**, 247

Abt H.A., 1986, ApJ, **309**, 260

Abt H.A., Brodzik D., Schaefer B., 1979, PASP, **91**, 176

Bidelman W.P., MacConnell D.J., 1973, AJ, **78**, 687

Bond H.E., 1970, ApJS, **22**, 117

Bond H.E., 1980, ApJS, **44**, 517

Cannon A.J., Pickering E.C., 1918-1924, Harvard Ann. 91-99 (HD)

Corbally C.J., 1987, AJ, **94**, 161

Crawford D.L., 1958, ApJ, **128**, 185

Crawford D.L., 1960, ApJ, **132**, 66

Crawford D.L., 1975, AJ, **80**, 955

Crawford D.L., 1978, AJ, **83**, 48

Crawford D.L., 1979, AJ, **84**, 1858

Crawford D.L., 1984, in: Garrison R.F. (ed.) *The MK Process and Stellar Classification*, David Dunlap Obs., Toronto, p. 191

Crawford D.L., Barnes J.V., 1970, AJ, **75**, 978

Crawford D.L., Mander J., 1966, AJ, **71**, 114

Edvardsson B., Andersen J., Gustafsson B., Lambert D.L., Nissen P.E., Tomkin J., 1993, A&A, **275**, 101

Garrison R.F., 1984, in: Garrison R.F. (ed.) *The MK Process and Stellar Classification*, David Dunlap Obs., Toronto, p. 415

Garrison R.F., 1985, in: Hayes D.S., Pasinetti L.E., Davis Philip A.G. (eds.) Proc. IAU Symp. 111, *Calibration of Fundamental Stellar Quantities*, Reidel, Dordrecht, p. 17

Garrison R.F., and Gray R.O., 1994, AJ, in press

Garrison R.F., Olsen E., 1992, JRASC, **86**, 278

Gray R.O., 1988, AJ, **95**, 220

Gray R.O., 1989, AJ, **98**, 1049

Gray R.O., 1991, A&A, **252**, 237

Gray R.O., 1992, A&A, **265**, 704

Gray R.O., Garrison R.F., 1987, ApJS, **65**, 581

Gray R.O., Garrison R.F., 1989a, ApJS, **69**, 301

Gray R.O., Garrison R.F., 1989b, ApJS, **70**, 623

Gyldenkerne K., 1993, private communication

Houk N., 1978, *Michigan Catalogue of Two-Dimensional Spectral Types for the HD Stars*, Vol. 2, Univ. of Michigan, Ann Arbor

Houk N., 1982, *Michigan Catalogue of Two-Dimensional Spectral Types for the HD Stars*, Vol. 3, Univ. of Michigan, Ann Arbor

Houk N., Cowley A.P., 1975, *University of Michigan Catalogue of Two-Dimensional Spectral Types for the HD Stars*, Vol. 1, Univ. of Michigan, Ann Arbor

Jaschek M., Andrillat Y., Jaschek C., 1989, A&A, **218**, 180

Johnson H.L., Knuckles C.F., 1955, ApJ, **122**, 209

Johnson H.L., Morgan W.W., 1953, ApJ, **117**, 313

Keenan P.C., 1984, in: Garrison R.F. (ed.) *The MK Process and Stellar Classification*, David Dunlap Obs., Toronto, p. 405

Keenan P.C., Morgan W.W., Münch G., 1948, AJ, **53**, 194

Morgan W.W., 1938, ApJ, **87**, 460

Morgan W.W., 1958, in: O'Connell D.J.K. (ed.) *Semaine d'Étude sur le Problème des Populations Stellaires*, Ricerche Astronomiche, Specola Vaticana 5, 323

Morgan W.W., 1960, PASP, **72**, 153

Morgan W.W., Keenan P.C., 1939, PASP, **51**, 355

Morgan W.W., Keenan P.C., Kellman E., 1943, *An Atlas of Stellar Spectra, with an outline of spectral classification*, Univ. of Chicago Press, Chicago

Morgan W.W., Sharpless S., Osterbrock D., 1952, AJ, **57**, 3

Morgan W.W., Whitford A.E., Code A.D., 1953, ApJ, **118**, 318

Nassau J.J., Morgan W.W. 1949, AJ, **54**, 192

Nassau J.J., Morgan W.W. 1950, ApJ, **113**, 141

Nielsen H.E.P., 1984, in: Garrison R.F. (ed.) *The MK Process and Stellar Classification*, David Dunlap Obs., Toronto, p. 230

Nissen P.E., Schuster W.J., 1991, A&A, **251**, 457

Nordström B., Andersen J., 1989, in: Davis Philip A.G., Lü P.K. (eds.) *The Gravitational Force Perpendicular to the Galactic Plane*, L. Davis Press, Schenectady, p. 153

Olsen E.H., 1978, in: Reiz A., Andersen T. (eds.) *Astronomical Papers Dedicated to Bengt Strömgren*, Copenhagen Univ. Obs., Copenhagen, p. 137

Olsen E.H., 1979, A&AS, **37**, 367

Olsen E.H., 1980, A&AS, **39**, 205

Olsen E.H., 1983, A&AS, **54**, 55

Olsen E.H., 1984, A&AS, **57**, 443

Olsen E.H., 1993, A&AS, **102**, 89

Olsen E.H., 1994, A&AS, in press

Perry C.L., 1969, AJ, **74**, 705

Roman N.G., 1950, ApJ, **112**, 554

Roman N.G., 1954, AJ, **59**, 307

Sandage A.R., 1964, ApJ, **139**, 442

Schmidt M., 1963, ApJ, **137**, 758

Strömgren B., 1937, in: Strömgren B. (ed.) *Handbuch der Experimentalphysik* 26, Akademische Verlagsgesellschaft, Leipzig, p. 321

Strömgren B., 1938, Nordisk Astron. Tidsskrift 19, 91

Strömgren B., 1948, Nordisk Astron. Tidsskrift 1948, 126

Strömgren B., 1949, Nordisk Astron. Tidsskrift 1949, 64

Strömgren B., 1950, Transactions I.A.U. 7, 404

Strömgren B., 1951a, AJ, **56**, 142

Strömgren B., 1951b, AJ, **56**, 174

Strömgren B., 1952a, AJ, **57**, 65

Strömgren B., 1952b, AJ, **57**, 196

Strömgren B., 1953, AJ, **58**, 272

Strömgren B., 1954, AJ, **59**, 193

Strömgren B., 1955a, *Universets udforskning*, 2. rev. ed., Hirschsprung, Copenhagen, pp. 99-100

Strömgren B., 1955b, AJ, **60**, 47

Strömgren B., 1956a, Vistas in Astronomy 2, 1336

Strömgren B., 1956b, in: Neyman J. (ed.) *Proc. Third Berkeley Symposium on Mathematical Statistics and Probability*, Vol. 3, Univ. California Press, Berkeley and Los Angeles, p. 49

Strömgren B., 1956c, AJ, **61**, 354

Strömgren B., 1957, AJ, **62**, 319

Strömgren B., 1958a, AJ, **63**, 199

Strömgren B., 1958b, in: O'Connell D.J.K. (ed.) *Semaine d'Étude sur le Problème des Populations Stellaires*, Ricerche Astronomiche, Specola Vaticana 5, 245

Strömgren B., 1958c, in: O'Connell D.J.K. (ed.) *Semaine d'Étude sur le Problème des Populations Stellaires*, Ricerche Astronomiche, Specola Vaticana 5, 385

Strömgren B., 1958d, Observatory 78, 137

Strömgren B., 1962, in: Woltjer L. (ed.) *The Distribution and Motion of Interstellar Matter in Galaxies*, W.A. Benjamin Inc., New York, p. 274

Strömgren B., 1963a, QJRAS, **4**, 8

Strömgren B., 1963b, in: Strand, K.A. (ed.) *Basic Astronomical Data*, Univ. of Chicago Press, Chicago and London, p. 123

Strömgren B., 1964, Rev. Mod. Phys. 36, 532

Strömgren B., 1966a, ARA&A, **4**, 433

Strömgren B., 1966b, Vistas in Astronomy 8, 159

Strömgren B., 1967, in: Woerden H. van (ed.) *Proc. IAU Symp. 31, Radio Astronomy and the Galactic System*, Academic Press, London and New York, p. 323

Strömgren B., 1983, ARA&A, **21**, 1

Strömgren B., Gyldenkerne K., 1955, ApJ, **121**, 43

Strömgren B., Perry C., 1962, *Photoelectric uvby Photometry for 1217 Stars Brighter than V=6.5, mostly of spectral classes A, F and G*, Institute for Advanced Study, Princeton, New Jersey, preprint

Strömgren E., Strömgren B., 1945, *Lærebog i Astronomi*, 2. rev. ed., Gyldendal Norsk Forlag, Oslo

Struve O., 1938, Nordisk Astron. Tidsskrift 19, 36

Discussion

Garrison: I have a couple of comments.

For a long time I have encouraged photometrists and classifiers to interact with one another in a complementary way. I first met Erik in 1984 or so and enjoyed our interactions so much that when Richard Gray finished his thesis in 1986, I encouraged (pushed) him to take a NATO fellowship and spend a few years with Erik in Denmark, which he did.

As an example of complementarity, I can use HD 204848, which was isolated by Erik as a potentially interesting star. He wanted to find out what the MK type was, so I took several spectra. It turns out to be a very metal-weak early K giant with very strong G band (CH), but no CN – i.e., a metal weak carbon star.

Olsen: To the first comment, I can only add that I enjoyed immensely to work together with Richard Gray, and benefitted greatly from the cooperation, both scientifically and otherwise.

To the second comment: I have been using HD 204848 as a secondary *uvby* standard for G and K dwarfs, because the combination of indices suggested that it might be an extremely weak-lined K2 dwarf. Despite Bob's classification, it can still be used as a secondary standard.

Abt: You emphasized, and that was appropriate for this meeting, the interaction of the Strömgren and MK systems, but I suspect that you have larger goals in your huge program. For instance, the goal of the Strömgren-Perry study was to plot the local interstellar gas. What were the goals of your programs?

Olsen: The goal of the F-star catalogue is to study the correlations between age, metallicity *and* kinematics for a very large sample of local F stars (10,000). The study by Edvardsson *et al.* (1993) shows the tremendous steps forward in our understanding of the evolution of the galactic disk, which can be achieved.

The goal of the two G-star programs is to get a large ($\approx$ 1500), complete sample of stars that live "forever" on the main sequence. Their metallicity distribution can be predicted by all models of galactic evolution, and it is therefore a powerful, observational constraint on these models, especially since the predicted abundance distribution is very sensitive to the model parameters.

The MK Process at 50 Years
ASP Conference Series, Vol. 60, 1994
C.J. Corbally, R.O. Gray, and R.F. Garrison (eds.)

Photometric Classification and the MK Process Revisited

David L. Crawford

Kitt Peak National Observatory/NOAO[1], *PO Box 26732, Tucson, AZ 85726 USA*

Abstract. As with spectral classification, there has been progress in photometric classification over recent years. The advances that are the best fully utilize the MK Process, as defined by Morgan and used by him and others to make "classification" the powerful tool that it is. In the case of photometry, existing photometric systems have been refined, and several new systems have been proposed. This paper reviews the MK Process philosophy as used in photometry, and it discusses specifically one new system in some detail: the Stromvil system recently proposed by Straižys, Crawford, and Philip.

1. Introduction

As I discussed in some detail in my paper (Crawford 1984) at the MK Process meeting in Toronto Canada ten years ago, MK classification and photometry go hand in hand. It is hard to imagine doing research without using both effectively. In fact, the title of the Toronto meeting was The MK Process and Stellar Classification, implying nicely that it is not just spectral classification that benefits but all kinds of classification, including photometry. I certainly agree. MK classification and photometric classification complement each other very well. I affirmed this fact in the paper I gave at the symposium in Toronto, but let me reaffirm it here and offer some additional comments.

Both techniques are using the information we receive at the telescope from the star (or other object), splitting the light up into a spectrum, really. With a spectrograph, we form the spectrum, with some spectral resolution, and measure or estimate things like line strengths or ratios. With photometry, we use filters (usually, anyway) to get the spectral resolution, and measure magnitudes, color indices, color differences, and the strengths of strong lines or features. With such techniques, we are looking at different things, and so the complementarity is great, as we might expect. The "confrontation" is also great, and it is not really confrontation but comparison. Looking at any differences in the stellar classification helps us a lot to understand both of the classifications.

The spectral range we cover with either method depends on our interests and needs, and on the detector used. The spectral resolution (or filter width, or bandwidth) we use depends on our needs and the method we are using. In

[1]Operated by AURA, Inc., under cooperative agreement with the National Science Foundation.

both cases, it is important to match closely the instrumental parameters with which the "system" was established. In both cases, it is essential to observe an adequate set of well established standard stars.

2. The MK System

The MK System was established in the blue region of the spectrum, for that is where the photographic plates used were sensitive. In addition, it is a region filled with much of the information that we want to observe, for "ordinary" stars anyway. We will hear much at this meeting about its history and its value, both of which are rich and well deserved.

3. Photometric Systems

The *UBV* system includes the spectral region where the 1P21 photomultiplier is sensitive. Note that the *UBV* filters essentially fill that spectral region, indeed with some overlap of the filter bandpasses. As such, it is a "natural system" to use, in the sense of the MK Process. It gives "all" the information that is there with that spectral resolution. It is a "wide band photometric system," with the filters having bandwidths of more than 100 nm each. It is based on much previous work, just as the MK system was. It offers a magnitude, V (for "visual"), a measure of the apparent visual brightness of the star. It offers two color indices, (B-V) and (U-B). (B-V) is quite similar to the earlier visual minus photographic magnitudes. (U-B) adds the ultraviolet region, to which the photomultiplier is sensitive. (B-V) is a good parameter to estimate temperature, subject to the effects of interstellar reddening (and other things). (U-B) is a good measure of the strength of the Balmer discontinuity. All this is to "first order," of course, as many other things arise to muck up the works. However, in spite of the observational problems inherent in a wide band system and the lack of information resolution, the *UBV* system is a most powerful tool, and history confirms this fact. Its strength is due in great measure to the fact that the MK Process was used in establishing the system. It is a natural system, empirically based, stable, much used and understood, with an excellent (at least to the accuracy with which the system is defined) set of standard stars. It is independent (mostly) of theory. It is empirical. That is its strength.

The *uvby* system was also based on much previous work, including the MK System and Process, and on the *UBV* and other photometric systems. It also is based on the spectral sensitivity of the 1P21 photomultiplier, but it uses four filters to fill the region to which the detector is sensitive. They do not overlap. Nor do they overlap the Balmer discontinuity. The parameters have been described in detail in the literature, but let me summarize some of that here. The y filter has the same central wavelength as V of *UBV* (actually, we must be careful of such "definitions"), and it is used to give a V magnitude. The V's in the two systems can be related to each other with high accuracy and precision. The (b-y) color index is a measure of temperature, but since the b band is located above the region in the spectrum where line blanketing becomes important, it is (to first order) independent of the blanketing, which (B-V) is not. The color index (v-b) is used with (b-y) to form a color difference,

m_1, a measure of the "metallic" line blanketing in the v region relative to a gradient defined by $(b\text{-}y)$. In a similar way, $(u\text{-}v)$ is used with $(v\text{-}b)$ to give c_1, another color difference, which is a good measure of the Balmer discontinuity. All of these are "cleaner" measures of the parameter of interest than are the ones in the *UBV* system, and, as such, offer better "information resolution." In addition, some of the observational difficulties that wide band systems have are minimized with the use of the intermediate band system. Still, great care is needed to insure that accurate and precise photometry is done. The literature is filled with papers where that is not the case. I discussed some of these problems in a recent paper (Crawford 1993) given in Cape Town, South Africa, at the conference honoring Alan Cousins, and I also gave a few examples of such.

With the advent of detectors giving good sensitivity in the near infrared, the *UBV* system was extended to include the R and I bands, giving a new color index $(R\text{-}I)$, one that is essentially independent of any blanketing effects and hence a clean measure of temperature. Other detectors more sensitive to the further infrared allowed the addition of other alphabetically named bands, ones of particular use to the classification of cooler stars and other applications.

The *uvby* system was supplemented with a narrow band system, β, measuring the strength of the Hβ hydrogen line. This added a useful parameter to the *uvby* system, as it is a measure of luminosity for stars hotter than the Balmer maximum and of temperature for those cooler than the maximum near spectral type A2. As it is defined with two filters having the same central wavelength, it is insensitive to interstellar reddening (and other things). It can be used with the *uvby* parameters to give good estimates of temperature, interstellar reddening, blanketing, and luminosity for all stars earlier than say G5 (it is pretty good even for later ones). The *uvby*,β systems offer a good deal of information resolution for such a few observational parameters. It is an excellent natural system in the MK Process philosophy, strictly empirical, and it has an excellent base of standards stars, also an essential part of the MK Process. It is an excellent complement to the MK System, of course, and the systems fit very well together. Most of us love them both.

Naturally, there are other excellent and useful photometric systems as well, some of them discussed in this symposium. Some are based rather well on the MK Process, others less so. I personally think that those that follow the philosophy the closest have turned out to be the best. No one system is ideal for all types of stars (or for all spectral types or for all regions of the HR diagram). One must use what one needs.

These systems, both photometric and spectral, are powerful tools in gaining astrophysical insight to the nature of stars and other objects in the universe, and also to insight in astronomical areas, such as galactic structure. They have enabled us to do many interesting things over the 50 years since the MKK Atlas, and they will continue to do so for many more. The applications are many, and they are growing as our technological tools and our understanding of the systems improve.

4. The Stromvil System

Naturally, things are not stagnant, even for long established and much used systems. There are new things to do, there is new technology, there are new ideas. Let me mention briefly one such expansion. We have proposed combining some of the best aspects of the Vilnius photometric system with the Strömgren *uvby* system, adding three of the Vilnius bands to the four bands of the *uvby* system (and to the two filters giving the β system, in some cases). These bands are:

1. P, at 375 nm, which is placed on the crowding of the higher members of the Balmer series near its limit. It is therefore quite sensitive to luminosity for early type stars.

2. Z, at 516 nm, placed on a wide and deep absorption feature in the spectrum of G, K, and M stars made by a crowding of metallic lines, of which the Mg I triplet is the strongest. In late K dwarfs and in M dwarfs, the bandpass also contains strong MgH molecular bands. These Mg features show a strong negative luminosity effect, making a color index containing the Z band a good luminosity parameter for late type stars.

3. S, at 656 nm, placed on the Hα line of hydrogen. It has the ability to show emission in this line, amongst other potential uses.

Both modeling studies and plots of data for standard stars show the potential power of the "new" system. An advantage is that much is already known about both systems, and excellent catalogs of standard stars already exist. The system should be of great value in classification studies of stars of all spectral types, including those with interstellar reddening present. Since all the bands (if we don't use the β system supplement) are intermediate width (about 20 to 30 nm half width), the system should be of value for CCD photometry, even to faint limits. The concept is discussed in more detail in a paper now in press by Straižys, Crawford, and Philip.

5. What Next?

Finally, let me offer a few comments about recent changes in technology and a bit about the future. Two things have made a major impact, and the future of photometry will be much different than the past. Present work already confirms this.

First, the increasing wavelength coverage. While much physical information is in the visual, blue, and UV regions of the spectrum, and we can never leave them out of future work, the expansion of detector sensitivity and space observations into the further UV and into the IR and beyond have opened vast opportunities, especially for the hottest and the coolest objects.

Second, the advent of the CCD detectors has allowed us to do panoramic photometry (imaging photometry), and the added quantum efficiency allows us to work fainter for the same time and accuracy. This is dynamite for many research programs. It also has turned quality small telescopes into much larger ones for many kinds of fundamental and frontier research.

However, it also greatly increases the complexity of the observations and of the reductions. Along with the increasing excitement of astronomical research, increasing the demand for telescope time, we are getting further removed from a complete understanding of what we are doing, probably leading to more "Second Class, First Class Astronomy," in the sense of what I discussed in the paper given recently at the conference in Cape Town (Crawford 1993). I hope not, but some present evidence and imagining what can happen do not bode well. We must do better. It is essential that the MK Process philosophy be used!

I wish to thank those who have helped greatly in my understanding of these things, such as it is, with special thanks to Morgan and to Strömgren and the many photometrists with whom I have discussed the issues over the years.

References

Crawford, D. L. 1993, Thoughts on Precision Photometry, in *Precision Photometry*, Editors: D. Kilkenney, E. Lastovica, J. Menzies, South African Astronomical Observatory, Cape Town, South Africa, pp. 18-24

Crawford, D. L., 1984, The Relationship Between the MK System and Photometric Classification, in The MK Process and Stellar Classification, Editor: R. Garrison, David Dunlap Observatory, Toronto, Canada, pp. 191-212

Discussion

Drilling: A lot of the problem with *U-B* for O and B stars lies in the assumption by many observers that the extinction coefficient is a linear function of color, whereas Johnson showed that this was not the case when he set up the standards for the *UBV* system. It is easy to see, though, that if the assumed linear extinction coefficient is determined from, say, A and F stars, and then extrapolated to the O and B stars, why different observers can disagree by 0.1 or 0.2 magnitudes in *U-B* for the latter.

Crawford: Certainly, but there are many other potential problems too. I could list a lot. Lack of adequate standards is high on the list.

Roman: How do you get enough standards if you have only two or three nights, not all of which are photometric?

Crawford: It's difficult, but you must do it. Otherwise, what you get is of much less value. It is even harder to get enough "utensils" to see what specific "system" you are on!

Weaver: R. I. Mitchel used 13-color photometry to try to span the separation between spectroscopy and photometry. Why hasn't it been more fully used?

Crawford: Perhaps lots of reasons, but one is that it did not offer enough added information beyond that existing in *UBVRI* or *uvby*,β photometry.

Garrison: Referring to your comment about spectroscopy and photometry converging towards spectrophotometry, I wondered if you could tell us your feeling about the number of filters one can use practically for single-star photometry.

That is, in adding more and more filters, when do you reach the point of diminishing returns so that it is better to do spectrophotometry?

Crawford: After about six, I think one is getting diminishing returns. For my natural group, this is about the amount of information content that will be there in wide band photometry. However, I think all this (three bands to however many) is really spectrophotometry anyway. We should ask Eric Olsen also.

Olsen (to Garrison): If you want to use many filters, and if they are not overlapping, you build a spectrograph, isolate the bands in question and send the light through filters. You get a fast filter-defined system, and all adverse effects, which are nearly grey (cirrus clouds, variable seeing) will not affect your data in any significant way, except in the magnitudes themselves, of course. Color indices are unaffected.

Crawford: Certainly, I agree. One can do it too, probably, by using a CCD and a low dispersion spectrograph to do the same thing. We need an $I(\lambda)$ machine, with "adjustments" for the λ range on the detector and the bandwidth (spectral resolution).

Penprase: I was wondering whether you could say something about the availability of CCD Strömgren photometry filter sets, and some of the technical issues in using them.

Crawford: It is a complex issue: identifying and obtaining excellent filters for a system. The IAU has and is discussing it at length; so are a few of us individually. We are now testing a new order of filters, and if okay, we will announce plans widely for a large order of matched filters.

Olsen (to Penprase): The Danish 1.5 m telescope on Cerro La Silla, ESO Chile, is equipped with large *uvby* filters to be used with its CCD camera. The same is the case with the Nordic Optical Telescope (NOT, 2.5 m) on La Palma, the Canary Islands.

Crawford: At KPNO and at CTIO, we have, and people use, *uvby* filters for CCD work. Still, as with photo multipliers, good matching to the standard systems is essential, and more discussion is needed to insure that we all are (or will be) on the same systems. "Be careful" must be the watch word.

The MK Process at 50 Years
ASP Conference Series, Vol. 60, 1994
C.J. Corbally, R.O. Gray, and R.F. Garrison (eds.)

UBVRIJHKL Photometry, Stellar Systems and the Carbon Stars

T. Lloyd Evans

South African Astronomical Observatory, P.O.Box 9, Observatory 7935, South Africa

Abstract. Carbon stars fall into several major groups on the basis of their spectral energy distributions (colors) as well as on details of their spectra. The metal-poor stars of the Small Magellanic Cloud have weaker CN than typical galactic carbon stars, as do some high velocity stars in the galactic field including the dwarf carbon star LHS 1075. The C/O ratio, the H abundance, the N abundance and the carbon isotope ratio all have a profound effect on the appearance of the spectra.

1. Introduction

The MK System aims to classify stars, from their spectra, in a multi-dimensional system. These empirical dimensions should correspond with physical parameters of which effective temperature and surface gravity are the most important. An abundance parameter (metals or CN) is often added. The carbon stars have proved more difficult to deal with. A temperature criterion is only available for the hotter R stars and there is no generally accepted luminosity criterion. The problem lies in the overlapping bands of carbon-bearing molecules (Keenan 1984). These cause severe difficulties in intercomparing the spectra of even "mild" carbon stars with those of normal stars of similar temperature (e.g. Lloyd Evans 1986a). The rarity of carbon stars in both the general field of the Galaxy and its star clusters eliminates one of the main sources of checks on luminosity classification. The presence of large numbers of carbon stars in the Magellanic Clouds and the recently recognised dwarf carbon stars provide suitable test objects which are now being exploited, although faint apparent magnitudes are a problem.

2. Observations

A large sample of both R and N stars in the galactic field and some N stars in the Small Magellanic Cloud have been observed on digital spectra of 1-3 Å resolution in the 5600-6700 Å region (all) and in several violet-green wavebands (galactic stars only). The R stars were observed in the $UBV(RI)_c$ system and separately in the *JHKL* system; most of the infrared photometry was by Robin Catchpole and other colleagues at SAAO. Published photometry, which in some cases needed to be transformed onto the system in use now, is available for many of the cooler stars (Catchpole *et al.* 1979; Walker 1979).

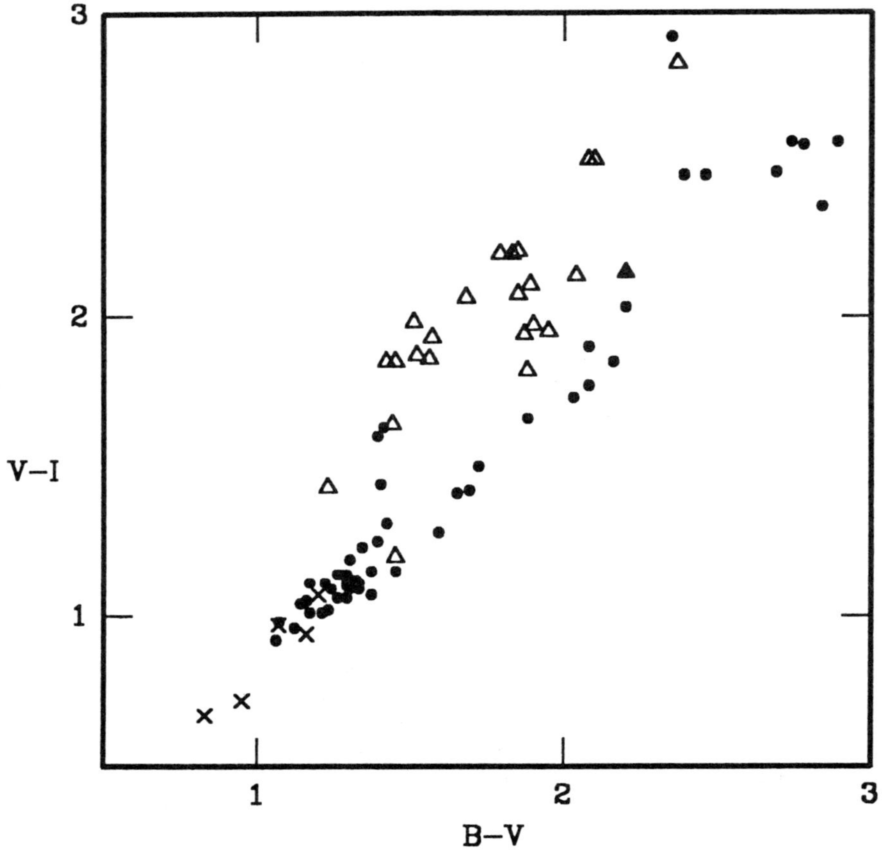

Figure 1. R stars in the *B-V*, *V-I* diagram. Ordinary R stars, filled circles; J stars, open triangles; HdC stars, crosses; V Ari, solid triangle.

We present some of the overall results based on measurements of a few of the spectral features. The G band of CH has been used to isolate the H-deficient stars and the isotopic bands of C_2 at 4737-44-52 Å and 6191-68Å and of CN at 6206-60 Å have been used to isolate stars with enhanced ^{13}C, the J subgroup of the carbon stars.

3. The Major Groups among the R Stars

There are three principal photometric groups among the R stars, as seen in the plot of *V-I* against *B-V* in Figure 1. Ordinary R stars, which are neither H-deficient nor ^{13}C-enhanced but which in this context do include the CH stars as well as some misclassified N stars at the red (cool) end of the distribution, follow a linear trend from which a few stars deviate in the direction of greater *V-I*. The J stars nearly all deviate in this direction, typically by 0.2 mag. This bifurcation

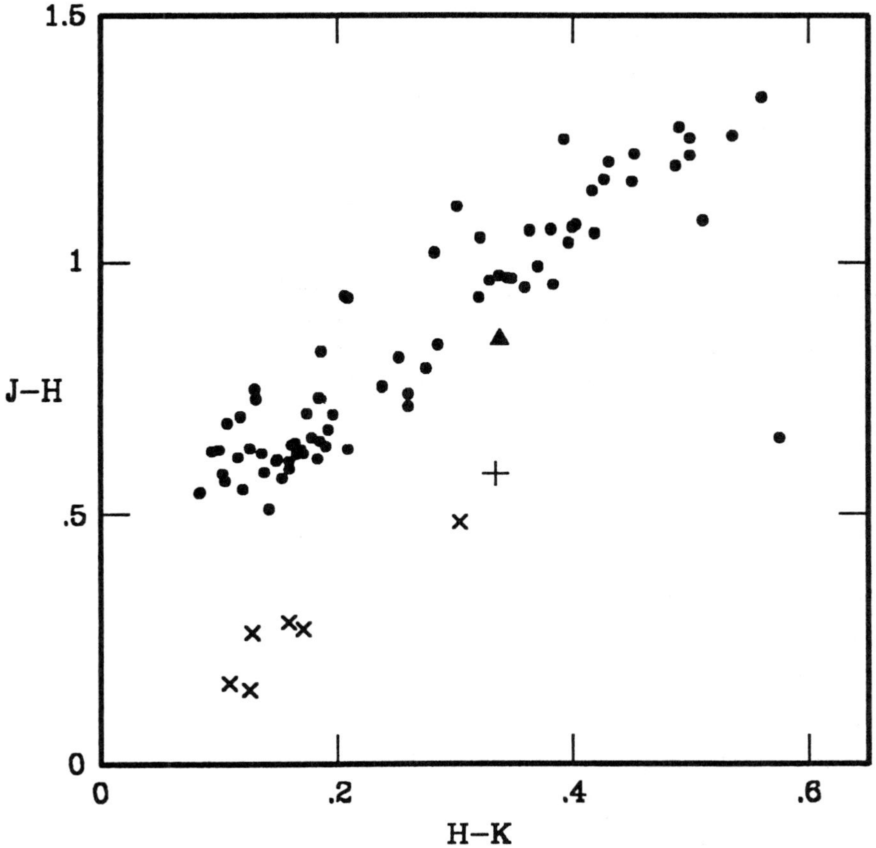

Figure 2. The *J-H*, *H-K* plot for the R stars. The J stars have not been distinguished from the ordinary R stars in this case, while the dwarf carbon star LHS 1075 is represented by a plus sign.

appears rather abruptly near $B\text{-}V = 1.3$. Only one J star falls to the blue of this but hotter ^{13}C-enhanced stars, identified from high resolution spectra by Dominy (1984), fall on the normal sequence with bluer *B-V*. The H-deficient stars also fall on the normal line, with one exception. They are amongst the bluest of the R stars and indeed two of them have *B-V* colors which are bluer than any normal carbon star, defined as a star with conspicuous C_2 bandheads. The marginal appearance of C_2 in an HdC star is well established in the pulsating R Coronae Borealis star RY Sgr at *B-V* of just less than 0.7 (Lloyd Evans 1986b). The absence of bluer normal R stars is probably not a selection effect as a few of the stars observed have very weak C_2 bands. The H-deficient stars differ from most R stars in their moderate metal deficiency (Warner 1967) as well as in their H-deficiency so that the atmospheric opacity caused by metallic line blanketing as well as by H^- absorption will be below normal.

The plot of *J-H* against *H-K* is shown for the same stars in Figure 2. We have not distinguished between J and ordinary R stars in this case as they follow almost identical loci (Lloyd Evans 1986c, 1990). V Ari shows a small displacement while HD 100764, which has enhanced ^{13}C according to Dominy (1984), has a large excess at *K* (Dominy *et al.* 1986). The HdC stars are 0.3 mag bluer than the rest in *J-H* at given *H-K*. The dwarf carbon star LHS 1075 lies on the same line as the HdC stars, using the infrared photometry by Bothun *et al.* (1991). Green *et al.* (1992) found that this and other dwarf carbon stars for which *JHK* photometry is available fall well below the general run of carbon stars in this diagram but it is surprising that they should deviate to the same extent as the HdC stars. Both groups are probably metal deficient. The deviations of both groups may arise from the absence of the flux excess in the H band where the H^- opacity minimum occurs and enhances the relative intensity in the spectral-energy distribution of normal stars. However, whereas this arises because of an actual lack of hydrogen in the HdC stars, the high gravity in the atmosphere of the dwarf star encourages the association of hydrogen (Dearborn *et al.* 1986). An additional factor is the weakness of CN, which is a conspicuous feature of the spectra of the HdC stars (see Figure 2 of Lloyd Evans *et al.* 1991) as well as of the dwarf carbon stars (Green *et al.* 1992).

4. Comparison of N stars in the Galaxy and the SMC

The strengths of ^{12}CN at 6206 Å and ^{12}C^{12}C at 6191 Å have been measured on spectra of N stars in the Galaxy and the Small Magellanic Cloud. J stars were omitted because they have weak CN 6206 Å ; they are uncommon in the SMC in any case (Lloyd Evans 1986c). The digital data confirm the earlier result from photographic spectroscopy that the Magellanic Cloud N stars have CN bands which are systematically too weak for their C_2 bands when compared to N stars in our own Galaxy (Lloyd Evans 1980). The weakness of CN was ascribed to a primordial deficiency of N along with other heavy elements in the Magellanic Clouds, whereas the overabundance of C is manufactured by the star from triple-alpha burning of the plentiful helium. The galactic high velocity star V Ari, noted above as falling below the general run of galactic stars in the *J-H*, *H-K* diagram, has similarly weak CN. The N stars in the Magellanic Clouds also fall below their galactic counterparts in the *J-H*, *H-K* diagram; this is explicable in terms of the reduced CN blanketing in the *J* band in these stars (Cohen *et al.* 1981).

5. Discussion

The limited use of the new material which has been made in this discussion has enabled us to show the existence of differences in the spectral energy distributions which correspond well with the differences in the optical spectra which lead to the recognition of the different groups in the first place. The J-type R stars are apparently faint in *V* relative to *B* and *I*, in agreement with the depression seen on our spectra in the 5800-6400 Å region where C_2 and CN are strong. The isotopic rotational lines provide absorption in addition to that caused by the saturated ^{12}C features. Nitrogen deficiency accounts for some of the anomalies seen in the spectra of the HdC stars and dwarf carbon stars: the unusual strength of C_2 at 6191 Å in the HdC stars may result from the relative weakness of the usual blanketing by CN in this region. The same effect may occur in the dwarf carbon stars where it would help account for the strength of the 6190 Å feature suggested by Green *et al.* (1992) as a possible luminosity criterion. This feature is evidently not the result of a low $^{12}C/^{13}C$ as not all the dwarf carbon stars have enhanced ^{13}C (Green and Margon, preprint). Much of the weakness of CN in the dwarf stars may result from high surface gravity rather than from a deficiency of nitrogen; the detection of dwarf carbon stars by criteria which are independent of space velocity and hence of metal content will permit a more detailed assessment of the factors involved.

There is also qualitative agreement between the CN strength in the red spectra of the N type stars and their *JHK* colors as seen in the comparison of the N stars in the Galaxy and the SMC. Only the CH stars among the well-known groups are not distinguished in the data presented here but these as well as lesser known groups among the carbon stars do appear well separated in other plots constructed from our data.

A major conclusion is that the main types of carbon star show differences in molecular blanketing and hence in spectral energy distribution. This points to the necessity of setting up separate spectral classification systems, as has indeed been done for several of these groups by Keenan (1993).

6. Acknowledgments

I am indebted to several colleagues, notably Robin Catchpole, for the infrared photometry used here.

References

Bothun, G., Elias, J.H., MacAlpine, G., Matthews, K., Mould, J.R., Neugebauer, G., and Reid, I.N. 1991, AJ, **101**, 2220

Catchpole, R.M., Robertson, B.S.C., Lloyd Evans, T.H.H., Feast, M.W., Glass, I.S., and Carter, B.S. 1979, S.A.A.O.Circ., **1**, 61

Cohen, J.G., Frogel, J.A., Persson, S.E., and Elias, J.H. 1981, ApJ, **249**, 481

Dearborn, D.S.P., Liebert, J., Aaronson, M., Dahn, C.C., Harrington, R., Mould, J., and Greenstein, J.L. 1986, ApJ, **300**, 314

Dominy, J.F. 1984, ApJS, **55**, 27

Dominy, J.F., Lambert, D.L., Gehrz, R.D., and Mozurkewich, D. 1986, AJ, **91**, 951

Green, P.J., Margon, B., Anderson, S.F., and MacConnell, D.J. 1992, ApJ, **400**, 659

Keenan, P.C. 1984, in *The MK Process and Stellar Classification*, R.F. Garrison, David Dunlap Observatory, 29

Keenan, P.C. 1993, PASP, **105**, 905

Lloyd Evans, T. 1980, MNRAS, **193**, 97

Lloyd Evans, T. 1986a, S.A.A.O. Circ., **10**, 1

Lloyd Evans, T. 1986b, MNRAS, **219**, 191
Lloyd Evans, T. 1986c, MNRAS, **220**, 723
Lloyd Evans, T. 1990, MNRAS, **243**, 336
Lloyd Evans, T., Kilkenny, D., and van Wyk, F. 1991, *Observatory*, **111**, 244
Walker, A.R. 1979, S.A.A.O.Circ., **1**, 112
Warner, B. 1967, MNRAS, **137**, 119

Discussion

Keenan: This is a great program! How many of those late-R or early-N stars were you able to include in your program?

Lloyd Evans: There are rather few late R (i.e., R8) stars or the R stars identified by Sanford from high resolution spectra; not more than ten such stars were found and most are hotter N stars with normal high $^{12}C/^{13}C$, whereas nearly all R5 stars have enhanced ^{13}C. There are many hotter N stars among those classified as N stars in the HD. Three R stars have weak C_2 and strong Na D; one is the Li-rich T ArA. Such stars are very rare among stars classified as N, although they are among the reddest (coolest?) R stars.

Liebert: I should like to point out that the dwarf carbon stars are now really a new spectral class and not a few isolated freaks. Jack MacConnell, Paul Green and Bruce Margon discovered several among known stars in catalogues, and now several more have been found recently in quasar and other surveys. Green and Margon (1993 *Ap.J.*, in press) summarize salient properties. These include: (1) enhancements in barium, (2) a wide range of $^{13}C/^{12}C$, but often as high as the J stars, (3) they seem to belong to both the halo and disk populations. The abundance patterns are reminiscent of the CH stars. Two recently have been found with visible white dwarf companions, and a mass-transfer, binary origin of the dwarf's peculiar abundances is strongly favored.

Drilling: Were any of the H-deficient carbon stars which you observed R CrB stars?

Lloyd Evans: Known R CrB stars were excluded (being studied by others at SAAO), but V1783 Sgr may be a new example while V1157 Sgr is almost certainly one, being very faint in 1991-92.

McCarthy: What differences do you see between carbon stars in the LMC and SMC? I note that in the CN/C_2 plot (Lloyd Evans 1980), the carbon stars in LMC and SMC seem to inhabit approximately the same domain.

Lloyd Evans: The LMC and SMC stars are not separated in the CN/C_2 plot, although one might have expected those in the SMC would be more different from those in the galaxy. I have more spectra of LMC stars but only at lower resolution.

McCarthy: Are there yet any signs of relating numbers of R and N stars which might permit us to see differences between the clouds? At much lower dispersions the ESO group (Azzopardi, Rebeirot, Westerlund) and CTIO group (Blanco, McCarthy) have observations of carbon stars in the LMC and SMC whose intercomparisons may be of interest when stars in common can be studied and differences noted.

Lloyd Evans: Searches for R stars are probably not complete enough in the clouds to see differences.

Parthasarathy: Do you find in all your R stars strong ^{13}C lines (i.e., low $^{12}C/^{13}C$)? Do you find any early R star with over abundance of s-process elements?

Lloyd Evans: Many R stars are too hot to show strong isotopic bands at my dispersion; in some cases high resolution spectroscopy has revealed the isotopic lines in the rotational spectra. A minority of the cooler R stars show no enhancement of ^{13}C. Some early R stars show enhanced s-process elements, in particular there are a few CH stars with strong Ba II. HD 24281 which Professor Keenan mentioned yesterday is one of them.

Garrison: With your measurement of spectral features by using the local continuum, do you think that veiling effects might lead to some misclassifications? If so, you might consider using line ratios.

Lloyd Evans: Veiling is undoubtedly important in the Mira variables with extended atmospheres in which dust and shock waves are important. However their spectra are very peculiar, in that they do not correspond instantaneously to those of the static spectra of constant stars, and it is doubtful that they would be meaningful in terms of temperature or luminosity. Estimates of the strengths of any features in the spectra of N stars are difficult because of overlapping features.

The MK Process at 50 Years
ASP Conference Series, Vol. 60, 1994
C.J. Corbally, R.O. Gray, and R.F. Garrison (eds.)

CCD Four-Color Observations of BHB and FHB Stars: An Example of Photometry Complementing Spectroscopy in Stellar Classification

A. G. Davis Philip

Van Vleck Observatory and Union College

Abstract. CCD photometric observations enable astronomers to make much more precise determinations of color indices of faint stars. Color-magnitude diagrams of globular clusters now allow more accurate assignments of stars to classes, such as main-sequence, giant branch, asymptotic giant branch and horizontal branch. Four-color CCD observations of globular clusters show that stars on the horizontal branch fall in two groups. One group of stars forms a tight lower envelope in the CM-Diagram, declining from the upper right to the lower left. The second group is a scattered one, the points falling about 0.2 mag above the lower distribution. The stars in each group fall together in other four-color diagrams, indicating that the photometry is selecting natural groups of stars.
The program, SPECTRUM, by Gray (1991) has been used to calculate a series of synthetic spectra, characteristic of low-metallicity stars. These spectra have been matched to spectra of some of the brighter FHB stars taken on the 6-meter telescope and some FHB stars in the Adelman *et al.* (1987) Atlas. The agreement is very good. Spectra of stars that fall in the different groups in the four-color diagrams can be investigated by comparison to the synthetic spectra. Differences in luminosity implied from the four-color photometric observations can be confirmed using the criteria identified in the synthetic spectra.

1. Introduction

CCD four-color photometry of blue horizontal branch (BHB) stars in NGC 1904 has been done with the 0.9-m telescope at CTIO. The stars on the HB form two main natural groups which have been identified with stars in different stages of evolution on the HB. Model spectra of field horizontal branch (FHB) stars, calculated with Richard Gray's SPECTRUM program, were used to construct grids of model spectra at temperatures from 7250 to 8250 K, $\log g$ values from 2.0 to 4.5 and [Fe/H] values from 0 to -3. The model spectra were searched for lines which seemed to be good indicators of luminosity and a region surrounding Hγ was found to have two possible indicators. These regions are compared with high dispersion spectra (from the Atlas by Adelman *et al.* 1987) for HD 161817 and HD 109995.

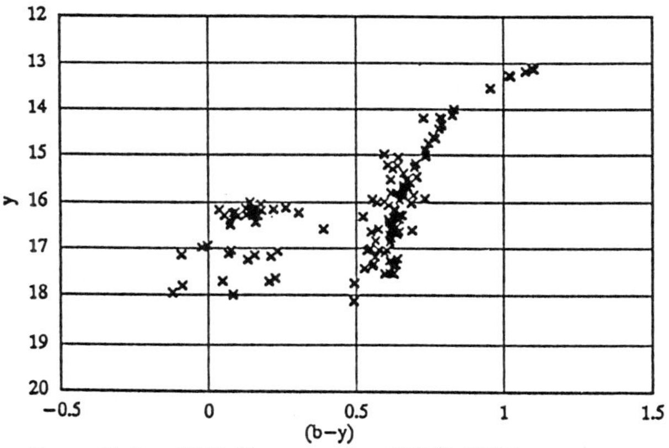

Figure 1. Four-Color CM diagram for NGC 1904.

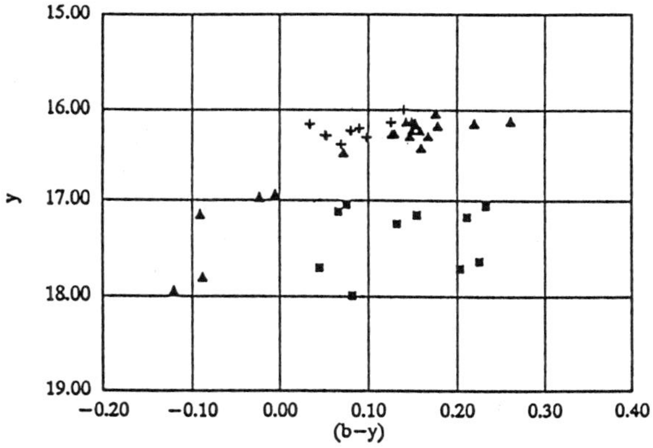

Figure 2. y vs $(b - y)$ for the horizontal branch of NGC 1904.

2. CCD Photometry

The globular cluster for which I have the greatest number of four-color measures of BHB stars is NGC 1904, a southern cluster which is located in the anticenter region of the Galaxy at an intermediate galactic latitude. There are few problems with field stars. The CCD frames were reduced at DAO using Stetson's (1987) DAOPHOT and ALLSTAR programs. The resulting CM diagram is shown in Figure 1. A well-developed horizontal branch can be seen and the giant and asymptotic giant branches are well defined. Figures 2 and 3 show enlargements of the horizontal branch. In Figure 2 y is plotted vs $(b-y)$ and in Figure 3 m_1 is plotted vs $(b - y)$. From previous work on other globular clusters (Philip 1988) I have found that in the four-color diagrams the BHB stars seem to fall in two groups. In the y vs $(b - y)$ diagram there is a tightly defined group (marked by solid triangles) that seems to follow the zero-age horizontal branch. There is a second group (marked by crosses) that lies a few tenths of a mag above the first.

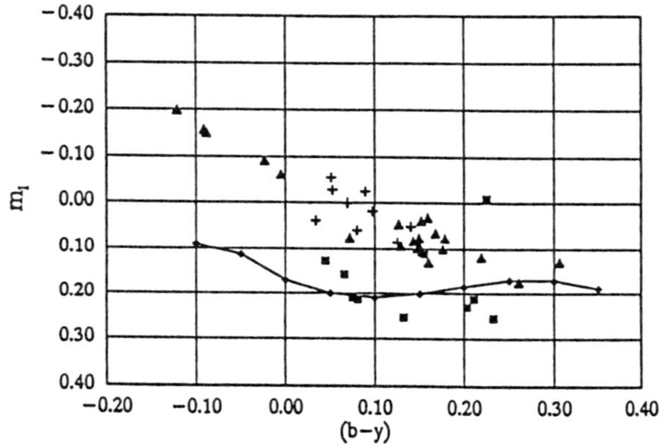

Figure 3. m_1 vs $(b-y)$ for the horizontal branch of NGC 1904.

In the m_1 vs $(b-y)$ diagram a line indicates the Pop I relation for A-type stars. The triangles and crosses fall in a region orthogonal to the line.

The RCA chip was fairly insensitive to the ultraviolet and although the frames that were taken looked well exposed, when the data were reduced the internal probable errors were very high and thus reliable c_1 indices could not be calculated.

In Figures 2 and 3 there are stars represented by squares. Each of the CCD frames was checked and each star in these plots is clean, i.e., has no pixel defects and is not crowded together with one or more stars. In the *UBV* plot by Stetson and Harris (1977) there were stars below the HB in a similar position as those found in the four-color CM diagram. One possible explanation for these stars is that they are blue stragglers. Since the cluster is at intermediate galactic latitudes, it is not likely that these stars (some of which are very close to the cluster center) are non-members.

3. Model Spectra

Richard Gray (1991) has written an IBM-PC-based stellar spectral synthesis program (SPECTRUM) using the new fully blanketed stellar models of Robert Kurucz (1991), calculated with a microturbulence of 2 km/s. The temperature range covered is 50,000 to 3,000 K (O to M in spectral type). There are models for stellar atmospheres of solar abundance and for [Fe/H] = -1, -2 and -3. On a 66 Mhz 486 PC a spectrum, calculated at a spacing of 0.02 Å, from 3900 Å to 4500 Å can be computed in about 17 minutes. SPECTRUM requires a Math coprocessor chip, otherwise the calculation time becomes too long. First I was interested in seeing how well SPECTRUM spectra matched the spectra of some of the classical FHB stars. A model spectrum ($T_{\rm eff}$ = 7750, $\log g$ = 3.0, [Fe/H] = -1.0, model 773010) was compared with a 6-m spectrum of HD 161817 (7 Å/mm) in the region 4300 Å to 4500 Å. Kodaira and Philip (1984) determined Teff = 7700 and $\log g$ = 2.9 for this star. The agreement was quite good.

3.1. The Hγ region

I computed a grid of SPECTRUM models with the 773010 model as the central model. Models at temperatures of 7250, 7500, 7750, 8000 and 8250 are shown in Figure 4, gravities of 4.0, 3.5, 3.0, 2.5 and 2.0 in Figure 5, metallicities of [Fe/H] = 0, -1, -2 and -3 are shown in Figure 6. With increasing temperature the lines generally weaken. With increasing luminosity most of the lines increase slightly in intensity. At λ = 4302 - 4304 there is a triplet which changes from the middle line being strongest ($\log g = 4$) to middle line being weakest ($\log g = 2.0$). At λ 4374 there is a doublet in which the shorter wavelength component becomes increasingly strong, relative to the longer wavelength component, with increasing luminosity.

Both HD 161817 and HD 109995 are on the high luminosity track, according to the four-color data. Sections from the Adelman *et al.* (1987) atlas are shown in Figure 7. The triplet shows the middle line as weaker than the two outer lines, in agreement with this classification. The doublet shows the shorter wavelength component and not the longer wavelength component. This also is in agreement with the higher luminosity track. To confirm the behavior of these lines as a function of luminosity in the spectra of FHB stars, spectra of additional FHB stars, on the lower luminosity and upper luminosity tracks should be obtained.

If the spectrum shows the higher luminosity features, the higher luminosity hypothesis can be checked by obtaining four-color photometry. Temperature affects the two selected sets of lines as follows; with increasing temperature the line ratios change somewhat in the same way as they do with increasing luminosity (or decreasing $\log g$ values). A temperature change of about 500 degrees would be needed to mimic the luminosity change and this difference is easily detected in the $(b - y)$ color. If the line ratio mimics a star of 8250 K and the color matches the color of a 7750 K star, then you can safely assume that the line ratio change is the result of the luminosity difference and not a temperature difference.

4. Future plans

I have a four-color and Hβ set of frames of M 92 taken with the 2k x 2k Tek chip at KPNO. On a 0.9-m telescope the field of view is 17 minutes of arc across, so the entire cluster is covered in each frame. Preliminary work done at DAO indicates that well over 100 BHB stars have been detected. Some of these are stars close to the cluster center so they have not been detected in previous single-channel photometric investigations. The u frames give good u magnitudes so for M 92 I will be able to calculate c_1 indices. These data will produce the most statistically significant set of BHB measures so far obtained in this series of investigations of globular cluster horizontal branches. It will be interesting to see if the patterns found in the NGC 1904 four-color diagrams are found also in the M 92 data.

At this time I have examples of high dispersion spectra of FHB stars in the Hγ region only for two FHB stars (from the Adelman Atlas). Both of these stars have been assigned to the upper, higher luminosity track, according to four-color photometry. I would like to obtain high dispersion spectra of HB stars, on both the upper and lower luminosity tracks and confirm, observationally, that

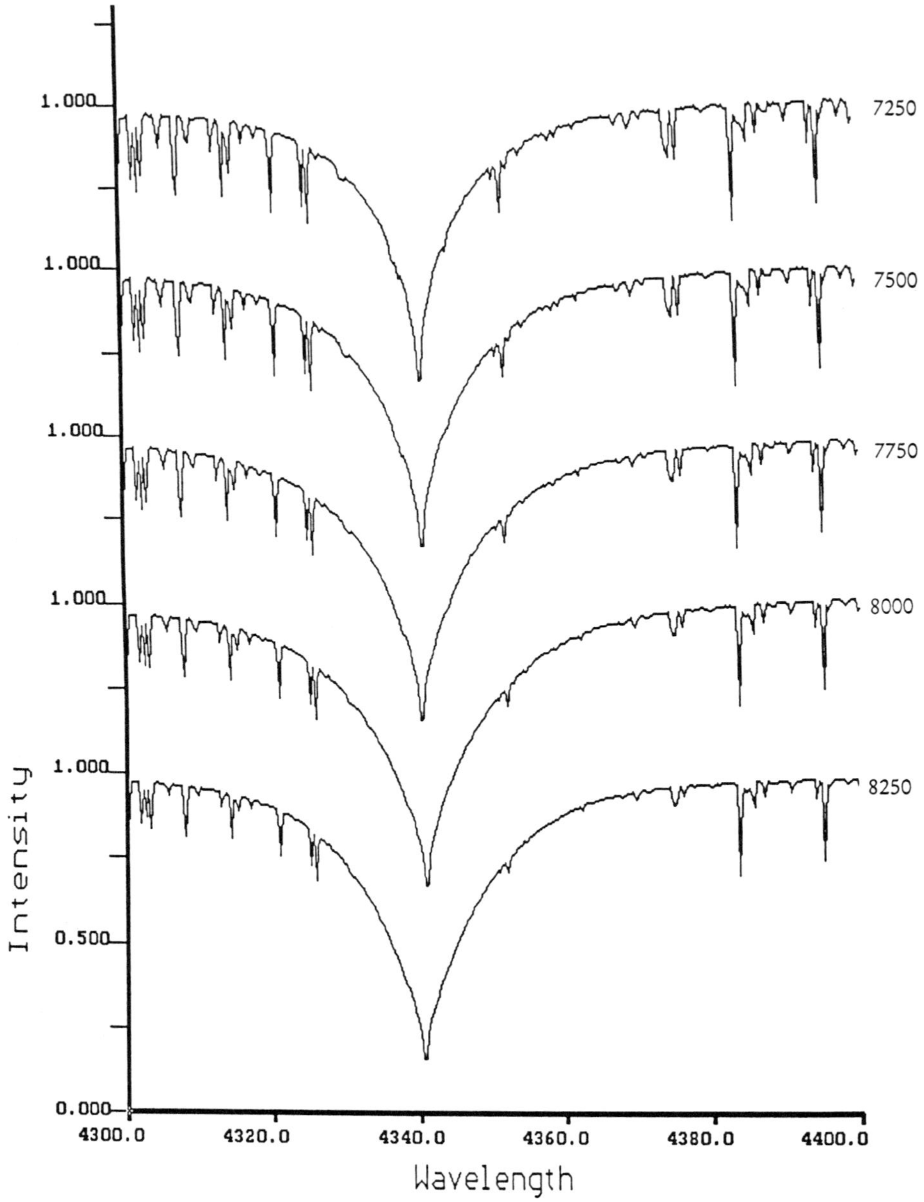

Figure 4. SPECTRUM Models at temperatures of 7250, 7500, 7750, 8000 and 8250.

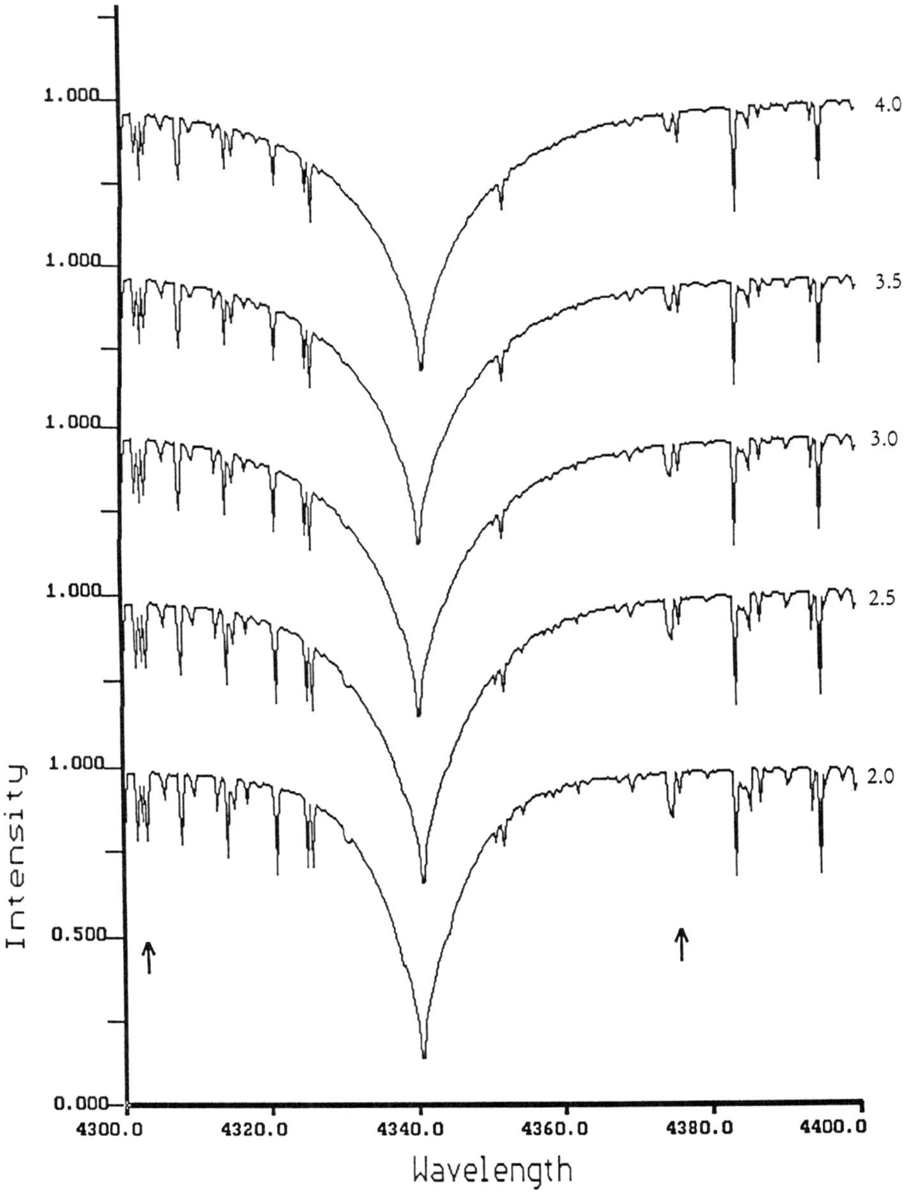

Figure 5. SPECTRUM Models at $\log g = 4$, 3.5, 3, 2.5 and 2.

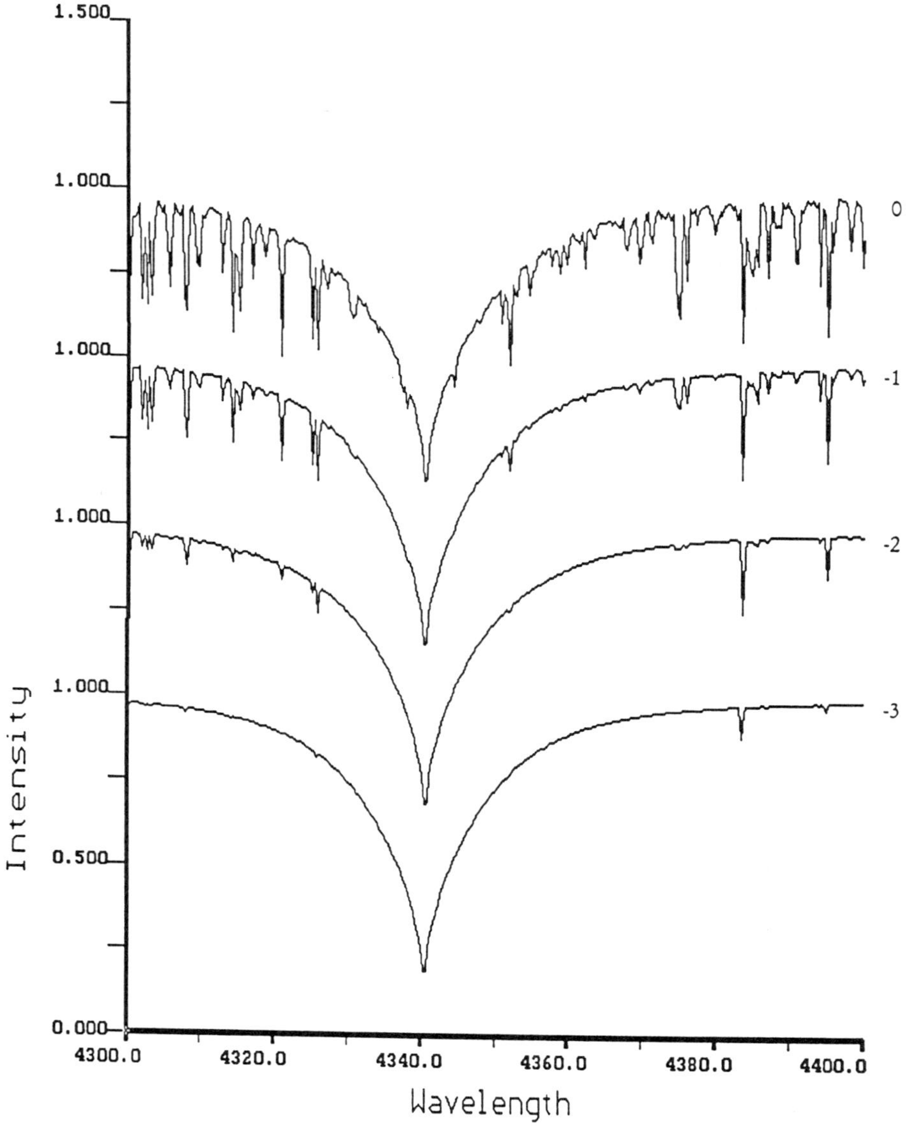

Figure 6. SPECTRUM Models at [Fe/H] = 0, -1, -2 and -3.

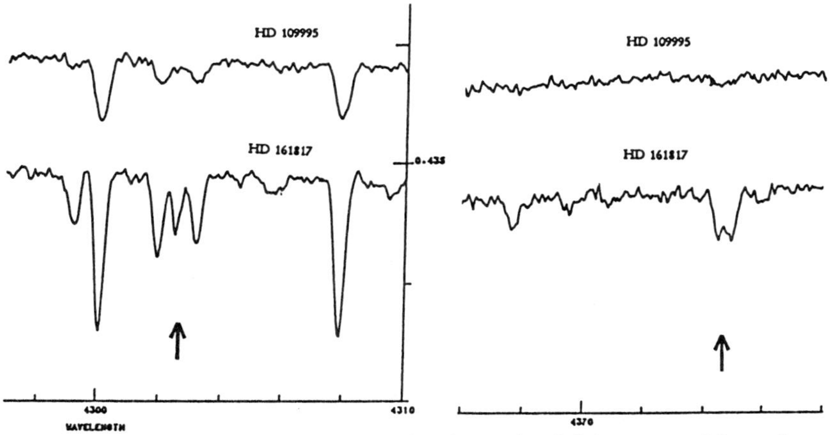

Figure 7. Section of spectra, near Hγ, from the Adelman *et al.*(1987) Atlas.

these luminosity indicators work for all FHB stars in this temperature region. If the luminosity indicators are confirmed then there will be a good spectroscopic indicator to confirm photometric assignments of HB stars to one or the other group on the horizontal branch.

References

Adelman, S. J., Fisher, W. A. and Hill, G. 1987, DAO Pub. XVI, 203

Gray, R. O. 1991, BAAS, **23**, 1382

Kodaira, K. and Philip, A. G. D. 1984, ApJ, **278**, 201

Kurucz, R. L. 1991, in IAU Symposium No. 149, The Stellar Populations of Galaxies, A. Reis, ed., Kluwer Academic Pub., Dordrecht

Philip. A. G. D. 1988, in Calibration of Stellar Ages, A. G. D. Philip, ed., L. Davis Press, Schenectady, p. 69

Stetson, P. B. 1987, PASP, **99**, 191

Stetson, P. B. and Harris, W. E. 1977, AJ, **82**, 954

Discussion

Drilling: Can you see these same evolutionary effects in c_1,β diagrams of horizontal branch stars?

Philip: The NGC 1904 frames were taken with the RCA CCD which did not give u magnitudes of sufficient accuracy to use. For the FHB stars, I have some Hβ observations but there are not sufficient data to make a good plot. In M92, I have u magnitudes of sufficient accuracy and a set of Hβ frames so in the near future, I will be able to make such plots.

Warren: Do you see an intrinsic width in your Zero Age Horizontal Branch (ZAHB) and, if so, how can you be sure that some of the spread is not due to small amounts of reddening?

Philip: The width of the lower track is about equal to the rms scatter of the observations. The upper track is about 0.2 mag or more brighter and shows greater scatter. The direction of the reddening vector in the y, $(b\text{-}y)$ diagram is two more points mainly to the right, there reddening changes would not account for the differences in position. In any case, in most of the clusters observed, the reddening is small (for M92 it is about $E_{B-V} = 0.01$).

Abt: You mentioned two lines: one is λ4374; is that the low-lying line of Fe I and λ4375? And what is λ4302?

Philip: In the Adelman, Fisher, Hill Atlas of FHB stars there is a blend at λ4374.4 and λ4374.8 (Sc II and Ti II, Y II) and a line at λ4375.9 (Fe I) which is the one you mentioned. There is a three line system near λ4302 composed of λ4301.9 (Ti II), λ4302.5 (Ca I) and λ4303.1 (Fe I).

The MK Process at 50 Years
ASP Conference Series, Vol. 60, 1994
C.J. Corbally, R.O. Gray, and R.F. Garrison (eds.)

Intrinsic Colors of MK Types in the Geneva Photometric System

B. Hauck

Institut d'Astronomie de l'Université de Lausanne 1290 Chavannes-des-Bois, Switzerland

1. Introduction

MK types furnish very important information on the physical parameters of a stellar atmosphere. We can also approach these parameters by photometric indices, for example, $B-V$, $R-I$, $b-y$, $B2-V1$ are among the various temperature indicators. It is important to know the intrinsic colors of MK types in the most useful photometric systems. With regard to the Geneva system, Meylan *et al.* (1980) have already derived these colors for the A to G stars, whilst Cramer (1993) has recently derived them for the B stars. In the present study we derive the intrinsic colors in the Geneva system using only the MK types found in the Michigan Spectral Catalogue. Taking $B2-V1$ as our temperature indicator, it will be compared to the spectral type. On the basis of new direct or semi-direct determinations of $T_{\rm eff}$ we first ascertain whether our previous calibration (Hauck 1985) of $B2-V1$ in terms of $T_{\rm eff}$ is still valid. The intrinsic colors for stars of luminosity class V and III were subsequently derived.

2. New calibration of $(B2-V1)$ in terms of $T_{\rm eff}$

The following publications were used in order to find $T_{\rm eff}$ values determined by a direct or semi-direct method:

1) Code *et al.*, 1976
2) Blackwell and Shallis, 1977
3) Saxner and Hammerbäck, 1985
4) Leggett *et al.*, 1986
5) Blackwell *et al.*, 1990
6) Malagnini and Morossi, 1990
7) Blackwell *et al.*, 1991

The source of the $B2-V1$ values is the Rufener (1988) catalogue.

Reddened stars have been excluded from the above list and only stars of luminosity class V were retained. Following a search for the best polynomial fit we found :

$$\Theta_{\rm eff} = 0.646 + 0.710(B2-V1) - 0.886(B2-V1)^2 + 1.533(B2-V1)^3. \quad (1)$$

If we consider two linear segments rather than a polynomial fitting, we obtain the following relations :

$$\Theta_{\rm eff} = 0.924 + 2.330(B2 - V1) \quad \text{for } B2 - V1 < -0.160, \qquad (2)$$
$$\pm\, 0.016 \pm\ 0.067$$

$$\text{and} \quad \Theta_{\rm eff} = 0.634 + 0.639(B2 - V1) \quad \text{for } B2 - V1 \geq -0.160. \qquad (3)$$
$$\pm\, 0.001 \pm\ 0.005$$

3. Test of two other calibrations

North & Nicolet (1990) and Kobi & North (1990) have proposed calibrations of some diagrams in the Geneva system in terms of $T_{\rm eff}$ and $\log g$. We can now compare the $T_{\rm eff}$ values given by these calibrations (hereafter called $T_{\rm GVA}$) and the data from the above publications. The following relations are obtained by linear regression :

$$T_{\rm GVA} = 343 + 0.958 T_1 \quad \text{r} = 0.997 \ \text{n} = 20,$$
$$\pm\, 252 \pm\ 0.015$$

$$T_{\rm GVA} = 30 + 0.996 T_3 \quad \text{r} = 0.966 \ \text{n} = 30,$$
$$\pm\, 227 \pm\ 0.037$$

$$T_{\rm GVA} = -23 + 1.009 T_4 \quad \text{r} = 0.997 \ \text{n} = 14,$$
$$\pm\, 209 \pm\ 0.019$$

$$T_{\rm GVA} = -185 + 1.029 T_5 \quad \text{r} = 0.990 \ \text{n} = 41,$$
$$\pm\, 252 \pm\ 0.015$$

$$T_{\rm GVA} = 152 + 0.964 T_6 \quad \text{r} = 0.973 \ \text{n} = 64,$$
$$\pm\, 170 \pm\ 0.021$$

$$T_{\rm GVA} = -235 + 1.034 T_7 \quad \text{r} = 0.991 \ \text{n} = 43,$$
$$\pm\, 91 \pm\ 0.014$$

and finally, if we merge the seven lists, we obtain :

$$T_{\rm GVA} = 252 + 0.963\ T_{\rm eff} \quad \text{r} = 0.996 \ \text{n} = 124.$$
$$\pm\, 58 \pm\ 0.006$$

4. Intrinsic colors of MK Types in the Geneva Photometric System

We present here a new determination of the intrinsic colors of MK types in the Geneva photometric system. Our source of MK types is the Michigan Spectral Catalogue (Houk & Cowley, 1975; Houk, 1978; Houk 1982; Houk & Smith-Moore, 1988). Our study has been restricted to unreddened stars, single stars, all visual unresolved binaries and all SB stars having been disregarded. In addition, all CP stars, stars known to be variable and stars with $\Delta m_2 \leq -0.010$

(metal-deficient stars) were excluded. It was necessary to limit our sample to stars having a solar metallicity so as to avoid a blanketing effect on $B2 - V1$ or an eventual effect on MK types due to the metal deficiency. Furthermore, stars that we found had obviously been misclassified were also excluded.

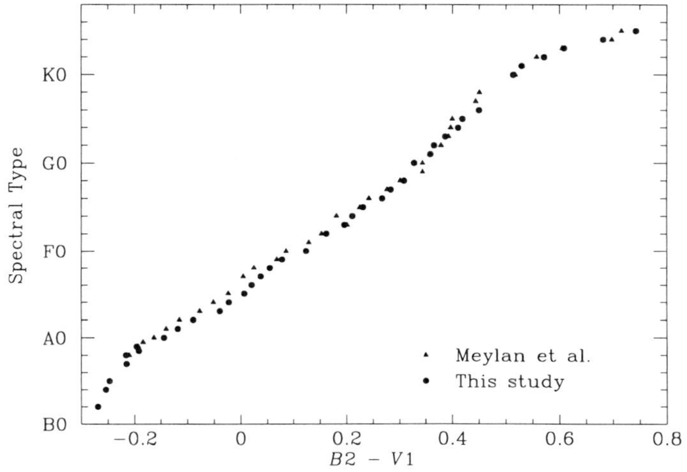

Figure 1. Spectral type vs $B2 - V1$

We have considered all Houk's class V stars with a value of $\Delta d \leq 0.100$ to be stars of luminosity class V, thus producing a sample of 947 stars. Table 1 shows the intrinsic colors for stars from B2V to K5V. A quick look at the same table in Meylan *et al.* (1980) shows that the intrinsic colors for the A-type stars derived from Houk are redder than those in Meylan *et al.*! This result should be analysed in more detail. Two factors could be responsible for this difference. First, despite our precaution to avoid reddened stars, we have nevertheless included some of them. Second, the rotational velocity affects the colors of A-type stars (Gray 1986, Gray & Garrison 1987, Hauck & Sletteback 1989). The mean $B2 - V1$ values derived by Hauck and Slettebak for stars with $0 < v \sin i < 100$ are in good agreement with those of Meylan *et al.* HD 205189 is a very interesting case. It is below the reference sequence in both a d vs $B2 - V1$ and c_1 vs $b - y$ diagrams. In a d vs Δ diagram this star is clearly in the area of metal-poor stars.

We have also examined the stars classified as class III by Houk. The same exclusion criteria were used, except that stars with $\Delta m_2 < -0.060$ were retained, thus giving a sample containing 364 stars. The results are given in Table 2. The last column of Tables 1 and 2 gives $T_{\rm eff}$ derived from the relations (2) or (3) of section 2.

Table 1. Luminosity Class V

MK type	*nb*	U	V	$B1$	$B2$	$V1$	G	$B2-V1$	$V1-G$	$B2-G$	d	Δ	g	m_2	T_{eff}
B2 V	8	0.518	1.176	0.794	1.592	1.861	2.402	-0.269	-0.541	-0.810	0.865	-0.400	-0.063	-0.675	16933
B4 V	7	0.653	1.143	0.804	1.577	1.831	2.366	-0.254	-0.535	-0.789	0.952	-0.268	-0.046	-0.656	15253
B5 V	4	0.724	1.134	0.805	1.572	1.819	2.352	-0.247	-0.533	-0.779	1.017	-0.200	-0.045	-0.655	14447
B7 V	13	0.900	1.082	0.823	1.555	1.770	2.288	-0.215	-0.518	-0.733	1.125	-0.045	-0.029	-0.634	11999
B8 V	23	1.010	1.078	0.827	1.551	1.767	2.286	-0.216	-0.519	-0.735	1.217	0.072	-0.019	-0.624	12044
B8/9 V	36	1.156	1.041	0.839	1.538	1.730	2.241	-0.192	-0.512	-0.703	1.316	0.203	-0.005	-0.611	10663
B9 V	10	1.262	1.040	0.842	1.534	1.730	2.241	-0.196	-0.511	-0.707	1.409	0.316	0.002	-0.602	10817
A0 V	33	1.464	0.953	0.878	1.503	1.647	2.144	-0.144	-0.497	-0.642	1.480	0.495	0.050	-0.559	9268
A1 V	18	1.539	0.908	0.895	1.487	1.605	2.093	-0.118	-0.488	-0.605	1.492	0.555	0.069	-0.539	9023
A2 V	14	1.545	0.859	0.911	1.469	1.559	2.035	-0.089	-0.477	-0.566	1.434	0.547	0.088	-0.518	8738
A3 V	7	1.504	0.775	0.936	1.442	1.481	1.936	-0.039	-0.454	-0.493	1.291	0.472	0.111	-0.488	8275
A4 V	2	1.534	0.762	0.938	1.444	1.466	1.911	-0.022	-0.446	-0.468	1.318	0.479	0.099	-0.495	8130
A5 V	5	1.528	0.716	0.947	1.431	1.424	1.860	0.007	-0.436	-0.429	1.273	0.454	0.108	-0.487	7895
A8 V	2	1.505	0.660	0.951	1.426	1.371	1.788	0.055	-0.417	-0.362	1.232	0.379	0.091	-0.500	7532
F0 V	8	1.348	0.566	0.962	1.410	1.287	1.672	0.124	-0.386	-0.262	1.026	0.155	0.075	-0.505	7069
F2 V	12	1.313	0.515	0.974	1.400	1.237	1.611	0.162	-0.374	-0.211	0.947	0.089	0.082	-0.499	6833
F3 V	39	1.285	0.467	0.984	1.390	1.194	1.554	0.196	-0.360	-0.164	0.882	0.032	0.083	-0.496	6639
F3/5 V	15	1.282	0.441	0.994	1.380	1.169	1.523	0.211	-0.354	-0.143	0.841	0.021	0.094	-0.483	6556
F5 V	67	1.281	0.415	1.000	1.375	1.144	1.493	0.231	-0.349	-0.118	0.818	0.004	0.098	-0.480	6452
F6 V	53	1.279	0.362	1.016	1.361	1.095	1.432	0.267	-0.337	-0.070	0.757	-0.024	0.112	-0.467	6267
F7 V	43	1.279	0.337	1.023	1.354	1.071	1.401	0.283	-0.330	-0.047	0.730	-0.035	0.117	-0.460	6187
F8 V	37	1.297	0.297	1.038	1.342	1.034	1.356	0.308	-0.322	-0.014	0.694	-0.033	0.133	-0.445	6066
G0 V	61	1.307	0.266	1.051	1.331	1.004	1.321	0.327	-0.317	0.010	0.656	-0.033	0.150	-0.429	5980
G1 V	36	1.330	0.219	1.067	1.318	0.960	1.265	0.358	-0.305	0.053	0.621	-0.031	0.163	-0.414	5845
G2 V	45	1.342	0.204	1.073	1.311	0.946	1.248	0.365	-0.301	0.063	0.609	-0.022	0.171	-0.404	5814
G3 V	51	1.366	0.171	1.088	1.300	0.915	1.208	0.386	-0.294	0.092	0.581	-0.011	0.186	-0.388	5726
G4 V	18	1.411	0.130	1.104	1.286	0.876	1.160	0.410	-0.284	0.126	0.568	0.020	0.204	-0.370	5625
G5 V	80	1.428	0.115	1.114	1.280	0.863	1.142	0.418	-0.279	0.139	0.553	0.033	0.212	-0.357	5596
G6 V	30	1.501	0.057	1.144	1.257	0.808	1.071	0.449	-0.264	0.186	0.519	0.089	0.244	-0.319	5472
K0 V	41	1.630	-0.061	1.200	1.213	0.700	0.924	0.513	-0.224	0.289	0.448	0.177	0.291	-0.247	5243
K1 V	35	1.684	-0.096	1.222	1.198	0.669	0.880	0.529	-0.210	0.318	0.429	0.221	0.309	-0.218	5187
K2 V	27	1.831	-0.183	1.277	1.161	0.589	0.762	0.571	-0.173	0.399	0.387	0.339	0.351	-0.144	5047
K3 V	34	1.955	-0.260	1.323	1.130	0.522	0.662	0.609	-0.141	0.468	0.356	0.435	0.384	-0.085	4929
K4 V	23	2.152	-0.404	1.409	1.076	0.394	0.468	0.682	-0.075	0.608	0.265	0.570	0.435	0.022	4713
K5 V	7	2.247	-0.501	1.457	1.049	0.306	0.341	0.743	-0.035	0.708	0.206	0.609	0.456	0.069	4549

Table 2. Luminosity Class III

MK type	*nb*	U	V	$B1$	$B2$	$V1$	G	$B2-V1$	$V1-G$	$B2-G$	d	Δ	g	m_2	$T_{\rm eff}$
B2 III	9	0.514	1.130	0.791	1.586	1.817	2.346	-0.231	-0.529	-0.760	0.859	-0.440	-0.077	-0.689	13388
B3 III	8	0.635	1.134	0.797	1.583	1.821	2.352	-0.238	-0.530	-0.769	0.962	-0.309	-0.066	-0.677	13850
B4 III	3	0.691	1.122	0.803	1.579	1.809	2.336	-0.230	-0.527	-0.757	0.997	-0.258	-0.061	-0.671	12981
B5 III	19	0.852	1.069	0.819	1.559	1.760	2.272	-0.200	-0.512	-0.712	1.093	-0.115	-0.046	-0.649	11380
B6 III	3	0.843	1.091	0.815	1.560	1.782	2.301	-0.222	-0.519	-0.741	1.094	-0.101	-0.041	-0.644	12392
B7 III	8	0.938	1.064	0.822	1.557	1.754	2.265	-0.198	-0.511	-0.709	1.166	-0.030	-0.041	-0.644	10936
B8 III	4	0.997	1.074	0.821	1.555	1.762	2.278	-0.207	-0.516	-0.723	1.226	0.043	-0.034	-0.640	11493
B9 III	8	1.231	1.019	0.837	1.542	1.714	2.211	-0.171	-0.497	-0.669	1.403	0.245	-0.031	-0.627	9796
A0 III	10	1.419	0.902	0.879	1.500	1.602	2.076	-0.102	-0.474	-0.576	1.427	0.398	0.022	-0.574	9054
A2 III	16	1.591	0.743	0.929	1.451	1.449	1.885	0.002	-0.436	-0.433	1.409	0.500	0.069	-0.523	7952
A3 III	23	1.582	0.749	0.936	1.443	1.457	1.898	-0.014	-0.441	-0.455	1.371	0.518	0.092	-0.500	8073
A5 III	4	1.560	0.680	0.955	1.427	1.392	1.817	0.034	-0.425	-0.391	1.279	0.459	0.105	-0.487	7696
A7 III	17	1.549	0.654	0.954	1.424	1.367	1.780	0.057	-0.413	-0.356	1.267	0.421	0.090	-0.496	7532
A8 III	12	1.533	0.631	0.961	1.418	1.347	1.752	0.071	-0.405	-0.333	1.226	0.392	0.092	-0.490	7421
A8/9 III	7	1.523	0.603	0.965	1.411	1.322	1.718	0.089	-0.396	-0.307	1.196	0.367	0.091	-0.487	7298
A9 III	16	1.487	0.582	0.969	1.409	1.300	1.692	0.108	-0.392	-0.284	1.148	0.314	0.092	-0.490	7172
F0 III	21	1.479	0.561	0.971	1.405	1.282	1.667	0.123	-0.385	-0.262	1.129	0.293	0.088	-0.490	7080
F2 III	9	1.496	0.510	0.985	1.393	1.236	1.605	0.157	-0.369	-0.212	1.094	0.279	0.093	-0.480	6869
F3 III	4	1.520	0.508	0.991	1.382	1.232	1.597	0.150	-0.365	-0.215	1.088	0.317	0.105	-0.459	6906
F3/5 III	6	1.477	0.451	1.008	1.375	1.179	1.538	0.196	-0.358	-0.163	0.993	0.238	0.120	-0.456	6646
F6 III	2	1.416	0.385	1.013	1.367	1.119	1.461	0.248	-0.341	-0.094	0.910	0.127	0.110	-0.468	6364
K0 III	28	2.157	-0.396	1.346	1.120	0.381	0.559	0.739	-0.178	0.561	0.488	0.571	0.468	-0.111	4558
K1 III	39	2.253	-0.448	1.377	1.101	0.333	0.496	0.768	-0.163	0.605	0.481	0.648	0.497	-0.075	4484
K2 III	36	2.468	-0.585	1.450	1.062	0.207	0.333	0.855	-0.126	0.729	0.462	0.799	0.560	-0.002	4277
K3 III	27	2.798	-0.805	1.565	1.008	0.004	0.075	1.004	-0.070	0.933	0.436	1.013	0.653	0.099	3955
K4 III	8	2.930	-0.882	1.613	0.986	-0.066	-0.019	1.053	-0.047	1.005	0.420	1.107	0.691	0.146	3860
K5 III	6	3.108	-1.013	1.665	0.963	-0.188	-0.166	1.152	-0.022	1.130	0.441	1.205	0.731	0.175	3680
M0 III	11	3.227	-1.105	1.695	0.952	-0.281	-0.259	1.233	-0.023	1.211	0.469	1.268	0.774	0.180	3545

5. Conclusions

Although we have more stars for which $T_{\rm eff}$ is determined by a semi-direct or a direct method, the relation between $\Theta_{\rm eff}$ and $B2 - V1$ is nearly the same as that determined a few years ago. In addition, we see that the calibration of North and Nicolet and that of Kobi and North give good estimated values of $T_{\rm eff}$. Concerning the intrinsic colors of MK types derived by Houk, we have seen that, for the A stars, they are redder than those we derived earlier. Before concluding that there is a real effect, we have to examine if the reddening effect is not more important than was assumed. More data is needed in this respect and we are awaiting with impatience further volumes of the Michigan Spectral Catalogue!

References

Blackwell D.E., Lynas-Gray A.E., Petford A.D., 1991, A&A, **245**, 567

Blackwell D.E., Petford A.D., Arrbas S., Haddock D.J., Selby M.J., 1990, A&A, **232**, 396

Blackwell D.E., Shallis M.J., 1977, MNRAS, **180**, 177

Code A.D., Davis J., Bless R.C., Hanbury-Brown R., 1976, ApJ, 203, 417

Cramer N., 1993, A&A, **269**, 457

Gray R.O., 1986, Thesis University of Toronto

Gray R.O., Garrison R.F., 1987, ApJS, 65, 581

Hauck B., 1985, in *Calibration of Fundamental Stellar Quantities, IAU Symp. 111,* eds. D.S. Hayes *et al.*, Kluwer, Dordrecht, pp. 271-284.

Hauck B., Slettebak A., 1989, A&A, **214**, 153

Houk N., 1978, Michigan Spectral Catalogue, Dept. of Astronomy, Univ. of Michigan, Ann Arbor, Michigan

Houk N., 1982, Michigan Spectral Catalogue, Dept. of Astronomy, Univ. of Michigan, Ann Arbor, Michigan

Houk N., Cowley A.P., 1975, Michigan Spectral Catalogue, Dept. of Astronomy, Univ. of Michigan, Ann Arbor, Michigan

Houk N., Smith-Moore M., 1988, Michigan Spectral Catalogue, Dept. of Astronomy, Univ. of Michigan, Ann Arbor, Michigan

Kobi D., North P., 1990, A&AS, **85**, 999

Leggett S.K., Mountain C.M., Selby M.J., Blackwell D.E., Booth A.J., Haddock D.-J., Petford A.D., 1986, A&A, **159**, 217

Malagnini M.C., Morossi C., 1990, A&AS, **85**, 1015

Meylan G., Python M., Hauck B., 1980, A&A, **90**, 83

North P., Nicolet B., 1990, A&A, **228**, 78

Rufener, F., 1988, *Catalogue of stars measured in the Geneva observatory photometric system* (fourth edition), Observatoire de Genève, Sauverny

Saxner M., Hammarbäck G., 1985, A&A, **151**, 372

Discussion

Houk: Is the catalogue by Mercedes Jaschek a compilation of spectral types from various classifiers?

Hauck: Yes, it is.

Garrison: How do you decide that something is "obviously misclassified" and so exclude it?

Hauck: A star can be considered as misclassified if its colors strongly disagree with the spectral type, or if its colors and MK type from another source disagree with the Houk type.

The MK Process at 50 Years
ASP Conference Series, Vol. 60, 1994
C.J. Corbally, R.O. Gray, and R.F. Garrison (eds.)

The Geneva Multicolor Photometric System as a Complementary Tool of Spectroscopic Spectral Classification

M. Golay

Observatoire de Genève, CH-1290 Sauverny, Switzerland

1. Historical review

G.B. Donati (1860, published 1862) introduced the concept of classification after having observed a few bright stars with the aid of a spectroscope. His classification criteria were first the color, then the line intensities. He noticed that there was some kind of relationship between a star's color and the appearance of the lines in its spectrum. During the 19th century, several attempts were made to classify stars by means of their colors. The notion of color index was introduced rather late in astronomical literature by C. Chandler (1888), and it was much used by Kapteyn, de Sitter and Scheiner between 1890 and 1900. We may consider that the first relation between spectral type and color index, in its modern sense, was probably that given in the article by J.A. Parkhurst (1912) titled "Yerkes Actinometry". The second historical relation between spectral classification and color indices is obviously that of H.L. Johnson and W.W. Morgan (1953) in their paper that refers to the Yerkes work in its title "Fundamental stellar photometry for standards of spectral type on the revised system of the Yerkes spectral atlas". The latter article marks the beginning of an uneasy but extremely interesting relationship between those who classify stellar spectra and those who derive color indices from multicolor photometry.

2. Intrinsic colors

For more than 50 years the authors of multicolor photometric systems have repeatedly attempted to establish intrinsic indices or parameters for each class and subclass. Nowadays, it is easy to make very accurate instantaneous photometric measurements. However, a great number of systematic errors occur during the reduction of the measurements outside the atmosphere, or during the computation of mean values over measurements made at different sites, with different photometers, at different dates (often covering long time spans). These adverse effects are strongly reduced if the photometric system is "isolated" or "closed," i.e., when all the operations that intervene between an observed color index and its published value, the construction of a photometer and the doing of the observations, etc., are controlled by the same group.

We know that, for a given spectral type, the color indices are affected by rotational velocity and the inclination of its axis, or by the spectral type and magnitude of a (generally undetected) companion. We can therefore only establish intrinsic colors for stars that are, as far as possible, single and non-variable, having rotational velocities close to the mean value for the spectral type considered, unreddened by interstellar matter and, of course, particularly well classified. The latter condition is in fact as difficult to achieve as the determination of accurate color indices reduced outside the atmosphere. An important critical effort is, however, being carried out by the specialists of the MK process (see Garrison, "A hierarchy of standards for the MK process", present publication), and the photometrists will likewise hasten to establish precise color indices for the "primary standards" and "secondary standards". For the current state of available spectral classifications, we have to select from a particularly homogeneous catalogue those stars that are destined to form a network of stars whose colors may be regarded as intrinsic. At the present, only the Michigan catalogue (in course of publication since 1975) fulfils the conditions that would characterise the spectroscopic counterpart of an "isolated" or "closed" system. Of the large number of stars classified therein, 20 000 are common with Geneva photometry.

Several favourable conditions prevail at the present: existence of a large and homogeneous catalogue of spectral classification, presence of a large and homogeneous catalogue of photometric data, the motivation to critically examine the lists of primary and secondary MK standards, and considerable technical progress related with the acquisition and reduction of photometric measurements. This situation leads us to question whether it is still relevant to publish tables of intrinsic colors for each spectral type. Such colors are mean values established over samples of stars that are assumed to belong to the same class, and they will fluctuate as the composition of each sample changes. Another way of looking at the problem would be to attribute the photometric colors and parameters of each star that has been selected as representative of a given spectral class, and that satisfies the criteria mentioned above, to its related intrinsic color. Here, it would be the spectroscopist who decides whether the star he has observed is of identical class to those of the list of stars with known intrinsic colors. This leads to an important question: what is the dispersion of the intrinsic colors of stars that share an identical MK classification? There is presently no satisfactory answer to that question at the level of the currently attainable photometric accuracy. Should one increase the number of spectroscopic parameters so as to improve the relation between spectral classes and intrinsic colors? This would be very useful for the study of interstellar extinction, for example.

Of Table 1, Table 2 and Table 3, the first intends to show the variety of spectral types of stars taken from the Michigan catalogue that would share the same colors in the Geneva 7-color photometry. It contains the stars that have colors corresponding to the intrinsic colors given by B. Hauck in this publication. The second (class V) and third (class III) tables illustrate the variety of spectral types attributed by 3 different sources to stars having almost the same colors as a reference star.

These three tables have been established by means of the technique of photometric boxes introduced by Golay (1969, 1977, 1978, 1991) and analysed by B. Nicolet in the present publication. The first selection is done In the color indices space U-B2, B1-B2, B2-V1, V1-G. We have selected all the stars of the Michigan catalogue that lie within a euclidian distance of 0.015 mag of the pseudo star having the intrinsic colors (Hauck, present paper) of each given spectral type. In this table, the Michigan stars are of quality classes 1 and 2, with quality 2 shown in brackets. The same process is carried out in table 2 and 3, where we

Table 1. Stars sharing the same colors in the Geneva system

HD	C_D_S	Michigan	R.O._Gray	Dist._(mag)	Rem.
27524	F5 V		F5 V m+0		
197692	F5 V	F5 V q=3	F6 V m+0	0.013	
211575	(F5)		F6 V m+0	0.019	
27808	F7 V		F8 V m+0		
142	G1 IV	F7 V q=1	F7 V m+0	0.016	
27383	F7 V		F7 V m+0		
3196	F8 V		F9 V m+0	0.019	
14802	G1 V	G2 V q=1	G0 V m+0		
2151	G2 IV	<G1 IV q=2	F9 V m+0	0.015	
91324	F6 V	F5 V q=1	F8 V m-1		D
5388	F6 V	F6 V q=1	F7 V m+0	0.006	
693	F6 V	F5 V q=1	F8 V m+0	0.017	
141103	(F5)		F8 V m-1.75	0.018	
223400	F7.5 V	F7/8 V q=1	F9 V m-1		
196378	F8 V	F7 V q=1	F9 V m-1.50	0.008	
48938	G2 V	F7 V q=1	F9 V m-1		
162396	F8 V	F8 IV/V q=2	F9 V m-1.25	0.014	
188815	F5 V	F5 V q=1	F8 V m-2		
61902	(F8)	F5/6wF2 q=2	F8 IV m-2	0.011	
25704	F7 V	F7 V q=1	F9 V m-2		
148816	F9 V		G0 V m-2.25	0.014	
22879	F9 V		G0 V m-2	0.019	
76932	F9 V	F7/8 IV/V q=2	F9 V m-2.5	0.02	
38510	(F8)	F5/6 V q=2	F9 IV-V m-2	0.02	
17072	(G2)	G2wF5 q=1	F9 III m-2		
208360	(G0)	F5/7 +(G) q=2	G0 III m-2	0.019	
63077	G0 VI	G0V q=1	G0 V m-2		
16784	G0 V	F8 V q=2	G0 V m-2	0.017	
56274	G1 V	G5 V q=1	G2 V m-2		
6434	G3 IV	G2/3 V q=2	G1 IV m-2	0.015	
150453	(F5)	F3 V q=1	F6 V m-2		D
7476	(F5)		F5 IV-V m+0	0.019	D
16784	G0 V	F8 V q=2	G0 V m-2		
79601	G2 V	G2 V q=1	G2 V m-2	0.006	
78747	(F8)	F8/G2 V q=3	G2 V m-2	0.013	
51929	F9 V	F8/G0 V q=2	G0 V m-2	0.014	
201099	(G0)		G0 V m-2	0.016	
63077	G0 VI	G0 V q=1	G0 V m-2	0.017	
17865	F8 V	F8 V q=1	G0 V m-2	0.02	
165401	G2 V		G1 V m-2		
20807	G2 V	G1 V q=1	G1 V m+0	0.017	D
113083	F9 V		G2 V m-3		
22879	F9 V		Go V m-2	0.019	

Table 2. Stars sharing almost the same colors as a reference star: class V

Sp.	Dist_ <=_0.005mag	Dist_ 0.006-0.010mag	Dist_ 0.011-0.015mag	Teff	Log g	Dist
B2V	B3 V	B2 V (B3 IV/V)	B3 V (B2 V)	18000	4.4	0.012
B4V	B3 IV	B5 V (B3 V to B5 Vn)	no	16000	4.4	0.017
B5V	B8 I I	B5 V (B3/5 V)	no	15000	4.1	0.018
B7V	B8 IV	B6 III to B7 III	no			
B8V	B8 V	B8 V (B9 IV)	B7/8 V	12500	4.1	0.016
B8/9V	B8/9 V to B9 V	(B9 V)	no	11500	4.25	0.01
B9V	no	no	B9 V (B9/9.5 V to B9 IV/VSi)	11000	4	0.02
A0V	A0 V	A0 V (A0/1 V)	A0 V	9750	4	0.011
A1V	<A0 V	A1 V	A1 V	9250	4	0.015
A2V	no	A1/2 V	ApSrEuCr to A2/3 V	8750	4	0.014
A3V	no	A3 V	A2/3 III to A3 V	8250	4.1	0.024
A4V	A4/5 IV	A3 V to A5 V	A2 V to A2 IV/V (A3/4 V)	8250	4.1	0.02
A5V	no	A5 II/III	A2mA5-A8 A7 III	8000	4	0.018
A8V	no	no	A3 V to A7 III	7500	3.9	0.02
F0V	F0 V to F2 V	no	F0 III (A9 III)	7250	3.9	0.018
F2V	F2 V	no	F3 V (F3 IV to F2 V)	7000	3.9	0.02
F3V	no	F3 IV to F3 V	F2/3 V to F3 V	6750	3.9	0.013
F3/5V	no	F3/5 V to F5 V (F3 V to F5 V)	F5 V	6750	4	0.01
F5V	F5 V	F3/5 V to F5 V	F5 V	6500	3.75	0.026
F6V	F6 V	F6 V to F7 V (F5 V to F6/7 V)	F5 V	6500	4	0.026
F7V	F7 V	F6 V to F8 V (G0 V)	F6/7 V (G0 V)	6250	4.1	0.026
F8V	no	F8 V	F7 V to F8V (F7 V)	6250	4	0.018
G0V	no	G0 V to G1 V (F8/G0 V to G0 V)	G0 V			
G1V	no	G2 V to G3 V	G1 V	6000	3.75	0.018
G2V	G3 V	G0V to G3 V (G2 V)	G1/2 V to G3 V	5900	3.9	0.026
G3V	G3 /5 V	G 0/1 V to G3 V	G1 V to G3 V	5900	4.1	0.014
G4V	no	no	G2 V to G6 IV (G5V)	5900	3.4	0.025
G5V	no	G3 V to G5 V (G3/5 IV)	G5 V	5750	3.4	0.019
G6V	no	G6 IV to K0 V (G8 V)	G8 V G8 IV/V	5750	3.5	0.016
K0V	no	(K0 V)	K1 III to G8/K0 V			
K1V	no	K0/1 V to K1 V	K0 V to K1 V (K2 III)			
K2V	no	no	K1 V to K3 V	5000	4.9	0.028
K3V	no	K 2/3 V to K3 V	K3 V (K1 V)	4900	4.9	0.024
K4V	no	no	K3/4 V to K4 V			
K5V	no	no	K4 V (K4 V to K4/5V)			

Table 3. Stars sharing almost the same colors as a reference star: class III

Sp.	Dist_ <=_0.005mag	Dist_ 0.006-0.010mag	Dist_ 0.011-0.015mag	Teff	Log g	Dist
B2III	no	no	B2 IV to B3 III (B3 III)			
B3III	no	no	B4 V to B3 III	17000	4.25	0.01
B4III	(B3 V)	B3 III to B3 IV (B4/5 III to B3 V)	B3/4 Vnn	15000	4.25	0.014
B5III	no	(B8 IV to B5/7 III)	B8 II/III			
B6III	B6 V	B8 II (B8 V pSi)	B8 II/III (B8 III)	14000	4.4	0.021
B7III	no	no	B7/8 II/III (B8 III)			
B8III	no	no	B8 IV/V to B8 V (B9 IV)	11500	4.25	0.019
B9III	no	B9 III	B8 V (pSi) (B8/9 III)			
A0III	no	A0 V (B9.5 V)	B9.5 IV to B9 V			
A2III	no	no	A5 V (A3 V)	8000	3.75	0.024
A3III	no	A3/5 IV/V to A4 IV	A2 III			
A5III	no	no	A9 V (A3/5mA5-A7 B9 V)	7750	3.9	0.027
A7III	no	A7 III	A7 III	7500	3.75	0.022
A8III	no	A8 III	(A9/F0 III)	7250	3.5	0.025
A8/9III	no	A7 III to F0 III (A9 III)	F0 III			
A9III	no	A9 III (F0/2 III)	A8 III to A5mA7-F0			
F0III	no	F0/2 III?IV to F2 III/IV (A3/5mF0-F2)	F0 II/III	7250	3.75	0.03
F2III	F3 III	F2 III	F0 IV/V to F2 V	6750	2.9	0.023
F3III	no	no	F3 IV/V	6750	2.9	0.026
F6III	no	F5m	(F7 IV/V)			
K1III	K2 III	K1 IIICN II to K1 III	K0 III			
K2III	no	K2 III	K2 III			
K3III	K2/3 III	K2/3 III (K3 III)	K3 III	4000	1.6	0.024
K4III	K4 III	no	K3 III to K3/4 III (K4 III)			
K5III	no	K4/5 III	K4/5 III to K5 III			
M0III	no	no	no			

have given the list of stars having distances ≤ 0.015 from those of a reference star (identified by bold characters in the first column). In this case, the stars and their classifications are taken from 3 different lists: CDS, Michigan (with indication of quality) and R.O. Gray 1989 (with indication of metallicity class).

These tables show that the more accurate is the classification, the less are the types dispersed in a photometric box. The classification by R.O. Gray, that includes an extra parameter, is very homogeneous within the boxes. The few observed deviations encourage a reappraisal of the spectral classification.

Photometric boxes can also be established in the D, Δ, g parameter space, thus making them less dependent on interstellar reddening. The stars in the "parameter" boxes that have the bluest color indices, and that are also members of the "color" boxes (containing at least 4 stars in the present context), are foreseen as potential members of the list of stars whose colors can be considered to represent very closely the relevant intrinsic colors.

3. Adjusting to stellar model atmospheres

The box method allows one to rapidly obtain the set of model atmospheres belonging to a grid such as that computed by Kurucz (1991 - 1993), that have synthetic colors or parameters that are identical (i.e. that lie within the same box of fixed dimensions) to those of a given observed star. As an example, table 2 lists for each set of intrinsic colors corresponding to a given spectral type the temperature and gravity of the closest solar composition model encountered in the colors space (its euclidian distance - in magnitudes - figures in the last column).

4. Conclusion

The Geneva photometric catalogue contains a great number of stars that are also measured in other photometric systems. It therefore provides a sufficiently large basis for applying discriminating tools, such as photometric boxes, that allow one to estimate the heterogeneity of other photometries or, conversely, to examine the residual heterogeneity of our own system. It is also an instrument that enables the objective assessment of the different types of information that are revealed by photometries having narrower bands, or by those that sample other regions of the spectral interval, by means of stars sharing the same Geneva photometric box. When applied to all the stars that are classified MK and that also have 7 color photometry, the box method allows the detection of stars with doubtful classification, or whose spectra present anomalies that should encourage the search for original and unforeseen solutions.

Acknowledgments. I would like to express my thanks to N. Cramer for the English version of this text and J. Guarinos for his assistance in the preparation of the paper.

References

Chandler, C. 1888, AJ, **8**, 137.

Donati, G.B. 1862, Annali del Museo Fiorentino.

Garrison, R.F. 1993, "A hierarchy of standards for the MK process", present publication.

Golay, M. Peytremann, E., Maeder, A. 1969, *Publ. Obs. Genève,Série A*, N°**76**, 44.

Golay, M. Mandwewala, N., Bartholdi, P. 1977, A&A, **60**, 181.

Golay, M. 1978, A&A, **62**, 189.

Golay, M. Nicolet, B., Cramer, N., 1991, "The photometric boxes in the UBVB1-B2V1G system", in *Precision Photometry: Astrophysics of the Galaxy*, L. Davis Press, Schenectady, N.Y., 303.

Gray, R.O. 1989 AJ**98**, 1049.

Hauck, B. 1993 " Intrinsic colors in the Geneva photometric system", present publication.

Houk, N., Cowley, A.P. 1975, "University of Michigan Catalogue of two dimensional spectral types for the HD stars", Vol 1, Dept. of Astronomy, University of Michigan, Ann Arbor.

Houk, N. 1982, Vol 2 et. seq.

Johnson, H.L., Morgan, W.W. 1953, ApJ, **117**, 313.

Kurucz, R. 1991, 1992, 1993, Private communication.

Nicolet, B. 1993, "Physical data from spectroscopy and high precision photometry", present publication.

Parkhurst, J.A. 1912, ApJ, **36**, 169.

Discussion

Philip: It is important to have the best spectral types when you make a comparison with stellar boxes. Some years ago, while I was at Strasbourg Observatory, Egret and I published a paper on photometric boxes in the Strömgren four-color system. We made tests of the stars in a photometric box by finding the spectral type of all stars in a given box. We found, in some cases, a large range of spectral types. We used a catalogue of spectral types compiled by M. Jaschek which took spectral types from the literature. I picked a number of stars, with divergent spectral types and gave the list to Pat Boeshaar who obtained good MK spectra from them. These I delivered to the Jaschek who then classified them. The scatter was greatly reduced.

Cramer - who read the paper in Golay's absence: Yes, it can also be reduced in the tables shown here.

Houk: In Tables 2 and 3 are the types in the body of the tables mine? (Answer: yes.) If so, then why is the first column, which is your "box," labeled "MK"? This is very confusing when you are using colors.

Cramer: You are absolutely right!

Hauck: The first column in those Tables is MK types. Golay has used as central stars of boxes the intrinsic colors I have derived on the basis of Dr. Houk's catalog.

Cramer: Yes. The header of the first column of Tables 2 and 3 is misleading. It should mean "box centered on the intrinsic colors corresponding to the MK type, as derived by B. Hauck." (Note: label changed in manuscript.)

Roman: I am puzzled why the physical parameters of A3 V and A4 V are identical in Table 2.

Cramer: The set of models, like the MK types, is a sequence of discrete values. The photometric box method crosses the model that is closest to the given MK location. If the photometric distance between two consecutive MK "points" is small, there is a chance that in both cases the same model is seen as the closest one.

The MK Process at 50 Years
ASP Conference Series, Vol. 60, 1994
C.J. Corbally, R.O. Gray, and R.F. Garrison (eds.)

The MK Classification of B-type Stars as seen by Geneva Photometry

N. Cramer

Observatoire de Genève, CH-1290 Sauverny, Switzerland

Abstract. The reddening-free color combinations X, Y and Z, defined in the Geneva system, are orthogonal to each other and have been specifically adjusted so as to optimise the analysis of B-type stars. X and Y estimate temperature and gravity effects, respectively, without ambiguity for O, B and the earliest A-types. These parameters are therefore narrowly correlated with the physical basis of the MK classification. Here, we examine the correlation with the MK system, and discuss the effects of reddening, rotation, binarity and chemical composition on the photometric classification.

1. Introduction

The Geneva photometric system was defined by M. Golay in 1958 and put into practice in 1960 by M. Golay and F. Rufener (Rufener, 1964). Great care has been taken to conserve the instrumental properties and reduction techniques of the system, with the result that the photometry has maintained a high degree of homogeneity up to this day (Rufener, 1985; Cramer, 1994). The system reproduces the general properties of the standard U,B,V system, though with a slightly higher feature sensitivity and different zero points for the color indices. The ability to classify stellar populations is considerably increased by the presence of the four intermediate bands B1, B2, V1 and G (see Rufener and Nicolet, 1988).

In the particular case of the B-type stars, it is possible to define a three-dimensional reddening-free representation (called X,Y,Z) based on linear combinations of the five intermediate pass-bands (see the sequence of papers by Cramer and Maeder, 1979, 1980 and Cramer, 1982, 1984a, 1984b, 1993). The parameters X and Y separate temperature and gravity effects, respectively, without ambiguity for types O, B and early A, while the Z parameter, which is orthogonal to the other two, shows very little spread for the normal stars but detects the magnetic Ap stars via their 5200 Å depletion and easily detects other peculiarities such as Wolf-Rayet features. The $T_{\rm eff}$ vs. $\log g$ separation in the X,Y plane (Figure 1, based on synthetic colors derived by B. Nicolet for Kurucz's 1991 models) reproduces the basic properties of the MK classification, and thus allows its correlation with the photometry to be carried out in a reddening-free environment. The advantage of doing so, particularly in the case of the early type stars that are virtually all significantly reddened, is that one can then make use of the *whole sample* of stars of a given subtype, thus deriving the mean colors

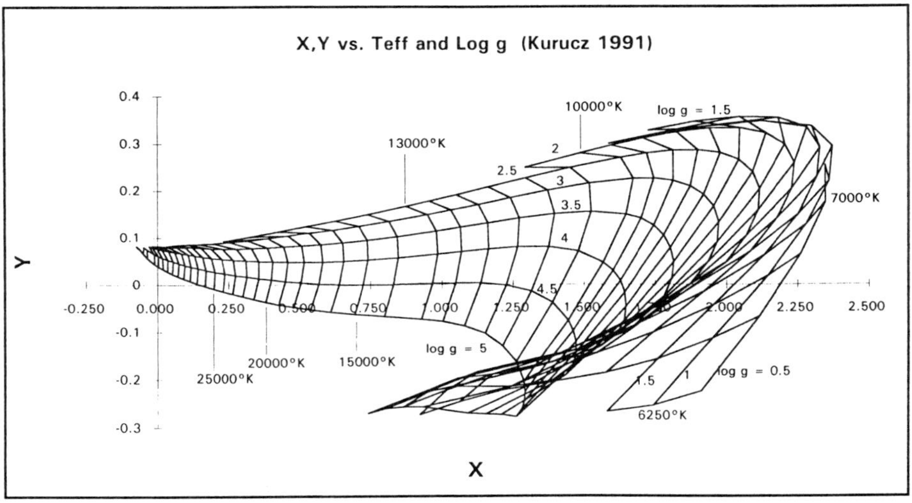

Figure 1.

in an objective manner. Otherwise, one has to consider for each sample a blue envelope that is probably affected by some residual reddening, and that also has a greater chance of being biased by untypical objects.

2. The Correlation

Geneva photometry is presently available for about 15,000 stars of types O, B and early A (Figure 2). Some 6,000 of these have MK types taken from various sources. The correlation presented here is based on a selection of 3,600 of the latter that are not mentioned as binaries, are not too heavily reddened and have been measured at least twice. The sources of the MK types are for about 80% the Michigan catalogue; the rest come mainly from earlier compilations based on selections by Jaschek and Jaschek or Cowley and Cowley. All subtypes and luminosity classes are represented, down to A5 for class V and to F5 for class Ia. The correlation was carried out as follows:

1. Each MK subtype was plotted in the X,Y plane. After removal of the obviously discrepant objects, the corresponding mean (X,Y) values were computed together with their standard deviations.

2. The data corresponding to each luminosity class were plotted in the X,Y plane. The values of each parameter were then slightly smoothed relatively to the successive subtypes, whereby the adjustments were kept well within the r.m.s. radius. Indeed, a smooth MK sequence must not necessarily imply a perfectly smooth photometric sequence since the classification criteria are different in both cases.

The resulting MK type calibration is shown in Figure 3. One notes the good

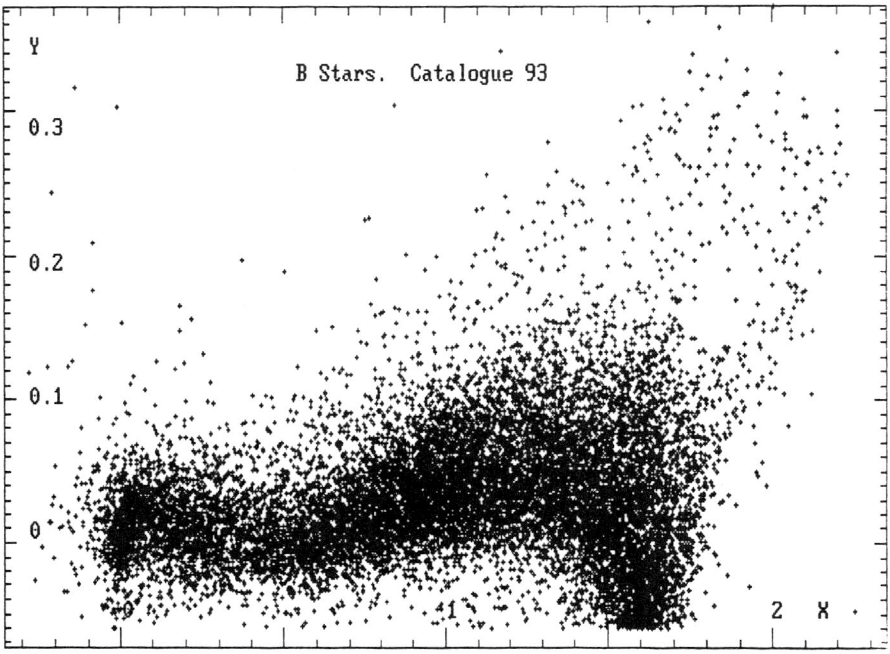

Figure 2.

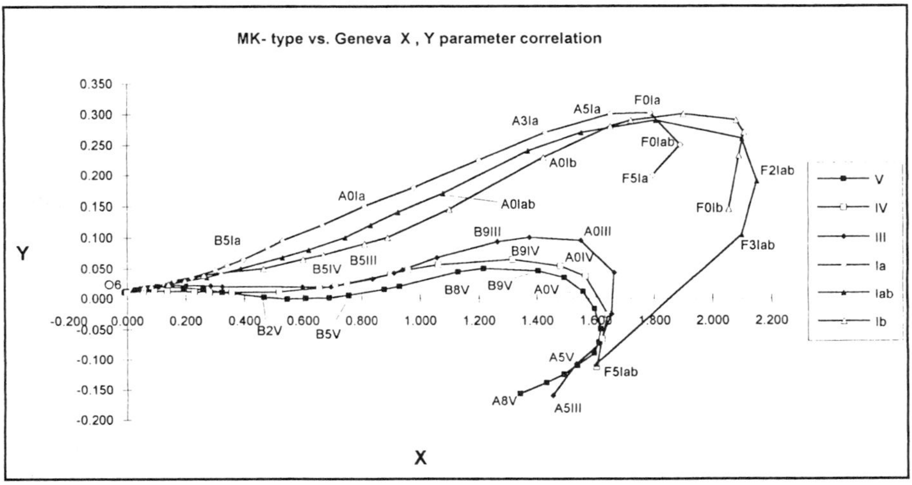

Figure 3.

overall separation of the subtypes, with the zones of ambiguity (predicted by the Kurucz models, Figure 1) for the late A- and early F type supergiants and for the late A type main sequence stars. One also notices that the B4 to B7 stars of classes IV and III are not separated in this correlation. The typical r.m.s. scatter of the correlations is shown for two areas of the relation. A calibration of the X,Y parameters in terms of intrinsic colors (Cramer, 1993) then allows the calculation of the corresponding intrinsic Geneva colors. An absolute magnitude calibration (Cramer, in preparation) based on cluster moduli derived by Meynet et al (1993) also provides an estimate of the relevant absolute magnitudes. These results are summarised in Table 1, Table 2 and Table 3.

3. Causes of scatter

Photometry is affected by reddening, rotation, binarity and chemical composition differences, and the effects on the indices can be predicted with a good degree of objectivity. The manner in which the attribution of an MK type is influenced by these parameters is, however, not well documented. Here, we restrict our discussion to their incidence on the photometric prediction of the MK type.

Reddening: The X,Y,Z parameters are corrected for reddening to the first degree. Some residual reddening is present for strongly reddened stars ($A_v \geq$ 1 mag) due to the non-linear nature of heterochromatic reddening lines, and to differences of interstellar extinction laws. The effects are negligible, however, in the present context since the typical deviations are of the order 10^{-2} mag per magnitude of visual absorption, which are small in comparison with the scatter of the correlations.

Rotation: The effects of rotation are estimated here on the basis of models computed by Maeder and Peytremann (1970) for different masses, inclinations and rotational velocities. For mid-B type stars, increasing rotation mimics a decrease of temperature and gravity in the X,Y plane. The effect is practically negligible for the equator-on aspect but becomes quite large for pole-on stars. The maximum variation, in terms of estimated MK type (in the B5V region), is an increase of two subclasses and a decrease of one luminosity class. In the region of the early A type stars, the effects may not be regarded as significant, except for the most extreme rotators which "appear" to be shifted about one subclass earlier. This is an artifact due to the overlapping of the photometric sequence in the X,Y diagram as shown by the models in Figure 1.

Binarity: The effects of duplicity can be estimated with the help of a ZAMS sequence established in the Geneva photometry. The variations within the ZAMS and along the class III sequence combined with ZAMS stars is shown in Figure 4. A comparison with Figure 3 shows that the general tendency is to increase the estimates of the luminosity class and the subtype.

Chemical composition: The B type stars distributed within the extended solar neighbourhood are a young population with a relatively homogeneous heavy element content. Also, their high effective temperatures decrease the importance of the spectral features produced by the latter in the visible. An estimate of the expected photometric effects can be made with the aid of Kurucz's models. Within the typical range of ± 0.5 around the solar value of the metallicity

Table 1. MK type versus X,Y,Z and related absolute magnitude and intrinsic color correlation: classes V and IV

Sp	X	Y	Z	Mv	[U-B]o	[B-V]o	[B1-B]o	[B2-B]o	[V1-B]o	[G-B]o
O6	-0.009	0.010	-0.020	-4.1	0.029	-1.349	0.739	1.644	2.009	2.578
O7	-0.006	0.011	-0.017	-4.1	0.032	-1.348	0.739	1.644	2.008	2.577
O8	-0.003	0.012	-0.012	-4.1	0.035	-1.347	0.739	1.644	2.007	2.576
O9V	0.034	0.015	-0.007	-4.0	0.069	-1.336	0.742	1.641	1.998	2.565
O9.5V	0.072	0.018	-0.004	-3.9	0.103	-1.325	0.745	1.638	1.989	2.554
B0V	0.103	0.020	-0.002	-3.8	0.131	-1.316	0.747	1.636	1.982	2.545
B0.5V	0.193	0.018	-0.002	-3.3	0.210	-1.290	0.755	1.628	1.962	2.523
B1V	0.260	0.015	-0.002	-2.9	0.269	-1.272	0.762	1.622	1.948	2.506
B1.5V	0.324	0.010	-0.002	-2.5	0.325	-1.255	0.768	1.615	1.934	2.490
B2V	0.468	0.002	-0.002	-1.8	0.452	-1.221	0.782	1.601	1.904	2.454
B2.5V	0.546	-0.001	-0.002	-1.4	0.521	-1.203	0.789	1.594	1.887	2.435
B3V	0.609	0.000	-0.002	-1.2	0.576	-1.191	0.794	1.589	1.875	2.420
B4V	0.689	0.001	-0.002	-0.9	0.645	-1.175	0.801	1.582	1.860	2.402
B5V	0.756	0.005	-0.002	-0.7	0.702	-1.163	0.805	1.578	1.849	2.388
B6V	0.878	0.015	-0.002	-0.4	0.805	-1.143	0.813	1.570	1.829	2.363
B7V	0.930	0.020	-0.002	-0.2	0.849	-1.134	0.816	1.567	1.821	2.353
B8V	1.129	0.044	-0.003	0.0	1.016	-1.106	0.826	1.556	1.791	2.315
B8.5V	1.217	0.050	-0.002	0.2	1.091	-1.091	0.831	1.549	1.776	2.296
B9V	1.404	0.046	-0.002	0.5	1.256	-1.051	0.848	1.532	1.736	2.249
B9.5V	1.493	0.034	-0.001	0.7	1.339	-1.023	0.859	1.521	1.709	2.217
A0V	1.557	0.011	-0.002	1.0	1.407	-0.991	0.872	1.508	1.678	2.182
A1V	1.595	-0.017	-0.004	1.3	1.457	-0.960	0.885	1.495	1.648	2.146
A2V	1.615	-0.049	-0.008	1.5	1.494	-0.929	0.898	1.482	1.616	2.109
A3V	1.609	-0.071	-0.012	1.6	1.501	-0.915	0.905	1.475	1.599	2.089
O9IV	-0.001	0.010	-0.012	-4.0	0.037	-1.346	0.739	1.643	2.007	2.576
O9.5IV	0.055	0.011	-0.010	-3.8	0.087	-1.328	0.744	1.638	1.993	2.560
B0IV	0.092	0.012	-0.007	-3.6	0.121	-1.317	0.748	1.635	1.985	2.550
B0.5IV	0.137	0.012	-0.005	-3.4	0.161	-1.304	0.752	1.632	1.975	2.538
B1IV	0.229	0.011	-0.002	-3.0	0.242	-1.279	0.760	1.624	1.954	2.514
B1.5IV	0.280	0.010	-0.001	-2.7	0.287	-1.266	0.764	1.619	1.943	2.501
B2IV	0.356	0.010	-0.001	-2.4	0.354	-1.247	0.771	1.613	1.928	2.483
B2.5IV	0.520	0.011	-0.001	-1.7	0.496	-1.211	0.785	1.599	1.896	2.444
B3IV	0.656	0.017	-0.001	-1.2	0.613	-1.185	0.794	1.589	1.871	2.414
B4IV	0.710	0.020	-0.001	-1.0	0.659	-1.175	0.798	1.585	1.862	2.403
B5IV	0.750	0.025	-0.001	-0.9	0.694	-1.169	0.800	1.583	1.855	2.394
B6IV	0.899	0.040	-0.002	-0.6	0.820	-1.146	0.808	1.574	1.832	2.364
B7IV	0.959	0.048	-0.002	-0.6	0.870	-1.138	0.811	1.571	1.823	2.352
B8IV	1.062	0.056	-0.002	-0.4	0.957	-1.122	0.818	1.564	1.806	2.331
B9IV	1.315	0.065	-0.002	0.1	1.172	-1.079	0.835	1.545	1.763	2.279
B9.5IV	1.480	0.054	-0.002	0.5	1.319	-1.038	0.852	1.527	1.723	2.233
A0IV	1.565	0.036	-0.003	0.7	1.403	-1.006	0.864	1.515	1.693	2.198
A1IV	1.640	-0.026	-0.007	1.4	1.505	-0.937	0.893	1.488	1.627	2.122
A2IV	1.621	-0.066	-0.011	1.6	1.509	-0.914	0.904	1.475	1.600	2.090

Table 2. MK type versus X,Y,Z and related absolute magnitude and intrinsic color correlation: classes III, II, and Ia

Sp	X	Y	Z	Mv	[U-B]o	[B-V]o	[B1-B]o	[B2-B]o	[V1-B]o	[G-B]o
O9III	0.022	0.015	-0.007	-4.1	0.058	-1.340	0.741	1.642	2.001	2.569
O9.5III	0.040	0.016	-0.005	-4.0	0.075	-1.334	0.742	1.641	1.997	2.563
B0III	0.044	0.017	-0.003	-4.0	0.078	-1.334	0.742	1.640	1.996	2.562
B0.5III	0.126	0.022	-0.003	-3.8	0.152	-1.310	0.749	1.634	1.977	2.539
B1III	0.201	0.022	-0.003	-3.4	0.218	-1.289	0.755	1.628	1.960	2.520
B1.5III	0.284	0.021	-0.002	-2.9	0.290	-1.267	0.763	1.621	1.943	2.500
B2III	0.324	0.020	-0.002	-2.7	0.326	-1.257	0.766	1.617	1.935	2.490
B3III	0.598	0.019	-0.002	-1.5	0.563	-1.197	0.789	1.594	1.882	2.427
B4III	0.694	0.019	-0.002	-1.1	0.646	-1.178	0.797	1.586	1.864	2.406
B5III	0.837	0.032	-0.002	-0.7	0.767	-1.155	0.805	1.577	1.841	2.376
B6III	0.912	0.042	-0.002	-0.6	0.830	-1.144	0.809	1.573	1.830	2.361
B7III	0.941	0.048	-0.002	-0.6	0.855	-1.141	0.810	1.572	1.826	2.356
B8III	1.057	0.068	-0.002	-0.6	0.952	-1.127	0.815	1.566	1.810	2.333
B9III	1.265	0.093	-0.002	-0.5	1.124	-1.100	0.825	1.555	1.780	2.295
B9.5III	1.375	0.100	-0.003	-0.4	1.216	-1.083	0.831	1.548	1.764	2.276
A0III	1.550	0.095	-0.003	-0.1	1.366	-1.046	0.846	1.533	1.729	2.237
A1III	1.663	0.043	-0.006	0.6	1.489	-0.984	0.871	1.507	1.672	2.172
A2III	1.653	-0.025	-0.011	1.4	1.516	-0.933	0.894	1.486	1.624	2.118
A3III	1.612	-0.074	-0.018	1.6	1.506	-0.912	0.906	1.473	1.595	2.085
O9II	0.017	0.014	-0.010	-4.1	0.053	-1.341	0.740	1.642	2.002	2.570
O9.5II	0.020	0.015	-0.008	-4.1	0.056	-1.341	0.740	1.642	2.002	2.569
B0II	0.045	0.016	-0.007	-4.0	0.079	-1.333	0.743	1.640	1.996	2.562
B0.5II	0.108	0.018	-0.005	-3.7	0.135	-1.314	0.748	1.635	1.981	2.544
B1II	0.154	0.020	-0.003	-3.6	0.177	-1.301	0.752	1.632	1.971	2.532
B1.5II	0.220	0.022	-0.002	-3.3	0.235	-1.284	0.757	1.626	1.956	2.515
B2II	0.253	0.023	-0.002	-3.1	0.264	-1.275	0.760	1.624	1.949	2.507
B3II	0.377	0.025	-0.002	-2.6	0.372	-1.245	0.770	1.614	1.925	2.477
B4II	0.580	0.037	-0.002	-1.9	0.547	-1.204	0.784	1.599	1.887	2.431
B5II	0.723	0.045	-0.002	-1.4	0.669	-1.179	0.794	1.589	1.863	2.400
B7II	0.880	0.054	-0.002	-1.0	0.803	-1.153	0.804	1.578	1.837	2.368
B8II	0.942	0.060	-0.004	-0.9	0.855	-1.144	0.808	1.574	1.827	2.356
B9II	1.004	0.070	-0.005	-0.8	0.908	-1.136	0.810	1.571	1.818	2.343
B0Ia	0.017	0.015	-0.005		0.053	-1.342	0.740	1.642	2.002	2.570
B0.5Ia	0.053	0.017	-0.005		0.086	-1.331	0.743	1.640	1.994	2.560
B1Ia	0.058	0.018	-0.005		0.091	-1.329	0.743	1.639	1.992	2.558
B1.5Ia	0.128	0.023	-0.005		0.153	-1.310	0.749	1.634	1.976	2.538
B2Ia	0.160	0.030	-0.005		0.183	-1.302	0.751	1.633	1.968	2.528
B2.5Ia	0.220	0.035	-0.005		0.236	-1.287	0.755	1.628	1.955	2.511
B3Ia	0.300	0.047	-0.005		0.308	-1.269	0.760	1.623	1.937	2.487
B4Ia	0.340	0.055	-0.005		0.345	-1.262	0.762	1.621	1.927	2.473
B5Ia	0.397	0.065	-0.005		0.397	-1.251	0.766	1.617	1.914	2.455
B6Ia	0.480	0.083	-0.005		0.474	-1.236	0.770	1.611	1.893	2.425
B8Ia	0.530	0.095	-0.005		0.521	-1.228	0.773	1.608	1.880	2.406
B9Ia	0.666	0.120	-0.005		0.644	-1.206	0.781	1.599	1.851	2.363
A0Ia	0.810	0.150	-0.005		0.776	-1.184	0.790	1.591	1.821	2.319
A1Ia	0.980	0.180	-0.005							
A2Ia	1.200	0.225	-0.005							
A3Ia	1.429	0.270	-0.005							
A5Ia	1.650	0.300	-0.005							
F0Ia	1.787	0.302	-0.005							
F2Ia	1.889	0.250	-0.002							
F5Ia	1.795	0.200	0.005							

Table 3. MK type versus X,Y,Z and related absolute magnitude and intrinsic color correlation: classes Iab and Ib

Sp	X	Y	Z	Mv	[U-B]o	[B-V]o	[B1-B]o	[B2-B]o	[V1-B]o	[G B]o
B0Iab	0.020	0.015	-0.005		0.056	-1.341	0.740	1.642	2.002	2.569
B0.5Iab	0.039	0.017	-0.005		0.074	-1.335	0.742	1.641	1.997	2.563
B1Iab	0.132	0.023	-0.005		0.157	-1.309	0.749	1.634	1.975	2.537
B2Iab	0.271	0.035	-0.005		0.280	-1.274	0.759	1.624	1.945	2.499
B3Iab	0.387	0.050	-0.005		0.384	-1.249	0.767	1.616	1.920	2.467
B5Iab	0.530	0.068	-0.005		0.510	-1.222	0.776	1.607	1.891	2.428
B6Iab	0.620	0.081	-0.005		0.589	-1.207	0.781	1.601	1.874	2.405
B7Iab	0.743	0.100	-0.005		0.697	-1.188	0.788	1.594	1.852	2.373
B8Iab	0.830	0.120	-0.005		0.775	-1.177	0.792	1.589	1.834	2.347
B9Iab	0.926	0.140	-0.005		0.859	-1.165	0.797	1.585	1.818	2.322
A0Iab	1.080	0.170	-0.005							
A2Iab	1.370	0.240	-0.005							
A3Iab	1.550	0.280	-0.005							
A7Iab	1.805	0.290	-0.005							
F0ab	2.100	0.260	-0.006							
F2Iab	2.150	0.190	-0.008							
F3Iab	2.100	0.104	-0.010							
F5Iab	1.599	-0.108	-0.012							
B0Ib	0.058	0.020	-0.004		0.091	-1.330	0.743	1.640	1.992	2.557
B0.5Ib	0.082	0.021	-0.004		0.112	-1.323	0.745	1.638	1.987	2.551
B1Ib	0.140	0.024	-0.004		0.164	-1.306	0.750	1.633	1.973	2.535
B1.5Ib	0.196	0.030	-0.004		0.214	-1.292	0.754	1.630	1.961	2.519
B2Ib	0.260	0.039	-0.004		0.272	-1.277	0.758	1.625	1.946	2.500
B2.5Ib	0.320	0.043	-0.004		0.325	-1.263	0.762	1.621	1.934	2.485
B3Ib	0.465	0.050	-0.004		0.451	-1.231	0.773	1.610	1.906	2.451
B5Ib	0.600	0.065	-0.004		0.568	-1.207	0.781	1.601	1.881	2.417
B6Ib	0.670	0.072	-0.004		0.628	-1.195	0.786	1.597	1.869	2.402
B7Ib	0.810	0.090	-0.004		0.749	-1.174	0.793	1.588	1.846	2.370
B8Ib	0.890	0.100	-0.004		0.816	-1.163	0.798	1.584	1.833	2.353
B9Ib	1.100	0.145	-0.004		0.997	-1.139	0.807	1.575	1.802	2.306
A0Ib	1.422	0.229	-0.004							
A1Ib	1.650	0.280	-0.004							
A2Ib	1.720	0.290	-0.004							
A3Ib	1.900	0.300	-0.004							
A4Ib	2.080	0.290	-0.004							
A5Ib	2.106	0.270	-0.004							
A7Ib	2.090	0.232	-0.004							
F0Ib	2.055	0.145	-0.004							

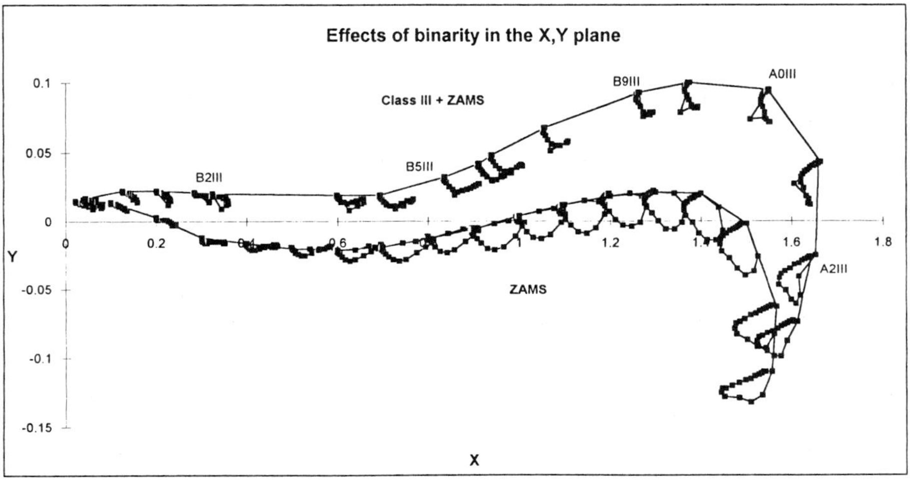

Figure 4.

parameter χ given by Kurucz, the X parameter varies by at most 0.05 mag, with X increasing as χ decreases. For Y, the maximum amplitude is 0.01 mag as Y increases with χ. In both cases, these effects certainly contribute to some extent to the scatter but are much less important than those of duplicity and rotation.

4. Conclusion

A reddening-free representation of O, B and early A type stars, that separates temperature and gravity effects in the sense of the MK classification, is possible in Geneva photometry. The clear separation of Ap, Wolf-Rayet and other peculiar stars by the photometry enhances its affinity with the MK Process. The photometric parameters are very well correlated with the MK criteria; they can be used to predict spectral types and serve to link the specific properties of both classifications with each other.

References

Cramer, N., Maeder, A. 1979, A&A, **78**, 305.

Cramer, N., Maeder, A. 1980, A&A, **88**, 135.

Cramer, N. 1982, A&A, **112**, 330.

Cramer, N. 1984a, A&A, **132**, 283.

Cramer, N. 1984b, A&A, **141**, 215.

Cramer, N. 1993, A&A, **269**, 457.

Cramer, N. 1994, in *The incidence of long term monitoring in variable star research*, NATO ARW, C. Sterken, M. De Groot, eds., Kluwer Acad. Publ., Dordrecht, to be published.

Maeder, A., Peytremann, E. 1970, A&A, **7**, 85.

Meynet, G., Mermilliod, J.-C., Maeder, A. 1993, A&A *Suppl. Ser.*, **98**, 477.

Rufener, F. 1964, *Publ. Obs. Genève, Série A*, **66**, 413.

Rufener, F. 1985, in *Calibration of fundamental Stellar quantities, IAU Coll. 111*, D.S. Hayes et al., Eds. Reidel Publ. Co., Dordrecht, 253.

Rufener, F., Nicolet, B. 1988, A&A, **206**, 357.

Discussion

Abt: What is the goal of your project: is it to determine the physical parameters appropriate for each spectral type or is it to replace getting MK types with photometry for faint stars?

Cramer: The object of this photometric classification is to get the same kind of information that one gets via MK types for stars that lack observational data. Incidentally, it also serves to correlate physical parameters with MK types.

Houk: If you could use a larger telescope, could you perhaps help with the problem Garrison was discussing, namely finding candidates of the rarer spectral types to check out as secondary standards between 14 and 16 magnitude?

Cramer: Yes, certainly. We will be setting up within two or three years two new 1.2 m telescopes, in collaboration with Lenven Observatory, one in each hemisphere. We are also testing the feasibility of reproducing the Geneva system using CCD detectors. This should enable the measurement of stars of that magnitude.

Hensberge: You told us that the quasi-simultaneous classical photometry in the Geneva system will be replaced in the future by CCD photometry. What will be, in your opinion, the consequence for the accuracy on the colors and the X, Y, Z parameters?

Cramer: There will still be one classical photometer using the quasi-simultaneous technique. We are attempting to reproduce the system using CCD detectors. First trials have been made a few months ago and preliminary results show that the accuracy of the color determinations is about half as good as with the classical aperture photometry. This is due to a large extent to the sequential integrations through the various filters. To reduce the scatter induced by that process, we hope to find some way of reproducing the quasi-simultaneous technique with the CCD photometer.

Garrison: Another problem will be changing detectors and thus having to define a new set of filters. Some stars may have different and possibly unpredictable values in the two systems, because some particular feature may fall in a different filter. This may only be a problem for a few types, but it is important to be aware of the problem.

Cramer: Yes. It is very important to reproduce as far as possible the shapes and locations of the passbands. For the CCD photometer that is being tested, we have used the same class filters as for the older classical photometers and done laboratory measurements of the sensitivity functions.

Osterbrock: Am I right in thinking that the stars you classified by Geneva photometry had previously been selected as B stars by a spectroscopic survey? If you push the photometric method to fainter stars, will you observe all stars and, if so, can you be sure that some of the stars that you select as B stars on the basis of this photometry are not actually stars from a very different region of the H-R diagram, or perhaps peculiar stars, or stars of a different population?

Cramer: The selection of the sample used for the present calibration was done by "cutting off" photometrically the region occupied by the B stars in the X, Y, Z space. The calibration was then carried out with the available MK types in that sub-sample. I do not think that the application of this photometric technique to fainter stars will reduce its ability to distinguish B stars from other types or "peculiar" objects; at least, if such surveys are carried out in our galaxy. I am curious, however, to see how B stars with lower abundances, such as those in the LMC, for example, would behave in this photometric representation.

Houk: I also have problems classifying stars in the B4, B5 range, telling whether they are giants or dwarfs. They look very similar on my plates.

Cramer: Yes, that is also what the photometric correlation with the MK types seems to show.

Abt: How useful is your system if you have difficulty distinguishing some Ap stars from supergiants?

Cramer: Actually, there is no difficulty in distinguishing Ap stars from supergiants. The supergiants are distributed in the X, Y plane in the direction of the Y parameter. The Ap stars stand out in the Z direction, which is normal to the X, Y plane. There is virtually no overlap of these two populations.

Weaver: How did you determine your coefficients? Did you use principal components?

Cramer: I did not use the method of principal components analysis to define the X, Y, Z parameters in 1979. They were derived by rotations in a three-dimensional space based on Q-type reddening-free color differences. Some years later, Egret from Strasbourg Observatory did a principal component analysis for the B-type stars, and found relations with coefficients that differed only by about 0.001 from mine. The parameters I am using were obtained "bare handed"!

The MK Process at 50 Years
ASP Conference Series, Vol. 60, 1994
C.J. Corbally, R.O. Gray, and R.F. Garrison (eds.)

Physical Data from Spectroscopy and High Precision Photometry

B. Nicolet

Observatoire de Genève, CH-1290 Sauverny, Switzerland

Abstract.
The Geneva photometric system offers an exceptionnally high degree of homogeneity. Intercomparison between stars (boxes technique) and between some stars and the new Kurucz models provides accurate information on $T_{\rm eff}$, log(g) and [M/H] for stars of various spectral types. Nevertheless the extinction or the binarity can deteriorate the calibrations, for example $T_{\rm eff}$ for A type stars and a good spectral classification provides then a very valuable help in most cases.

1. Introduction

According to the stellar internal structure and evolution theories, the most *fundamental* characteristics of a star, M_0, χ_0, t, v_{rot} (initial mass, initial chemical composition, age, and rotational velocity), lead to *atmospheric* properties like L, R, $T_{\rm eff}$, log(g), χ (luminosity, radius, gravity, superficial chemical composition) defining themselves a spectrum which may be expressed as a flux $E(\lambda)$.

The features of $E(\lambda)$ can be determined by a *spectrograph* having a sensitivity function S(1) and an instrumental profile $P(\lambda)$ giving $E'(\lambda)$ where $E' = P*(S{\cdot}E)$.

Photometers use filters having pass-bands $S_i(\lambda)$ and measure (after reduction outside terrestrial atmosphere) color indices

$$C_k = -2.5 \log_{10} \int S_{i_k}(\lambda)E(\lambda)d\lambda \;/\; \int S_{j_k}(\lambda)E(\lambda)d\lambda$$

For photometrists the whole chain gives a function F between fundamental properties and color indices (or parameters)

$$(M_0, \chi_0, t) \vec{F} \vec{C} = (\ C_k)$$

Estimating the fundamental values from color indices is equivalent to defining a reciprocal relation F^{-1}. To obtain reliable astrophysical data (M, $T_{\rm eff}$, log(g), χ, pecularities, extinction), it is indispensable that F^{-1} is univocal, continuous, and that the C_k are very accurate.

2. Geneva system and photometric boxes

The Geneva photometric system uses 7 filters: U, B, V, B_1, B_2, V_1, and G. The first three are more or less similar *but not identical* to those of the 3-color system

UBV. For a description of the properties of this system see a review by Golay (1980).

During more than 30 years, special care was devoted to the *accuracy* and *homogeneity* of this system (Cramer, 1994). The observational techniques are carefully defined, the filters of a given denomination come from the same piece of glass, some of the standard stars were monitored during the whole lifetime of the system.

The reduction itself is especially carefully carried out: the slope of the continuum and other 2 nd order terms are taken into account. Moreover, in view of a maximal coherence, it is worked out exclusively at the Geneva Observatory.

The last version (Rufener 1988) contains almost 30 000 stars. 10 000 more were measured since and are being reduced.

Given the high degree of homogeneity resulting from this strict methodology, we can expect that two stars S and S′having nearby colors C_k and C'_k would have similar physical properties. We will examine this point.

3. Boxes as a topology defined by a metric

Golay et al. (1969) defined as a *box around a star* S the set of all stars S'_1, S'_2,... having all $\Delta C_k = C'_k - C_k$ within 0.01 magnitude in absolute value. Here C_1, C_2, C_3, C_4, C_5 and C_6 are U, V, B_1, B_2, V_1, G in the relations U-B,V-B,... S is called the *central star* and the S'_i *member stars.*

We can consider $\vec{C} = (C_k)$ and $\vec{C'}$ as vectors in a 6-dimensional vectorial space or as points in a punctual space. The representative points of the star of a box are contained in a 6-dimensional *hypercube* centered on $\vec{C}$. Such boxes are now designed as *cubic*. At this epoch the Geneva catalogue contained N = 1400 stars. From then on N increased by a factor 30 and the number of "proximity relations" by the square 900. A more refined treatment became necessary. It is not very "reasonable" that a point close to a corner would belong to the box and, in a statistical work, costly information may be lost.

To increase the efficiency of this technique, Nicolet (1981a, 1993a,b) enlarged this definition. It is well-known that the errors on the colors indices are not equal. $\sigma(U)$ is greater than the other $\sigma(C_k)$. Moreover these errors may be correlated or anti-correlated. All these errors and correlations between them may be described by a covariance matrix $V = (V_{ij})$. Instead of simple error bars, *error ellipses* give a much better description of the errors and their correlations.

The *elliptical boxes* homothetical to the error ellipses were introduced for most applications.

Elliptical as well as cubic boxes have all characteristics of the mathematical concept of *metric* based on a *distance* δ. To each couple (P; P′) of points belonging to a space E corresponds a positive number δ (P; P′) having the following properties:

1) $\delta(P; P) = 0$ for each $P \in E$
2) $\delta(P; P') = \delta(P'; P)$ for each couple $(P; P') \in E \times E$
3) $\delta(P; P'') \leq \delta(P; P') + \delta(P'; P'')$ Triangle inequality

The distance between $\vec{C}$ and $\vec{C}'$ corresponding to the *cubic* boxes is given by: $\delta(\vec{C} ; \vec{C}') = \text{Sup}_k \mid \Delta C_k \mid$.

A box around S (having colors C) is the *set of the stars S'* such that $\delta(\vec{C} ; \vec{C}') \leq r$ where r is the *size* (or *radius*) of the box. Each star is the center of a box. Two or more boxes may overlap. Such boxes are *closed neighborhoods* in the sense of the mathematical *topology.*

This topology *isn't a classification* like the MK process: in a proper classification each element (star) of a set is put into a subset (class) and these subsets have no intersection.

For *elliptical boxes*, the only difference is the definition of the distance. Let V' be V^{-1}, the inverse of covariance matrix, and f a "convenient" multiplicative factor, then

$$\delta^2(\vec{C} ; \vec{C}') = f \sum \Delta C_j V'_{jk} \Delta C_k$$

4. Boxes on parameters

The differential extinction may *redden* the color indices. So similar stars differing only by their color excess do not fall into the same box as defined in the preceding section. But this problem disappears if we define boxes on *reddening-free* parameters. In the Geneva sytem they are:

$$d = (U - B_1) - 1.430\,(B_1 - B_2)$$
$$\Delta = (U - B_2) - 0.832\,(B_2 - G)$$
$$g = (B_1 - B_2) - 1.357\,(V_1 - G)$$

see Golay (1980) and references therein.

A FORTRAN software package as well as a FORTRAN program PHOTOM of internal use at the Geneva Observatory allows us to obtain boxes of any kind and size in a few tenths of a second. On output, member stars are sorted by increasing distances. In addition to 6-colors (U, V, B_1, B_2 , V_1 , G normalized to B) boxes, and d, Δ, g parameters boxes, an option (frequently used) allows us to build boxes involving only the 5 intermediate passbands, therefore 4 color indices: U - B_1, B_1 - B_2, B_2 - V_1 , and V_1 -G.

The purpose of this paper is not to describe applications of the box method, for instance together with MK classification (see Golay 1994), but to illustrate by a few examples the resolving power of the boxes and the necessity of complementary information. Let us see a box around HD 90470 (Cf. Table 1).

The central star (HD 90470) leads; the member stars such as HD 21912, HD 258755, etc. follow. Spectral types from Michigan catalogues (Houk and Cowley 1975, Houk 1978 and 1982, Houk and Smith-Moore 1988) are given with their qualities (Houk q); spectral types without quality come from other sources.

If HD 90470 were faint (say $V \simeq 12$), its measurement with a medium-sized telescope (diameter $\simeq$ 0.70 m) would need about 15 minutes and the photometric results would provide a *prediction* (not a determination!) of the spectral

Table 1.

Name	V Mag.	Sp. Houk	q	U -B_2	B_1-B_2	B_2 -V_1	B_2 -G	dist.
HD 90470	6.008	A2V		0.093	-0.495	-0.043	-0.498	
Ellipt. box on 4 colors								
HD 21912	5.807	A5m		0.093	-0.499	-0.046	-0.497	0.008
HD 258755	9.347	(A2)		0.102	-0.497	-0.039	-0.497	0.009
HD 98154	7.354	A3V		0.094	-0.501	-0.043	-0.503	0.010
HD 101549	7.965	A3V		0.095	-0.499	-0.042	-0.488	0.012
HD 133954	8.121	A2/3 III	1	0.078	-0.500	-0.035	-0.491	0.012
HD 76512	6.382	A4 V	1	0.081	-0.503	-0.035	-0.485	0.014

type, using either calibrations and diagrams, or the box method. We immediately see that the box method is *auto-calibrating* and *multidimensional* instead of 2-dimensional like the diagrams. It uses the totality of the available information.

For HD 90470, a reasonable prediction gives a spectral type A3V, which is rather correct. Nevertheless, despite its accuracy, photometric data may lead to error. Binarity, for instance, may bias the colors and the physical data about a star, if ignored.

More pernicious biases may arise if we are not watchful enough. Theoretical models illustrate this point.

5. Boxes involving models

Recently Kurucz produced models and fluxes from new opacity calculations (Kurucz 1991, 1993). Using the passband function of Rufener and Nicolet (1988), we easily deduce Geneva color indices for these models. The Kurucz color grid was enriched by cubic spline interpolations on $\Theta_{\rm eff} = 5040/T_{\rm eff}$, and on log(g) and put into the files read by PHOTOM. Table 2 gives the models falling in an elliptic box on 4 colors, constructed with HD 90470 as the central star.

Table 2.

Name	V Mag.	Sp. Houk	q	U -B_2	B_1-B_2	B_2 -V_1	B_2 -G	dist.
HD 90470	6.008	A2V		0.093	-0.495	0.043	-0.498	
$T_{\rm eff}$	Log(g)	[M|H]	E(B_2 -V_1)	U -B_2	B_1 -B_2	B_2 -V_1	B_2 -G	dist.
8250	4.10	0.00	0.000	0.094	-0.486	-0.043	-0.505	0.013
8250	4.00	0.00	0.000	0.119	-0.492	-0.050	-0.513	0.022
9500	4.40	0.00	0.100	0.068	-0.507	-0.034	-0.502	0.022
9000	4.40	0.00	0.060	0.077	-0.490	-0.042	-0.511	0.023
8500	4.25	0.00	0.030	0.097	-0.479	-0.034	-0.496	0.024

Despite the huge number of lines ($\simeq$ 58 000 000) taken into account in the elaboration of the models of Kurucz and the care of his work, it would be foolish to deduce $T_{\rm eff}$, log(g) and χ from the box in table 2 without further work. Even if the models were perfect, the passband function would have an error bar of about 3% larger than the box size! The reason to use boxes with models is to illustrate effects on colors arising from changes of atmospheric data or color excesses.

To simplify, let us write $\partial\vec{C}\ (\Theta_{\rm eff})$ for $\partial\vec{C}\ /\ \partial\Theta_{\rm eff}$, and then $\partial\vec{C}\,[\log(g)]$, $\partial\vec{C}\,(\chi)$ and $\partial\vec{C}\,[E(B\text{-}V)]$ with an analogous convention. These first derivatives are not affected by the systematic errors on the models and the passbands. We admit, as a work hypothesis in order to simplify the language, that such errors are null.

The third member of the box: ($T_{\rm eff}$ = 9 500, log(g) = 4.40 , E(B_2 -V_1)=0.100) must be discarded. The distance of a A2V type star with a magnitude of 6 is of the order of 150 pc and a color excess of 0.100 is impossible. But $\partial\vec{C}\ (\Theta_{\rm eff})$ and $\partial\vec{C}\,[E(B\text{-}V)]$ are nearly parallel and a reddened star having $T_{\rm eff}$ = 9500, log(g) = 4.40, χ = 0.00 has nearly the same colors as HD 90470. Complementary information (here: E(B_2 -V_1) = 0) is necessary to deduce $T_{\rm eff} \simeq 8300$ and log(g) $\simeq$ 4.10 for this star.

Let us examine another example: star 60 in the open cluster NGC 2547 (Cf. Table 3). The first model should be given the highest weight in a statistical treatment. Unfortunately it would be unrealistic since the cluster NGC 2547 is reddened by E(B_2- V_1) = 0.038 (Nicolet 1981b). Members 3 and 4 are more adequate since E(B_2- V_1) = 0.030 for these models. In this region of the photometric space a change in $T_{\rm eff}$ together with a change in log(g) mimic a reddening. Expanding in terms of vectors:

Table 3.

Name	V mag.	We	Sig	U -B2	B1-B2	B2-V1	B2-G
NGC 2547 60	10.904	2	4	-0.003	-0.401	0.172	-0.190

Teff	Log(g)	[M|H]	E(B2-V1)	U-B2	B1-B2	B2-V1	B2-G	dist.
6750	3.40	0.00	0.000	0.006	-0.395	0.181	-0.191	0.017
6750	3.50	0.00	0.000	-0.009	-0.394	0.183	-0.189	0.021
7000	3.75	0.00	0.030	-0.023	-0.403	0.177	-0.199	0.021
7000	3.60	0.00	0.030	0.011	-0.405	0.172	-0.205	0.021
7250	4.00	0.00	0.060	-0.010	-0.406	0.171	-0.210	0.022

$$\partial\vec{C}\,[E(B\text{-}V)] \simeq \alpha\ \partial\vec{C}\ (\Theta_{\rm eff}) + \beta\ \partial\vec{C}\,[\log(g)],$$

and, here again, complementary information is required.

6. Conclusion

The box method, which uses all the information provided by a very homogeneous photometric system, is a tool with a fantastic resolving power. For a given type of box, since the number of relations is proportional to N^2 (the square of the number of stars measured in this system), the statistical yield of this method will dramatically increase not only for spectral type predictions (as described in this paper), but also for determining physical parameters and cluster distances when the results of HIPPARCOS become available.

Nevertheless, photometry alone is not able to avoid confusions, especially those due to the extinction. A good 2-dimensional MK classification may provide complementary information, and more refined classifications (e.g., Gray 1989, 1994) are welcome.

References

Cramer, N. 1994, to be published in *The impact of long-term monitoring on variable star research* (Chr. Sterken and Maart de Groot, Eds), NATO ARW, Kluwer

Golay, M.: 1980, *Vistas in Astron.* **24**, 141

Golay, M. 1994, this volume

Golay, M., Peytremann, E., Maeder, A. 1969, *Publ. Obs. Genève, Série A*, No 76, 44

Gray, R. O. 1989, AJ, **98**, 1049

Gray, R. O. 1994, this volume

Guarinos, J., Golay, M. 1994, this volume

Houk, N., Cowley, A.P. 1975, *University of Michigan catalogue of two-dimensional spectral types for the HD stars*, vol. 1, University of Michigan, Ann Arbor, Michigan

Houk, N.: 1978, *University of Michigan catalogue of two-dimensional spectral types for the HD stars*, vol. 2, University of Michigan, Ann Arbor, Michigan

Houk, N.: 1982, *University of Michigan catalogue of two-dimensional spectral types for the HD stars*, vol. 3, University of Michigan, Ann Arbor, Michigan

Houk, N., Smith-Moore, M.: 1988, *University of Michigan catalogue of two-dimensional spectral types for the HD stars*, vol. 4, University of Michigan, Ann Arbor, Michigan

Kurucz, R. 1993, Private communication

Kurucz, R. 1991, in *Stellar atmospheres: Beyond classical models*, Crivellari, L, Hubeny, I., and Hummer, D. G. (Eds) NATO ASI, Kluwer, Dordrecht, p. 441

Nicolet, B. 1981a, A&A, **97**, 85

Nicolet, B. 1981b, A&A, **104**, 185

Nicolet, B. 1993a, in *IAU Coll. 136*, Elliott, I. and Butler, C. J. (Eds) Poster Papers on stellar photometry, p 14

Nicolet, B. 1993b, in *IAU Coll. 136*, Elliott, I. and Butler, C. J. (Eds) Poster Papers on stellar photometry, p 17

Rufener, F. 1988, *Catalogue of stars measured in the Geneva observatory photometric system* (fourth edition) Observatoire de Genève, CH-1290 Sauverny (Switzerland)

Rufener, F., Nicolet, B. 1988, A&A, **206**, 357

Discussion

Philip: With square photometric boxes you cover the photometric space completely. However, elliptical "boxes" do not tile the parameter space completely. If you are investigating photometric box space, how do you handle this problem?

Nicolet: The box technique isn't a *classification*, but a *topology*. That means, that each star (point in the photometric space) may be the center of a box (neighborhood) and that boxes may overlap. Thus, the elliptical shape is not a problem.

Abt: May I ask a stupid question? Your photometric parameters are continuous variables, but the MK types are compartmentalized. What is gained by putting the photometric variables into boxes?

Nicolet: There are no stupid questions. When a star varies, the center of its box does so. Consequently, also the contents (stars or synthetic models) vary. About this point, it must be mentioned that variable stars are not suitable member stars of boxes.

Hauck: I want to make a comment relative to the question of Dr. Abt. This technique of boxes allows us to obtain very quickly all the stars in our database having the same color as a star that is interesting to you. You have then many possible applications, and certainly the obtaining of a pseudo spectral type is not the best one, since you gain no more information. But you have many other successful applications.

Abt: Is, then, the point of the boxes a way of specifying the accuracies of the photometric variables?

Nicolet: The interesting point is the accuracy of physical variables. In the best case, it can be shown that absolute magnitudes agree to ± 0.10 within a box. Nevertheless, this technique doesn't aim at "automatic spectral classification." On the contrary, as shown in this meeting (e.g., in the paper by Guarinos and Golay using Information Theory), spectral classification and photometry mutually complement the information.

The MK Process at 50 Years
ASP Conference Series, Vol. 60, 1994
C.J. Corbally, R.O. Gray, and R.F. Garrison (eds.)

Information Theory as a Tool for Extracting the Best Information Relevant to Spectral Classification from Geneva Photometry

J. Guarinos and M. Golay

Observatoire de Genève, CH-1290 Sauverny, Switzerland

1. Introduction

Information Theory was developed by C.E. Shannon, originally for telecommunications. Nevertheless, it is also a powerful tool for any study in which one intends to measure the quantity of useful information lying in a set of measurements – or lacking in it. Therefore, Information Theory (IT) can be used for pointing out the situations in which the efficiency of a system could be improved and how it could be improved.

We began a study of the Geneva photometric system's resolving power regarding spectral classification, using IT. With this method, it is possible to take into account the dispersion around the true values of the variables, along with the redundant information lying in every system composed of dependant sub-systems. In the following, we will call a "system", any set of variables deduced from observations and allowing the description of a star in physical terms ($T_{\rm eff}$, $\log g$ and [M/H] for example), such as the spectral type and the luminosity class. As a test sample, we use 400 standard stars carefully selected for having well-defined spectroscopic characteristics and photometric colors.

Golay introduced the concept of "photometric box" (see Golay *et al.* 1969) which was extended by Nicolet (1981). It is based on topological neighbourhoods around the colors of a *central star* in the multidimensional space built either with the color indices or with the reddening free parameters of Geneva photometry. If we use the d, Δ, g reddening free parameters, the stars included in each box (the "box members") are supposed, in most cases, to have fundamental astrophysical parameters almost identical to those of the central star. Using our 400 standard stars as central stars of photometric boxes, we can get a measure of the photometric space's homogeneity around each central star. Roughly speaking, the larger this homogeneity is, the better our photometric system will be able to extract reliable information for spectral classification in the related area of the photometric space. In other words, the homogeneity of spectral characteristics within the box is a measure of our photometric system's resolving power.

Our ambition is to provide the means of learning which kind of information is lacking in various areas of the multidimensional photometric space, how big is this uncertainty and how to remove it. This can be done applying the Information Theory formalism (Weaver and Shannon 1949), which provides a means of measuring quantitatively and objectively the information content of the Geneva photometry as a function of the considered region of the photometric space. As a preliminary step, we present results of a study taking into account only the spectral types. The figures presented provide a first estimation of the ability of the photometric indices of our 400 standards to give reliable and rich information regarding the spectral type.

2. Method

Golay (1991) introduced empirical parameters called "Spectral Number" (N_S) and "Luminosity Number" (N_L). The first can be considered as a numerical scale which translates the spectroscopic difference between two spectral types in the MK system into an integer. The second establishes another numerical scale which expresses the difference between two luminosity classes in terms of an absolute magnitude difference. Each central star and each member of a box centered around it can be assigned a Spectral Number and a Luminosity Number. These two empirical parameters will be used for estimating the homogeneity of each photometric box regarding spectral classification.

2.1. Why use Information Theory

Nicolet (1982) and Jaschek and Frankel (1986) have shown how probabilities can help in estimating the information lying in a photometric system. But their studies were first attempts and did not take into account two important points.

Firstly, when several spectroscopists classify a star (even a standard star), they often come to different conclusions. Therefore, the result of their examination is not a *certain* spectral type but a *probable* spectral type. There is a dispersion of the measurements (i.e., the spectral classifications and, thus, the Spectral Number) around the true value. The photometric measurements which were used for building the photometric boxes must be interpreted the same way. IT takes into account these uncertainties.

The second point is the following: For classifying the stars (assigning a spectral type, a luminosity class and a chemical peculiarity), we use a "system." This system roughly consists in Geneva photometry associated with the MK system. IT can take into account the existence of redundant information, which might lead to an overestimation of the information brought by the system. Sometimes, a measurement of a physical quantity reduces the indetermination of another quantity, as we will see below. Observational biases also have much influence on this matter.

2.2. Information Theory applied to spectral classification

In IT, the information, I, brought by a system is considered as a reduction of the *a priori* indetermination, made possible thanks to the observations.

$$I = \left[\begin{array}{c} \text{A priori} \\ \text{indetermination} \end{array} \right] - \left[\begin{array}{c} \text{A posteriori} \\ \text{indetermination} \end{array} \right] \tag{1}$$

Let's consider a variable X and let's use the *entropy* as a measurement of its indetermination. We get :

$$I = < H(X_k) > - < H(X_k|X_l) > \tag{2}$$

where $H(X_k|X_l)$ is the entropy related to the probability that the real state be X_k given the fact that the observed state is X_l, and $< H >$ is a mean entropy.

Let's assume that the variable X is N_S. It can carry an amount of information directly linked to the number of different states it allows to distinguish. The $\log_2$ of the inverse probability of observing a value for N_S indicates a degree of indetermination, thus an *entropy*, if the concept introduced in statistical mechanics is used with a binary encoding. We get :

$$< H(X_k) > = -\Sigma_{k=1}^{n}[P(X_k)\ log_2 P(X_k)]. \quad (3)$$

$P(X_k)$ is the probability that the variable X have the value X_k. The index k refers to the *real* states, while the index l refers to the *observed* states. Now, let $P(X_k|X_l)$ be the *a posteriori* probability that the real state be X_k, given the fact that the observed state is X_l. Equation (2) can be expressed as follows:

$$I = -\Sigma_{k=1}^{n}[P(X_k)\ log_2 P(X_k)] + \Sigma_{l=1}^{m}\Sigma_{k=1}^{n}[P(X_l)\ P(X_k|X_l)\ log_2 P(X_k|X_l)] \quad (4)$$

$P(X_k|X_l)$ can be calculated thanks to the Bayes theorem:

$$P(X_k = i|X_l = j) = \frac{P(X_k = i)\ P(X_l = j|X_k = i)}{\Sigma_{k=1}^{n} P(X_k)\ P(X_l = j|X_k)} \quad (5)$$

This theorem, well known in statistics, uses *a priori* probabilities, or conditional probabilities. They represent the dispersion of the observed variables around their true values. One such probability, $P(X_l = j|X_k = i)$, corresponds to the probability of observing a state l, if the reality is state k. We calculated these probabilities assuming a gaussian dispersion law around the true values for N_S, as a first approximation.

Therefore, we get that the information carried by N_S is the reduction of the *a priori* indetermination to the *a posteriori* indetermination. Since there is a dispersion of the observations (which are the values of N_S computed from the *deduced* MK types in each photometric box) around the true N_S (computed from the *true* MK type), this reduction of indetermination, hence the corresponding information gain, is not maximal.

Now, if we consider the system composed of all the variables (N_S, N_L and a "Chemical Peculiarity Number" which we still have to define), we have to take into account the redundant information which exists because these three variables are not independent. The system carries less information than what could be computed by simply adding the indetermination reductions corresponding to each variable, because they have common or *mutual* information. For example, it is highly probable that a giant star be a cold red star rather than a hot blue one. Thus, N_L carries a piece of information regarding the possible states (or values) of N_S. A similar situation happens with the Ap, Bp, Am, Fm stars: The Peculiarity Number gives an information on N_S. The CN weak or strong stars are G-type stars; CNO stars are B-type stars... Observational and physical bias also (such as the Hertzsprung gap) must be taken into account, in order to fix "deduction rules" which are necessary for computing the *mutual* informations. The latter will be computed using entropies related to a posteriori probabilities: The mutual informations correspond to the entropy reduction of a variable (or the reduction of number of allowed values) caused by the information brought by another variable. Once all the mutual informations are computed, we have to substract them from the sum of the informations brought by each of the

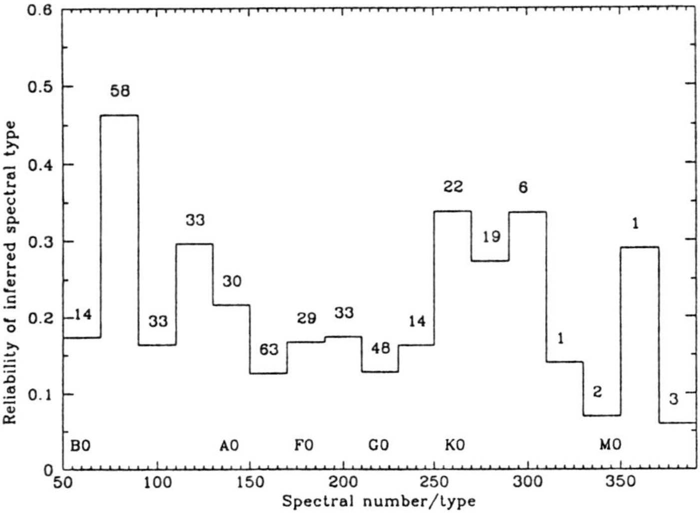

Figure 1. The reliability of the inferred spectral types of 400 standard stars as a function of the Spectral Numbers and their corresponding spectral types.

variables, in order to obtain the information quantity brought by the [N_S, N_L, Peculiarity Number] system.

For example, let's consider a second variable, Y here, which will be the Luminosity Number. The total information is not the sum of the informations gained by each variable, since there is mutual information. Between the two variables "Spectral Number" and "Luminosity Number", the mutual information, $I_{\rm mut}$, can be calculated as follows :

$$I_{\rm mut} = \begin{bmatrix} \text{Luminosity entropy} \\ \text{before} \\ \text{spectral type observation} \end{bmatrix} - \begin{bmatrix} \text{Reduction of luminosity entropy} \\ \text{thanks to} \\ \text{spectral type determination} \end{bmatrix} \quad (6)$$

This can be expressed this way:

$$I_{\rm mut} = < H(Y_k) > -H < (Y_k|X_l) > \quad (7)$$

3. Preliminary results

Figure 1 shows the reliability of the spectral types which were inferred for our 400 standard stars as a function of the Spectral Numbers and their corresponding spectral types. The average reliability has been computed for bins 20 units large in Spectral Numbers. The numbers written on top of each column are the number of stars taken into account in each bin. We can see that we lack high reliability for assigning a precise spectral type from the early A-type to late G-type stars. On the contrary, there is high confidence in the spectral types

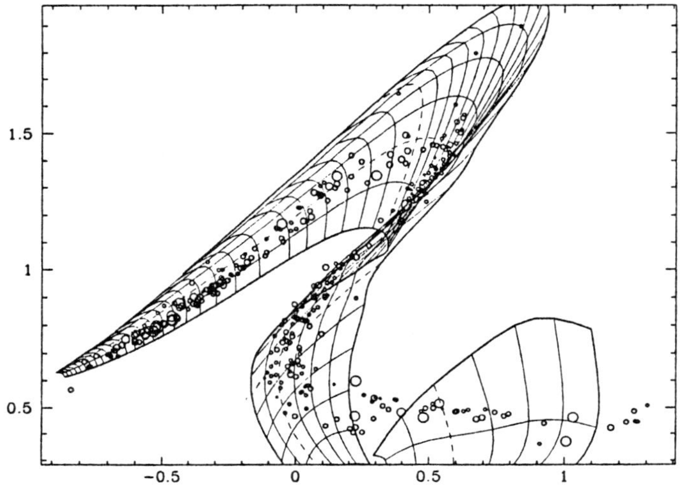

Figure 2. The distribution of our standard stars in d/Δ diagrams, with grids drawn from Kurucz's (1991) stellar atmosphere models for [M/H]=0.0.

inferred from our photometry for the early B-type to early A-type stars, except around B5, and for the late G-type to K5-type stars. Unfortunately, we face a selection effect in our list of standards after K5.

Figures 2, 3 and 4 show the distribution of our standard stars in d/Δ diagrams, with grids drawn from Kurucz's stellar atmosphere models (Kurucz 1991) for [M/H]=0.0 (Figure 2) and [M/H]=-0.5 (Figure 3), where the grids exhibit the loci of stars having $\log g$ = 2, 2.5, 3, 3.5, 4, 4.5 and 5, and $\log g$ = 4.0 (Figure 4), where the grid exhibits the loci of stars having [M/H] = -1.5, -1.0, -0.5, -0.2, 0.0, 0.2, 0.5. In these diagrams, the circles indicate the positions of standard stars. The larger a circle is, the more the star can be assigned a reliable spectral type with its reddening-free parameters, according to the photometric box content.

In Figures 2 and 3, the dotted line evolving with the temperature corresponds to gravity 4.0 while, in Figure 4, it corresponds to [M/H] = 0.0. In these three diagrams, the two other dotted lines correspond to $T_{\rm eff}$ = 6250 K (the lower one) and $T_{\rm eff}$ = 10000 K (the upper one). The latter is not visible in Figure 4. We can see that between $T_{\rm eff}$ = 10000 K (B9.5 spectral type) and $T_{\rm eff}$ = 6250 K (F8 spectral type), the grid drawn for [M/H] = 0.0 (Figure 2) overlaps in such a way that two stars lying in the same area of the photometric diagram can have very different temperatures, depending on their gravities. Figure 3 shows that for stars with [M/H] = -0.5 , the situation is even worse: The grid overlaps for stars ranging from $T_{\rm eff}$ = 6000 K to almost $T_{\rm eff}$ = 13000 K, thus for the spectral types between B7 and G0. On the contrary, Figure 4 shows that, for most of these stars, the temperatures are well separated and there cannot occur a mixing of different temperatures corresponding to different metallicities, provided the gravity is well known. We now can say that a large part of the

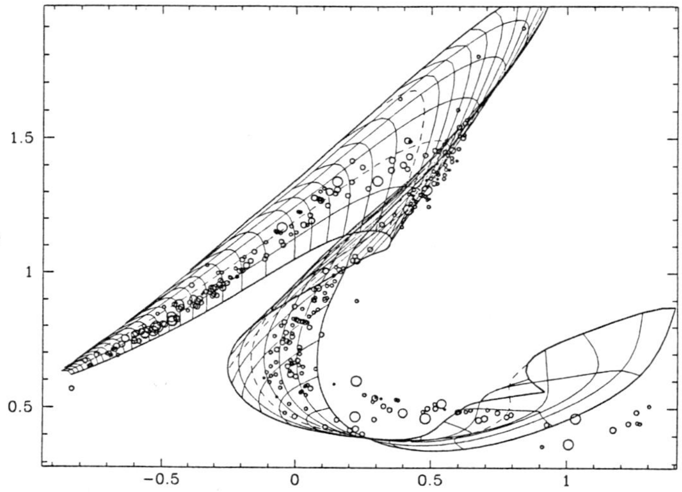

Figure 3. The distribution of our standard stars in d/Δ diagrams, with grids drawn from Kurucz's (1991) stellar atmosphere models for [M/H]=-0.5.

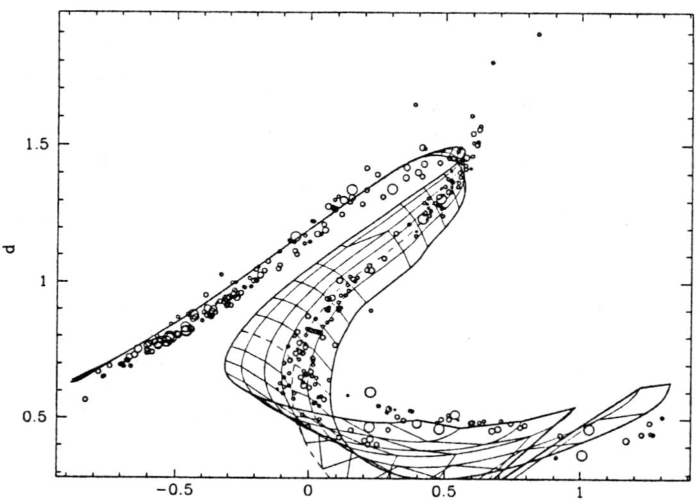

Figure 4. The distribution of our standard stars in d/Δ diagrams, with grids drawn from Kurucz's (1991) stellar atmosphere models for $\log g = 4.0$.

uncertainty in the spectral types of A- and F-type stars that is clearly visible in Figure 1 can be removed if a more precise determination of the gravity can be achieved (of course, there are also effects of chemical peculiarities).

The very good ability of Geneva photometry for providing a precise spectral type for the B-type stars is not a surprise if we take a look at the Kurucz's model-based grids. There is no overlapping, hence the temperatures are very well separated. Furthermore, in this area of the photometric space, the metallicity effects are nonexistent.

Even if the theoretical atmosphere models still have to be improved, especially regarding the cold spectral types, they are ideal tools for helping understand which kind of supplementary information could be added to Geneva photometry to perform accurate automatic stellar classification.

4. Conclusion and future prospects

The results are encouraging and show that our formalism and methodology can be applied to a larger set of stars (with checked spectroscopic and photometric data) and can take into account the luminosity class and the chemical peculiarity. If applied to a large number of selected stars, this process will indicate the variation of the Geneva photometry's resolving power regarding spectral classification as a function of the colors of stars. This will be very useful in the prospect of automatic spectral classification which could be performed on large samples of stars observed through the Geneva photometric system.

Acknowledgments. We gratefully acknowledge the great help provided by M. and C. Jaschek for establishing the list of MK standards. We would also like to acknowledge that without the program "Photom" developed by B. Nicolet, this work would not have been possible.

References

Golay, M. 1991, private communication.

Golay, M., Peytremann, E., Maeder, A. 1969, *Publ. Obs. Genève, Série A*, N°**76**, 44.

Jaschek, C., Frankel, S. 1986, A&A, **158**, 174.

Kurucz, R.L. 1991, in *Stellar atmospheres: beyond Classical Models*, ed. by L. Crivellari, I. Hubeny, and D.G. Hummer, NATO ASI Series, Kluwer, Dordrecht.

Nicolet, B. 1981, A&A, **97**, 85.

Nicolet, B. 1982, in *Les boîtes photométriques*, Ph.D. thesis, Observatoire de Genève.

Weaver, W. and Shannon, C.E. 1949, in *The mathematical theory of communication*, Board of Trustees of the University of Illinois.

PART V.

THE HISTORY OF MK CLASSIFICATION

The MK Process at 50 Years
ASP Conference Series, Vol. 60, 1994
C.J. Corbally, R.O. Gray, and R.F. Garrison (eds.)

Fifty Years Ago: Astronomy; Yerkes Observatory; Morgan, Keenan and Kellman

Donald E. Osterbrock

UCO/Lick Observatory, University of California, Santa Cruz, CA 95064

1. Introduction

When we think of William W. Morgan and Philip C. Keenan, we tend to think of them as they are now, or as they were when we first met them. But in fact, as we all realize, they were very different in 1943, the year of publication of the MKK Atlas, than they are today. Astronomy, and Yerkes Observatory were very different also, and in this paper we will begin with these two aspects of our story, and then work our way back to Morgan, Keenan, and Edith Kellman.

2. Astronomy

In 1943 America was locked in a great World War, which had followed closely on the heels of the Great Depression. Many American astronomers were working in wartime technical development programs, "the best damn second-rate physicists you can find," in Bart Bok's words. Nevertheless, they still considered themselves astronomers, and the membership of the American Astronomical Society continued to grow slowly. A far smaller percentage of its members were research astronomers than today: only the Mount Wilson, Yerkes, Harvard, Lick, Lowell and Naval Observatory staff members, and a few superstars like Henry Norris Russell. Most astronomers taught courses every quarter or semester in colleges and universities, and did research when they could in their spare time or on vacations. Even presidents of the American Astronomical Society, like Joel Stebbins at the University of Wisconsin, taught undergraduate elementary astronomy courses, with little, or in most cases, no help from teaching assistants.

In the five years ending in 1941 (after that wartime restrictions cut down on travel, and many astronomers could not get leave from their posts), the typical number of members at an AAS meeting ranged from 70 (the low, in far-off Berkeley, in 1939) to 150 (the high, in Philadelphia in 1940), plus 20 to 40 guests (graduate students, local mathematicians and physicists, and advanced amateur astronomers). The Society always held its winter meeting between Christmas and New Year, so that astronomers could get to it without missing their classes! There were no poster papers, no parallel sessions, and no registration fees.

There were far fewer astronomers than today. The AAS had approximately 650 members in 1943, compared with 5,000 fifty years later. There were women astronomers, but nearly all of them were teachers or research assistants. No woman astronomer had a faculty appointment in any of the premier research institutions mentioned above; even such high-level researchers as Annie J. Cannon and Cecilia Payne-Gaposchkin at Harvard, and Charlotte Moore Sitterly at Princeton, had less prestigious, less well-paid positions.

In 1943 there was almost no radio astronomy, no X-ray astronomy at all, no space astronomy, and very little infrared astronomy. Almost all astronomy was ground-based, optical astronomy. Almost all astronomical research was on stars, although some work was being done on gaseous nebulae. Practically no one was working on galaxies, which most astronomers called extragalactic nebulae.

There was no NASA, no NSF, no government funds, and little money of any kind for astronomical research. The Draper Fund of the National Academy of Sciences, and the Rumford Fund of the American Academy of Arts and Sciences could make grants of amounts like $500 or $1,000 for special items of new equipment, such as a photometer, a Schmidt corrector plate, or a sensitometer. The only source of postdoctoral fellowships was the National Research Council, which might or might not include one astronomer in the quota of fifteen to twenty physicists who received them. One or two senior astronomers a year might get Guggenheim fellowships for a year's sabbatical research. Otherwise the research support had to come from university funds. Under these conditions an observatory could only rarely get a new spectrograph, and long-term observational programs were the rule.

In the years before 1943, the *Astrophysical Journal* came out in ten issues per year, skipping two months. These issues made two volumes for the year, totaling approximately 1200 pages, more or less the equivalent of two ten-day issues of today's *Journal*, when its larger page size is taken into account. The *Journal* belonged to the University of Chicago, which published it, and the director of Yerkes Observatory was automatically its managing editor.

There were no national observatories, and few astronomers ever got the chance to obtain data anywhere but at their own observatory. The dominant astronomical research institution in the United States was the Mount Wilson Observatory, and its 100-inch reflector was the largest telescope in the world. Under its longtime director, the outstanding stellar spectroscopist Walter S. Adams, a small, highly skilled staff worked away, each at his own specialty, Paul W. Merrill on long-period variable stars, Alfred H. Joy on cepheid variables, Roscoe F. Sanford on carbon stars, etc. They could work at their own pace, secure in the knowledge that no one else anywhere could compete with them or scoop them on a new result. Edwin Hubble, a Yerkes Observatory alumnus, and Walter Baade were among the few astronomers doing observational research on galaxies.

Lick Observatory, under its long-time staff members William H. Wright (director 1935-42) and Joseph H. Moore (1942-1945), had an ongoing radial-velocity program, built around its old 36-inch refractor. Especially under Wright, it had moved into more imaginative astrophysical research, but its antiquated 36-inch Crossley reflector kept it far behind Mount Wilson.

Harvard College Observatory did not have a large telescope, nor a Mount Hopkins station. Its longtime director, Harlow Shapley, believed in spending the absolute minimum possible on telescopes, and during the Great Depression the minimum was very small indeed. As a result Harvard had a large collection of old, small photographic telescopes, most of them at its Southern Hemisphere station, where permanent observers took plates and shipped them back to Cambridge, for analysis and study there. Shapley himself had done important research on variable stars, globular clusters, galactic structure and galaxies in his earlier years as the "boy director" of HCO. A onetime newspaperman, he was more publicity conscious than most astronomers (or perhaps more successful in attracting publicity), and by 1943 he depended heavily on his scientific assistants, and had little time for research except to summarize the data they had collected and reduced.

Russell, at Princeton University, 66 years old in 1943, was the undisputed high priest of American astronomy. He had begun his career in classical astronomy, but had become a master in the theoretical analysis of laboratory spectra. With his brilliant mind, and wide knowledge of every field of astronomy and astrophysics, he was extremely prolific on his own, and gave freely of his advice to other astronomers. Russell was famous for commenting intelligently on practically every paper presented at the AAS meetings he attended. He always strove for theoretical understanding in his own research, and encouraged others to do the same. He made frequent working visits to Harvard, where Shapley and Donald H. Menzel, one of the few theorists in American astronomy, had both been his students, and where Payne-Gaposchkin had been his protégé since she had come to America in 1923. Russell was a Research Associate of the Carnegie Institution of Washington, which provided funds for his frequent long visits to Pasadena to work with the Mount Wilson astronomers and to gain access to their latest observational results.

This was the world of American astronomy which Morgan entered at Yerkes Observatory in 1926, and Keenan at the University of Arizona at nearly the same time.

3. Yerkes Observatory

Yerkes Observatory went into operation in 1897, built around the 40-inch refractor, the largest telescope in the world. Young George Ellery Hale, with President William Rainey Harper of the brand new University of Chicago, had persuaded predatory capitalist Charles T. Yerkes to provide the money to build it. The observatory had been erected in Williams Bay, Wisconsin, seventy miles northwest of the fog, mists, and smoke of Chicago. Hale, 30 years old in 1897, was an MIT graduate, who had done excellent solar research in his senior thesis there, and in his own private Kenwood Physical Observatory. It had been built for him by his wealthy father around an excellent 12-inch refractor, optimized for astrophysical work on the sun. At Yerkes Hale planned to continue research, and especially to compare the spectra of representative bright stars with the sun's. The two senior faculty members, S. W. Burnham, a visual double-star observer, and E. E. Barnard, a planetary visual observer and expert astronomical photographer, were sterling representatives of the old school of positional astronomy.

Hale, always more of a restless organizer than steady researcher, soon tired of the 40-inch refractor, especially after he found that its flexure, together with the flexure of a heavy spectrograph mounted at its eyepiece end, made it impossible to take long-exposure high-dispersion spectra of even the brightest stars. He knew that reflectors, not refractors, would be the monster telescopes of the future, necessary for the astrophysical research that was his life's dream. In 1902 he went to California to set up a remote observing station of Yerkes Observatory on Mount Wilson. By 1904 he was able to secure a grant from the new Carnegie Institution of Washington, which enabled him to break away from the University of Chicago and found the new, independent Mount Wilson Solar Observatory. He took with him two solar telescopes from Yerkes, and his personal research group: George Willis Ritchey, who had already begun working on the mirror for the 60-inch reflector in the Yerkes optical shop in 1897, Walter S. Adams, Hale's graduate student and scientific assistant, and Ferdinand Ellerman, his observing assistant. (Wright 1966; Osterbrock 1993). Edwin B. Frost was left at Yerkes Observatory as its acting director, and he succeeded Hale as director in 1904. Then 38 years old, Frost had been born in Vermont, and grew up in Hanover, New Hampshire, where his father was a professor and then dean of the Dartmouth Medical School. Young Frost received his bachelor's degree at Dartmouth in 1886, then took some graduate work there and at Princeton, where his teacher was Charles A. Young, the pioneer American astrophysicist, and a friend of Frost's father. Next he went to Europe for two years as a graduate student, the second mostly at Potsdam, where H. C. Vogel and Julius Scheiner were opening up photographic stellar spectroscopy. Frost returned to Dartmouth as the assistant professor of astronomy in 1892. He translated Scheiner's new book on the spectral analysis of stars into English, and it became the standard text on the subject for twenty years. Frost settled easily into the comfortable life of a New England college professor.

But in 1898, when Hale raised the money to hire a stellar spectroscopist on the Yerkes faculty, James E. Keeler of Allegheny and Lick Observatories, the outstanding American researcher in the field, declined the post, and then Frost, Hale's second choice, accepted it. He brought Adams from Dartmouth with him, and they began a radial-velocity program with the Yerkes 40-inch refractor. After Adams departed for Pasadena, Frost continued it with various faculty members and graduate student assistants. It was important work, but W. W. Campbell and his highly organized team at Lick Observatory were doing it much better with its 36-inch refractor at a far superior observing site. Frost could not match Campbell's engineering approach to designing and building spectrographs optimized for the single purpose of obtaining radial-velocity spectrograms, nor his systematic, factory-like organization for measuring and reducing them. Frost, unlike Hale or Otto Struve, the third director of Yerkes, was not in the least innovative or imaginative scientifically. He was content to continue the observatory, much as he had inherited it from Hale, for nearly thirty years.

Frost had inherited severe near-sightedness, and in 1915 he suffered a detached retina in his right eye, while observing. He soon lost nearly completely his sight in that eye, and had to stop observing. Within a few years he developed a very rapidly advancing cataract in his other eye. It led to a hemorrhage in 1921, and Frost became completely blind. Nevertheless, he continued as director for eleven years, helped by a succession of secretaries, who read letters to him and took dictation from him, and assistants who read scientific papers to him. Frost was an extremely courageous person. He was a warm, friendly human being. He deserves great sympathy. He wrote charming little essays on nature, and an autobiography that recreated his New England boyhood and his friendships with his college chums. But he was not a good director for what should have been a leading scientific research institution.

After he became blind, Frost hardly ever went to an astronomical meeting. Although he corresponded frequently with Hale, he had almost no contacts with the younger generation of astronomers and astrophysicists at other institutions. He did not seek (and in fact actively avoided) outside advice on new appointments and promotions for the Yerkes staff. When Burnham and Barnard reached the ends of their careers, Frost replaced them with successors as nearly like them as he could, double-star observer George van Biesbroeck and wide-field astronomical photographer Frank E. Ross. Both were excellent in their specialties, but Yerkes Observatory could not grow into new fields under Frost's policies. (Frost 1933; Struve 1938)

Yerkes Observatory was the base of most of the faculty and graduate students of the Department of Astronomy, although there were also two or three classical astronomers on the campus, who taught celestial mechanics and mathematics. Chicago and California were the two main sources of astronomy Ph.D.'s in the United States. Harvard had no graduate program in astronomy at that time, but Michigan, Yale and Columbia did. At Yerkes there were typically five to ten graduate students, who took courses, actually chiefly reading and practical observing experience, and worked on their theses. In summer, several more would usually appear, nearly all of them teachers at Midwestern colleges and universities , attracted as much by the charms of the Lake Geneva area as by the opportunity to work slowly toward the Ph.D. degree that would bring a raise in status and pay at their institutions. Few of the graduates of Frost's era made much of an impression in their later careers in astronomy, but five who did were Edwin Hubble, Otto Struve, Nicholas T. Bobrovnikoff, William W. Morgan, and Philip C. Keenan.

4. William W. Morgan

Morgan was born in 1906 in a very small town in Tennessee. His parents were both home missionaries, and his father became a minister who, according to Morgan's account, moved frequently within the South. During World War I he served as an army chaplain, and afterward was an itinerant inspirational lecturer. Morgan attended a small private high school for two years, which was attached to "a tiny church college" in Missouri, where his mother was in charge of a women's dormitory. Then they moved to Washington, D.C., where he graduated from Central High School. He was a very good student, especially interested in English literature, but with an awakening interest in science. In 1923 he entered Washington and Lee University, in Lexington, Virginia, as an English major (Morgan 1988). There the professor of physics was Benjamin A. Wooten, and it was through him that Morgan got his chance to begin the career at Yerkes Observatory which led to the MKK spectral atlas and the discovery of the spiral arms in our Galaxy.

The direct cause of this opportunity was the acrimonious resignation of Oliver J. Lee from the Yerkes faculty, in the summer of 1926. Lee, a singularly unproductive research worker who was a Yerkes Ph.D. himself, felt that after thirteen years on the staff he deserved a promotion to associate professor and the salary raise that would go with it. He did not get the promotion, and in a hot-tempered letter to the Board of Trustees, threatened to resign. Lee's wife was Frost's secretary, and he probably thought that the director could not get along without her. However, Frost, angered by the long personal harangue to which Lee subjected him, held firm, and Dean Henry G. Gale, in Chicago, was glad to get rid of a malcontent, unproductive assistant professor. To Lee's surprise they accepted his resignation, and he became an insurance salesman.

One of Lee's jobs had been to take the routine, daily direct photographs of the sun which Frost had continued ever since Hale had departed for Mount Wilson. This program had to go on, in the director's opinion, so he had to find a replacement for Lee. Frost's first idea was to offer the job to Julius Arthur Brown, a fellow Dartmouth graduate, who was the professor of astronomy at the American University in Beirut. Brown was to return to the United States for a year's leave, and Frost thought he would make an excellent temporary replacement. Brown had a nice wife and five children and would need a job. This plan fell through, however, and Frost began looking for "some young graduates" to "train up for the solar, astrometric and perhaps also for the photometric work."

Wooten, Morgan's professor at Washington and Lee, was at Yerkes Observatory that summer, and had mentioned his young student who was so interested in astronomy. Frost needed someone fast, and was entranced by Wooten's description of Morgan, as nearly 21 years old and "very strong and healthy." Morgan himself wrote Frost that he had "always been very anxious to take up some form of astronomical work as a profession," and that although he had one more year's work to do at Washington and Lee, he would be glad to "postpone graduation" to take the job. Frost offered him an assistantship at $900 a year plus tuition, and Morgan was at work at Yerkes by mid-September. He could not officially register as a graduate student until he had his bachelor's degree, which Frost arranged for him to get at the end of his first academic year's residence at Yerkes, without ever having taken a course on the campus or indeed visiting it until a few weeks before the commencement in June 1927.

Morgan was an extremely successful assistant, intelligent, hard working, very gifted in photographic techniques and dedicated to astronomy. He loved observing. After two years at Williams Bay, in June 1928 he married Helen Barrett, daughter of Storrs Barrett, the longtime secretary at Yerkes Observatory. After another year at the observatory taking graduate courses, Morgan was granted a fellowship and the young couple moved to Chicago for two quarters, so that he could take the required beginning graduate courses in mathematics, celestial mechanics, and physics. He later claimed that he learned "nothing" from them, which of course in fact was far from the case. But he certainly had much less training in these subjects than any of his fellow astronomy graduate students, few of whom were ever heard of again in astronomy.

In the spring quarter of 1930, Morgan returned to Yerkes, to begin a thesis with Struve on the spectra of the peculiar A stars. By this time he was committed to stellar spectroscopy as his life's work. He did all the observing for his thesis with the 40-inch refractor and its radial-velocity spectrograph. In spite of his lack of formal training in physics, Morgan had made himself an expert in recognizing subtle patterns, similarities and differences in spectra, what he later called "the thing itself." He classified and arranged, almost purely in observational spectroscopic terms, these stars and described their spectral peculiarities. It was a good thesis (Morgan 1931a,b, 1932). Morgan completed it and took his final examination in December 1931, when the Great Depression was at its worst. The University of Chicago was operating under deficit conditions and the administration was applying severe pressure to reduce spending. However, Frost was able to keep Morgan on the payroll, as an assistant at $1500 per year.

5. Philip C. Keenan

Keenan had a more conventional astronomical education. Born in Bellvue, Pennsylvania, a suburb of Pittsburgh, he had grown up and attended high school in Ojai, California. Then he entered the University of Arizona, where he took many advanced courses in physics and mathematics, along with seminars and reading courses in astronomy. Keenan completed his B.S. at Tucson in 1929, and then continued with a master's thesis under Edwin F. Carpenter, working on lunar photographic photometry with Steward Observatory's 36-inch reflector, which was then on the campus in Tucson (Keenan 1931).

Keenan received his M.S. in 1930 and entered Yerkes Observatory that same summer. (He had spent the previous summer there as an assistant.) He took more graduate courses, and held the same solar assistantship which Morgan had had. Life was not easy for a graduate student in the Great Depression. Keenan, as Morgan had before him, slept in a large room in the attic of Yerkes Observatory with the other students. It was called "the battleship" because of its small, widely separated, round porthole windows, but the roof leaked and the room became a cold, dark tomb on winter weekends, when the electric power (except to the telescope) and heat were turned off in the building. Water dripped in during summer thunderstorms, and snow drifted in during the winter. The only bathtub in the building was in the middle of the men's room in the basement, adjoining the instrument, optical, and woodwork shops. There were no showers. The students took their meals at a nearby boarding house, where the more affluent visitors to the observatory stayed.

Keenan did his thesis on an astrophysical study of the chromosphere, based on spectra which he took with the 40-inch refractor. His advisers were Struve and Christian T. Elvey. This thesis was a noble effort, but the main thing it proved was that a site with better daytime seeing, and less scattering in the atmosphere above it (such as Mount Wilson) was necessary to get good spectrograms of the chromosphere (Keenan 1932). Nevertheless Keenan received his Ph.D. in June 1932. He continued on the Yerkes staff as an assistant, while Morgan was promoted to instructor, effective July 1, 1932.

That same day Otto Struve succeeded Frost as director of Yerkes Observatory. Keenan never forgot it. He began, that very morning, the daily inspections that all Yerkes faculty members soon came to expect. Walking down the marble hall, his measured footsteps announcing his presence, he checked each office to see who was working and who was not. An astronomer who had observed the second half of the previous night could sleep until noon; everyone else was expected to be there, at work, by the time Struve arrived at the observatory.

6. Otto Struve

Struve, one of the hardest working directors the world has ever seen, was determined to restore Yerkes Observatory to the astrophysical greatness it had held under Hale. Struve represented the fourth generation of a famous astronomical family. His great-grandfather, grandfather, and father had all been directors of observatories in Russia, and his uncle in Germany. Struve had done his undergraduate work in mathematics and classical astronomy at Kharkov. He served as an officer in the Russian army in World War I, and then in the White (loyalist) army fighting the Bolsheviks in the Russian Revolution. He barely escaped alive after the final defeat of the Whites in 1920, and ended up in a refugee camp near Constantinople. Frost, a great humanitarian, learned of Struve's plight through German astronomers, and managed to pull enough strings to get him out of Turkey and into the United States. Struve reached Yerkes Observatory in October 1921, became a graduate student immediately, perfected his English enough to pass all his courses, and did his thesis on short-period spectroscopic binaries as part of the radial velocity program on the 40-inch refractor. He received his Ph.D. at the end of 1923, and became an instructor on the Yerkes faculty (Goldberg 1964; Krisciunas 1992).

Struve quickly began grinding out papers at a prodigious rate. He realized that the Yerkes plate files contained hundreds of spectrograms showing interstellar absorption lines, and used them to become a pioneer in this field. Reviewing Cecilia Payne's (later Payne-Gaposchkin's) book on Stellar Atmospheres for the *Astrophysical Journal* converted Struve to trying to understand the nature of the stars, and those were the topics he worked on, prodigiously, for the rest of his life (Struve 1926). Frost, for all his faults, recognized Struve's greatness from the beginning, and fought to keep him at Yerkes, warding off outside job offers with raises, promotions, and promises of research leaves. Struve traveled widely, attended meetings, gave papers and colloquia at the drop of a hat, discussed research with everyone, and generally did all the things that Frost had not done. When Frost reached the normal retirement age, the University of Chicago administration was glad to appoint Struve in his place, in spite of his youth (he was then 34) and the anti-foreign sentiment that was rampant in American astronomy. (Struve had become a U. S. citizen in 1927, and had married an American.)

Struve, however, knew by this time that he and Yerkes Observatory needed a larger telescope in a better climate. Simultaneously with the Yerkes directorship, he was offered a professorship and the assistant directorship at Harvard, then second only to Mount Wilson in prestige in the United States. Struve used this offer as leverage to get Chicago to promise the bigger telescope, and then miraculously brought it about himself through a joint agreement with the University of Texas, to operate an observatory which it would build with money which it had received under the will of William J. McDonald. Once again, though Frost had not foreseen this possibility at all, he had opened the way for it years before by his work "for the good of astronomy." He had provided reams of material for the Texas lawyers to use in defeating McDonald's relatives' efforts to break his will, on the grounds that he must have been insane if he left his money to support research in astronomy. Struve provided the astronomical expertise which went into building the McDonald Observatory reflector in the mountains of west Texas. It went into operation in 1939, and after that he did most of his observing there, as joint director of Yerkes and McDonald Observatories (Evans and Mulholland 1986).

7. Morgan and Keenan

In 1932, when Struve took over as director, the Great Depression was hitting the University of Chicago very hard. There was constant pressure to reduce costs. Salary raises and promotions were infrequent, and obtaining new positions was unthinkable. Struve wanted to keep Morgan and Keenan on the faculty, and made it his first order of business to get rid of two of Frost's unproductive favorites, Clifford C. Crump, an associate professor, and Arthur S. Fairley, an instructor, so that he could.

Morgan continued working on peculiar A stars. He experimented with new types of photographic developers, and with new red-sensitive emulsions. He reviewed Payne's second book, The Stars of High Luminosity, no doubt at Struve's behest, but rather than awakening his interest in astrophysics, it repelled him (Morgan 1933). Mostly Morgan worked alone, but when Nova Herculis blazed up at the end of 1934, he collaborated with Struve in obtaining nightly spectrograms of it through December 27. Struve showed slides of these spectra and discussed them in their joint paper at the AAS meeting in Philadelphia, just two days later.

Struve thought highly of Morgan's research, and recommended him for "a small raise" after he had been at $1,800 a year for three years. To Dean Henry G. Gale, Struve wrote that in astronomical spectroscopy his young staff member was "the best investigator in America." But the director's own private judgement was that Morgan needed to learn more theoretical astrophysics, and he urged him to study on his own to remedy the deficiencies of his Washington and Lee, and Yerkes training. Struve thought Morgan should learn more about effective temperatures and electron pressures, dilute radiation and multiplet structure, as he had done himself. Morgan made some half-hearted attempts in this direction, but true to his own nature, instead read more and more about spectral classification.

The Henry Draper catalogue, produced at Harvard by Annie J. Cannon and her assistants, was the standard catalogue of spectral classification. Based on objective-prism spectra, it was a one-dimensional system and, as Morgan and others came to understand, quite inhomogeneous. The brightest stars were overexposed even on the short-exposure plates, and had been classified on higher-dispersion spectrograms, while the fainter stars were underexposed even on the long-exposure plates, and were not in reality classified in the same way as the stars of intermediate brightness.

Furthermore, although Antonia C. Maury had recognized the c stars long ago, and they had been understood as supergiants from the time of Hertzsprung, the Draper catalogue did not differentiate between dwarfs, giants and supergiants.

Walter S. Adams had pioneered in two-dimensional spectral classification, giving the luminosity of the star as well as its spectral type. He, and the Mount Wilson astronomers who worked with him, used slit spectrograms taken primarily for measurement of radial velocities. These tended to be narrow, and developed to high contrast, which made them ideal for their primary purpose, but not for classification. Furthermore, although the Mount Wilson observers gave the spectral type of a star "on the Draper system," they listed a numerical value of the absolute magnitude as the second coordinate. It was derived from a plot of one or more estimated line ratios against the absolute magnitudes of stars at "known" distances.

Morgan, with his unconventional, unphysical way of thinking, came to realize that this was not a good system to use. Each time the methods of distance determination were improved, the absolute magnitudes of all the calibrating stars changed, as did the whole system, and all the derived absolute magnitudes with it. In a paper published in 1937, he showed specific examples, culled from Mount Wilson publications, in which this had happened. He stated that if it had been traditional to give spectral type in terms of a numerical value of the effective temperature of the star, so that the Sun for example were described as a 5730 K star rather than G2, then any change in the effective temperature calibration would be reflected in a change in all the "effective temperature types" as well. Just as it was more logical to give the spectral type, it was more logical to give the spectral "luminosity class," for which Morgan invented the symbols ranging from Ia and Ib for supergiants through III for giants to V for main-sequence stars, so that the Sun's full spectral classification was G2 V. Then the calibration process could be clearly separated from the classification process. Thus although the effective temperature and gravity (or electron pressure) were the primary physical parameters defining the stellar atmosphere, and the luminosity or the absolute magnitude and the effective temperature or spectral type were the primary parameters for the Hertzsprung-Russell diagram, the spectral type and luminosity class were the natural parameters for spectral classification. Furthermore, as Morgan emphasized, these different descriptions are not simply one-parameter functions of one another; for instance, stars of the same spectral type but different luminosity classes do not all have the same effective temperature. However, each pair of coordinates is, in principle, a transformation of any of the other pairs of coordinates. In his first paper on the subject, Morgan made these arguments in somewhat awkward numerical form, no doubt the way in which he had first grasped them. They certainly could be criticized, as they were, but the overall conclusions are correct, and proved extremely useful (Morgan 1937).

Morgan also had come to understand that for accurate spectral classification it was essential to have homogeneous data with (in modern terms) a very high signal-to-noise ratio. This could be achieved by using the same spectrograph, dispersion, plates and development (the "same system") for bright stars and faint, and exposing them all to the same level. Furthermore, it meant using unusually low-contrast, fine-grain development, and widening the spectra greatly during the exposure, to get the maximum signal, and to store it effectively on photographic plates. Finally, Morgan realized that the "system" of classification should be defined by a relatively few "standard stars" of each type and luminosity class, rather than by supposed values, equalities, or appearances or disappearances of specified line ratios, as in the Draper and Mount Wilson systems. In his mind, although he did not emphasize it in his papers, spectral type and luminosity class were discrete coordinates, much like quantum numbers (though not for the same reason), not continuous variables. Thus in the MKK Spectral Atlas, there is no standard of spectral type G3.5 V, or F2 IV.V, or even F1 V, and hence there are no stars of these types in the MKK system. This bewildered many of Morgan's contemporaries who thought in more physical terms, but it emphasized that "the" spectral classification of a star depends on the amount of light available, and the instrument used, as much as on the properties of the star itself. He was to give several examples of this general principle in his paper on "natural groups" at the symposium in 1950 for the dedication of the Curtis Schmidt telescope (Morgan 1951).

In his second paper on spectral classification, Morgan spelled out the power of accurate two-dimensional spectral classification for studies of galactic structure. The observed parameters, spectral type and luminosity class, can be calibrated to yield both absolute magnitude and intrinsic color index. Then, from the observed color index of the star, the extinction can be calculated (using the empirical result that the form of the extinction curve is to the first approximation independent of the position in the sky or the distance of the star, already reasonably well established at that time). Thus the distance of the star can be established (to the accuracy of the classification) and it can be plotted on a three-dimensional map of the Galaxy. In this paper Morgan (1938) used the early accurate photoelectric colors measured by the pioneer K. Bottlinger at Potsdam, but only 70 miles from Williams Bay, at the University of Wisconsin's Washburn Observatory (and on visits to Mount Wilson), Joel Stebbins, and his collaborators C. Morse Huffer and Albert E. Whitford, were beginning to make mass-production photoelectric measurements of early-type stars. They were using spectral types from the Draper catalogue, and were finding puzzlingly large scatter in the derived extinctions and distances to stars which appeared to be close to one another in space (Stebbins, Huffer, and Whitford 1940). With Morgan's types these discrepancies disappeared, and a new era of accurate distance determinations, or "spectroscopic parallaxes," began.

Keenan, with much broader training in astronomy and physics, worked on many subjects, including prominences, the chromosphere, solar granules and limb darkening, photometry of diffuse nebulae and elliptical "nebulae," and an integration of a model stellar interior. In 1940, when Grote Reber came from nearby Wheaton, Illinois, to show Struve the records of "cosmic static" which he had taken with his homemade antenna in his backyard, the director assigned Louis G. Henyey and Keenan to investigate the phenomenon theoretically. They calculated the free-free emission from interstellar ionized gas, the mechanism Reber (1940) had suggested, and found that although it could be stretched to fit his 80-cm observations, it was entirely insufficient to match Carl Jansky's earlier measurements at longer wavelengths (Henyey & Keenan 1940). The concepts of interstellar magnetic fields, relativistic electrons, and synchrotron radiation were still in the future.

In 1935 Keenan, under pressure from Struve, who did not feel he could pay him more at Yerkes, accepted an instructorship at Ohio Wesleyan University on a trial basis. Its Perkins Observatory, with a 69-inch reflector equipped with a Yerkes spectrograph, had almost no money but provided some observing opportunities. The Perkins director, Nicholas T. Bobrovnikoff, a Yerkes Ph.D., was a refugee from Russia and good friend of Struve, and he had persuaded Keenan to take the job for a year to see if he liked it. He quickly found that the research opportunities were very limited, and returned to Yerkes as soon as the year was up. Keenan was promoted to instructor in 1936, and sent to the campus in Chicago for several quarters of teaching, but he always came back to Williams Bay as soon as he could. His heart was in research, and ultimately Struve replaced him on the campus, first with Bengt Strömgren (in 1936), then with Thornton L. Page (in 1938).

Around 1939 Keenan began working closely with Morgan on spectral classification, with Struve's encouragement. Their first joint publication was on the spectral type of the sun, and the observational confirmation of the predicted difference between the types of the center of the disk, the limb, and the integrated light (Morgan & Keenan 1939).

The first problem Kennan did on his own, suggested by Morgan, was the spectral classification of the stars in the North Polar Sequence. Stebbins and Whitford had measured photoelectrically their magnitudes and colors, which with the Draper and Mount Wilson spectral types seemed to show unexpectedly large interstellar extinction in that region of the sky, with considerable scatter. Morgan knew that the colors were accurate, and suspected that the classifications of the faint stars were systematically too late. Keenan (1941a), with the help of Daniel M. Popper and Horace W. Babcock, was able to obtain good slit spectrograms of the NPS stars down to photographic magnitude 11, with the 82-inch reflector at McDonald Observatory, which had just gone into operation. His data showed that the earlier classifications of the fainter stars were incorrect, and that the extinction at the pole was smaller and better determined than had previously appeared. Then Keenan (1941b) went on to a careful study of the usefulness of the CN $\lambda\lambda$3882, 4215 bands as luminosity criteria. Although they discussed all their classification work with one another, Morgan tended to concentrate on the earlier-type stars, and Keenan on the later.

Struve authorized the construction of a small, light spectrograph, optimized for the classification program, in the Yerkes shop. It was much easier to mount and dismount from the telescope than the large, heavy Bruce spectrograph. The new spectrograph had a long slit, a fast camera which used the typically poor-seeing images effectively, and did away (in ordinary use) with a comparison source altogether, so that many greatly widened spectra of stars could be recorded on a single plate. This "classification spectrograph" was completed in the summer of 1940.

By then Morgan had the concept of the Atlas of Stellar Spectra firmly in mind, and he and Keenan began taking the spectrograms for it with the new instrument. The Atlas was to be one of the series of astrophysical monographs, published by the University of Chicago Press. Struve had conceived of this project and persuaded the university administration to subsidize it; the first volume by Bart J. Bok, on the distribution of the stars in space (as revealed by star counts) had appeared in 1937. Struve's plan was to publish one volume a year, and subsequent books were those by S. Chandrasekhar, on stellar interiors (Morgan read the entire manuscript for him); by Paul W. Merrill, on long-period variables; and by Russell and Moore, on stellar masses. Struve was eager to publish another volume by a Yerkes author, and the Atlas fitted the bill.

Struve and Morgan knew well, from experience with the *Astrophysical Journal*, how hard it was to reproduce satisfactorily the full range of gradation in astronomical photographs by any printing process then available. Hence Morgan's idea from the beginning was that the main part of the Atlas would be a set of photographic prints, prepared directly under his supervision at the observatory. They would be accompanied by a short text, outlining the principles of the system, describing the spectral features used in each range of types and luminosity classes, and listing the standard stars which defined the system. This publication scheme was exactly the one which Frank E. Ross had adopted for his Atlas of the Northern Milky Way, published by the University of Chicago Press in 1934.

Morgan and Keenan, assisted by night assistant Frank R. Sullivan and (at various times) by graduate students Frances Sherman, John A. O'Keefe, Louis R. Henrich, and William P. Bidelman, began taking the spectrograms in the summer of 1940. Morgan already had long ago outlined the Atlas, but preparing the illustrations and writing the text were still tedious tasks.

8. Edith Kellman

Edith Kellman was born on a farm near Walworth, Wisconsin, five miles from Yerkes Observatory. She went to a country school until her parents moved to Williams Bay, where she started in seventh grade, and went through high school. Then she entered Wheaton College, west of Chicago. She was very good in mathematics and science, and she took a B.S. degree with an education minor, to train to be a high school mathematics teacher. She graduated from Wheaton in June, 1933, but because of the Depression there were no jobs to be had, even for mathematics teachers. Kellman was unemployed for a year, but then she got a job as a computer at Yerkes Observatory. George van Biesbroeck, the double-star expert, was a longtime member of the Williams Bay School Board and had a special concern for good science students. He learned about her plight through his older daughter Simone, and recommended Kellman for the observatory job.

She began working as Morgan's assistant in the summer of 1934. She did all his photographic work, including developing the spectroscopic plates which he took at the telescope, and left in the darkroom when he went home, as her first task for the morning. She made hundreds of observing cards for him and his graduate students, tracing the star fields from the Bonner Durchmusterung. When Morgan and Keenan decided on the spectrograms which would go into the Atlas, Kellman made enlargements of them onto paper prints, then cut, trimmed and mounted them. Morgan then printed the text on the mounts, and Kellman photographed the resulting Atlas pages on 8×10 plates, made copy negatives, and printed all the copies of the first edition of the Atlas.

There were several other computers in the observatory; all women. They worked five and a half days a week, including Saturday mornings, like the men in the shop and on the buildings and grounds crew. The senior computer was Mary Calvert, Barnard's niece who had been his assistant until he died, and afterward became a general high-level assistant and photographic technician. She was "very nice" in Kellman's opinion, as was Marguerite van Biesbroeck, the astronomer's sister, who was the observatory's half-time librarian.

Kellman got along well with everyone at Yerkes, but she wanted to be a teacher. By the end of December 1942 she had made all the prints for the Atlas.

The war created many new openings for teachers, as men and women left for the services or for better paying jobs in industry. She registered with a teachers' agency for a job anywhere in Wisconsin, but she got a position exactly where she wanted to be, at the Williams Bay High School. Van Biesbroeck and the other School Board members already knew her, and they thought she would be as good if not better than anyone else they might be able to hire from out of town. Even though she left the regular observatory staff, Kellman came back to work at Yerkes during her summer vacations for several more years, and recommended her best students for jobs of their own as assistants.

9. The MKK Atlas

Morgan had originally planned that the Atlas would include 35 plates, and expected to finish it in 1941, but like most research projects it grew. Ultimately there were 55 plates in the set, and he completed the manuscript in August, 1942. In its printed form the text was 35 pages long, and it is still an admirable description of the philosophy of photographic spectral classification, and of the practical means of carrying it out . It concentrated heavily on normal stars, but included short sections on peculiar A stars, metallic-line stars, and, a year before Walter Baade published his epochal paper on the two stellar populations, a paragraph on the "abnormal weakness" of the CN absorption in high-velocity giants, and another on the weakness of all the lines in two "subdwarfs" (Morgan, Keenan and Kellman 1943). Kellman had produced about 7000 prints, and the original edition of only 100 copies of the Atlas came out in January, 1943, with the other sets reserved for the Yerkes and McDonald astronomers. The list price was $10. Because of World War II, few observatories outside North America even ordered copies. Morgan was naturally concerned about what the response would be, particularly from Mount Wilson Observatory, as the Atlas could certainly be interpreted as a criticism of its methods.

He had discussed the monograph as a work-in-progress with senior Mount Wilson spectroscopist Merrill at the AAS meeting held at Yerkes Observatory in September, 1941, the last big meeting before the United States entered the war. It featured a symposium on astronomical spectroscopy (almost entirely on stars), organized by Struve. In it, Morgan gave an invited paper on spectral classification, and Merrill one on the spectra of symbiotic stars. In their discussions Merrill encouraged Morgan to go ahead with the Atlas, and so did Adams, after the Yerkes astronomer wrote him and described in detail his plans for the monograph. Alfred H. Joy, a third, highly respected, Mount Wilson stellar spectroscopist, was equally encouraging.

When the Atlas came out, Joy (1943) wrote an enthusiastic review of it, which Morgan always believed got him his promotion to an associate professorship in the summer of 1943. He had been the local product, the American who stayed home and watched Struve import Gerard P. Kuiper, Chandrasekhar, and Strömgren, at ranks and salaries higher than his own. Now he had arrived, and he knew that his future was secure.

Keenan was not at Yerkes Observatory when the Atlas was published. A bachelor, with a good training in physics, he had gone on leave to work on wartime weapons development at the Navy Department Bureau of Ordnance in

Washington. Morgan, with far less competence in sciences related to causing death and destruction, and with two young children, was over 35 when Pearl Harbor was attacked. He, like Struve, remained at Yerkes during World War II. They were in the minority; when Morgan (1946) presented a paper on the spectral classification of the B stars at the AAS meeting at nearby Northwestern University in December, 1942, it was one of the smallest meetings on record in that era, with only 28 members present, most of them from Chicago, Yerkes and Madison.

After the war ended, Keenan accepted a position at Ohio State University, which by then played the major role in operating Perkins Observatory. He had a long and successful career in stellar spectroscopy, concentrating on later type stars, and including astrophysical analyses as well as spectral classification. He made many observing visits to Mount Wilson, and later to the national observatories. Morgan stayed on the Yerkes staff all his life, and had an outstanding career. He pioneered in adapting his methods to finding faint OB stars on Schmidt objective-prism surveys, and brilliantly applied the methods of spectral classification to delineating the nearby spiral arms in our Galaxy. Later, first as a visitor at Pasadena and Lick Observatory, he broadened his field of expertise to include the spectra of globular clusters and galaxies, and the forms of galaxies. Always he trusted his eyes and "the thing itself" as he saw it. He made many discoveries that way.

Acknowledgments. This paper is based in part on a book in progress, *Yerkes Observatory: The Birth, Near-Death and Resurrection of a Scientific Research Institution, 1892-1950.* It will contain full references to the many news items, letters, books, reports, and papers which are its sources. Hence only a few of the most important papers and books are listed here. This paper is also based on written notes of my interviews of William W. Morgan, Philip C. Keenan, and Edith Kellman over the years since 1986 (although I have known all three of them for more than forty years).

References

Evans, D. S., and Mulholland, D. J. 1986, *Big and Bright: A History of McDonald Observatory*

Frost, E. B. 1933, *An Astronomer's Life*

Goldberg, L. 1964, QJRAS, **5**, 284

Henyey, L. G. and Keenan, P. C. 1940, ApJ, **91**, 625

Joy, A. H. 1943, ApJ, **98**, 240

Keenan, P. C. 1931, PASP, **43**, 203

—— 1932, ApJ, **75**, 277

—— 1941a, ApJ, **93**, 64

—— 1941b, ApJ, **93**, 475

Krisciunas, K. 1992, Bio. Mem. NAS, **61**, 351

Morgan, W. W. 1931a, ApJ, **73**, 104

—— 1931b, ApJ, **74**, 24

—— 1932, ApJ, **75**, 46

—— 1933, ApJ, **77**, 72

—— 1937, ApJ, **85**, 380

—— 1938, ApJ, **87**, 460

—— 1946, Pub.AAS, **10**, 310

—— 1951, Pub. Univ. Michigan Obs., **10**, 33

—— 1988, ARA&A, **26**, 1

Morgan, W. W. and Keenan, P. C. 1939, PASP, **51**, 355

Morgan, W. W., Keenan, P. C., and Kellman, E. 1943, *An Atlas of Stellar Spectra: With an Outline of Spectral Classification*

Osterbrock, D. E. 1993, *Pauper & Prince: Ritchey, Hale and Big American Telescopes*

Reber. G. 1940, ApJ, **91**, 623

Stebbins, J. S., Huffer, C. M., and Whitford, A. E. 1940, ApJ, **92**, 193

Struve, O. 1926, ApJ, **64**, 204

—— 1938, Bio. Mem. NAS, **19**, 25

Wright, H. 1966, *Explorer of the Universe: A Biography of George Ellery Hale*

The MK Process at 50 Years
ASP Conference Series, Vol. 60, 1994
C.J. Corbally, R.O. Gray, and R.F. Garrison (eds.)

Reminiscences on Antonia Maury and the *c*-Characteristic

Dorrit Hoffleit

Yale University, New Haven, Connecticut 06511

Abstract. In 1905 Hertzsprung discovered that stars of the same color differed widely in absolute magnitude. He inferred that such differences should be reflected in their spectra. The work of Antonia Maury confirmed his expectations. Mistakenly, E.C. Pickering at Harvard disagreed.

1. Early Maury and Cannon Classifications

In 1905 Ejnar Hertzsprung (1905,1907, 1908) discovered that stars of the same color differed widely in absolute magnitude and he inferred that such differences should be reflected in their spectra. However, he found no clues in the Harvard spectral classifications of southern stars by Annie J. Cannon (1901) on the system established by E.C. Pickering and Williamina Fleming at Harvard (Pickering 1890a). But Antonia Maury's catalogue of northern stars (1897) happily revealed just what he hoped for. Her *c*-characteristic differentiated the high luminosity stars from the rest. But Pickering, miffed that she had set up an independent classification system of her own rather than blindly accepting the Pickering-Fleming system, steadfastly maintained (Jones and Boyd 1971) that the spectra she examined were not of good enough quality to reveal such distinctions! Cannon did not incorporate such distinctions into her classes but she did give remarks calling attention to the stars with sharp narrow lines.

It was also Maury who pointed out that in the Pickering-Fleming classifications (arranged alphabetically in order of increasing complexity of appearance) class B should logically precede A, and the absorption O, called Oe5, should precede B. The Oe5 stars are the only instance in the Henry Draper Catalogue (Cannon 1918-24) where the letter e does not stand for emission. Emission O stars, classified Oa through Od on the HD system, Maury assigned to Group XXII following the red stars with molecular spectra. Figure 1 shows how well her standard stars for her groups I through XX correlate with modern MK temperature classes for the same stars. The symbols differentiate the various MK luminosity classes. If this graph is replotted with different symbols for Maury's subdivisions a, b, and c , it is clear that the small dispersion in Figure 1 is not correlated with either her divisions or the MK luminosity classes, confirming that both the HD classification and Maury's are strictly correlated with temperature and not with luminosity. Her groups XXI (C types, or carbon stars originally classed Na) and XXII (Wolf-Rayet stars) do not follow the major temperature sequence, but represent distinct groups. Group XXII would best have preceded Group I, the absorption Oe5 types. The Oe5 designation has now been

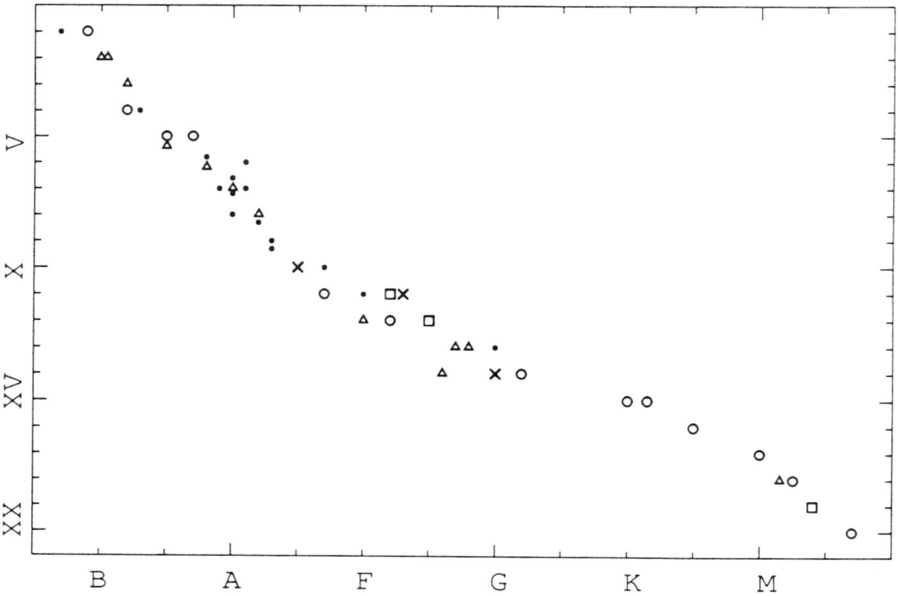

Figure 1. Maury's Groups correlated with MK classes: dots for luminosity V, crosses IV, open circles III, squares II, and triangles I.

replaced by O6-O9; the Wolf-Rayet, or Oa-d, are now reclassified as WR or WC. Emission in other classes of spectra Maury noted only in her *Remarks*.

Maury's catalogue of 681 bright stars north of -30° (Maury 1897) included eighteen *c*-stars. Her spectral classes are compared in Table1 with Miss Cannon's HD classes, and with modern MK classes. The latter have been converted to absolute magnitudes on the basis of calibration tables presented by Corbally and Garrison (1984) in the *Proceedings* of the first MK Process meeting. The correlations between these absolute magnitudes and Maury's spectral classes are shown in Figure 2. All of the stars are supergiants, but Maury's *c*-designation does not distinguish between luminosity classes 0, Ia or Ib. The open circles and triangles in Figure 2 are for the stars she classified *ac* (Table 2), spectra showing sharp lines but less sharp than those classified *c*. But for Pickering's disapproval of her work, Maury would surely have found the means for subdividing her *c* and *ac* divisions to reveal further subdivisions in their luminosities. With but one exception, μ UMa (Table 2), all of Maury's *c* and *ac* stars are supergiants. She classified μ UMa XI*ac*; the recent MK type is F2IV, corresponding to luminosity +1.9V, a δ Scuti type variable.Trigonometric parallaxes are now available for thirteen of the eighteen *c*-stars, but only two of the parallaxes are greater than twice their standard deviations, and the apparent magnitudes of these are strongly affected by interstellar absorption. Among the *ac*-stars the only one with a parallax significantly larger than $0''.020$ is μ UMa with $0''.039\pm.007$, yielding an absolute magnitude +1.7, consistent with the spectroscopic.

Annie J. Cannon, in her *Remarks* on the 1122 southern stars she had classified, noted nineteen stars as having spectra showing abnormally sharp lines.

Table 1. Stars with Sharp Spectral Lines

Name	Maury	Cannon	MK	M_v	Var
χ^2 Ori	III*c*	B2p	B2Iave	-6.8	-
o^2 CMa	V*c*	B5p	B3Iab	-6.2	-
η CMa	V*c*	B5p	B5Ia	-7.0	-
67 Oph	V*c*	B5p	B5Ib	-5.7	-
+59° 660	VI*c*	B9p	B9Ia	-7.1	NSV 1152
+58° 607	VI*c*	A0p	A0Iae	-7.1	-
β Ori	VI*c*	B8p	B8Iae	-7.1	-
μ Sgr	VI*c*	B8p	B8Iaep	-7.1	EA+α Cyg
15 Sgr	VI*c*	B0	B0Ia	-6.2	-
4 Lac	VI*c*	B8p	B9Iab	-6.3	-
η Leo	VII*c*	A0p	A0Ib	-5.2	-
α Cyg	VIII*c*	A2p	A2Iae	-7.3	α Cyg
ϵ Aur	XII*c*	F5p	F0Iae	-8.5	EA
35 Cyg	XII*c*	F5p	F5Ib	-4.5	NSV 12994
δ CMa	XIII*c*	F8p	F8Ia	-8.0	-
45 Dra	XIII*c*	F8p	F7Ib	-4.4	-
γ Cyg	XIII*c*	F8p	F8Ib	-4.3	-
ρ Cas	XIII*c*	F8p	G20e	-8±	SRd

She had classified all as B1 to F8, and with but one exception (β^2 Tuc) all have now been assigned MK lumimosity classes 0 (zero) to Ib. (β^2 Tuc comprises components CD of a visual quadruple system, A2V+A7V.) Thus, with few exceptions, both Maury and Cannon were correct in noting unusual sharpness of lines that later proved to be indicative of high luminosity.

The first seven Groups in Maury's classification system contained lines of unknown origin at the time of her compilation.These lines were called Orion lines in view of the fact that many of the stars showing them are in Orion. Maury meticulously measured and tabulated these lines, later identified with helium. P.J.C.Janssen (1868) had discovered a line of unknown origin during the total eclipse of 1868. This he reported to N. Lockyer who named the element helium (Asimov 1982). It was not until 1895 that W. Ramsay discovered the element in the mineral cleveite; and H.C.Vogel in December 1895, and C.Runge and F. Paschen in January 1896, published wave lengths. Although Maury's catalogue was not published until February 1897, it had been completed well before these papers appeared, but too late to warrant the long delays that would have been necessary for her to revise her manuscritpt accordingly. Hence she simply added an appendix comparing her measurements of the stellar spectra with the wave lengths of the laboratory sources. One series of helium lines was tentatively ascribed to an unknown element called parhelium. Maury noted that the lines of both helium and parhelium occurred in the stellar spectra and she discussed how they varied from group to group. She commented, "This fact, when the more precise behavior of the spectrum of helium under varying conditions of temperature and pressure has been investigated, may furnish a clue to the physical condition of the atmospheres of these stars." Moreover, she predicted that "the marked narrowness of the lines of hydrogen characterizing Division *c*, may some

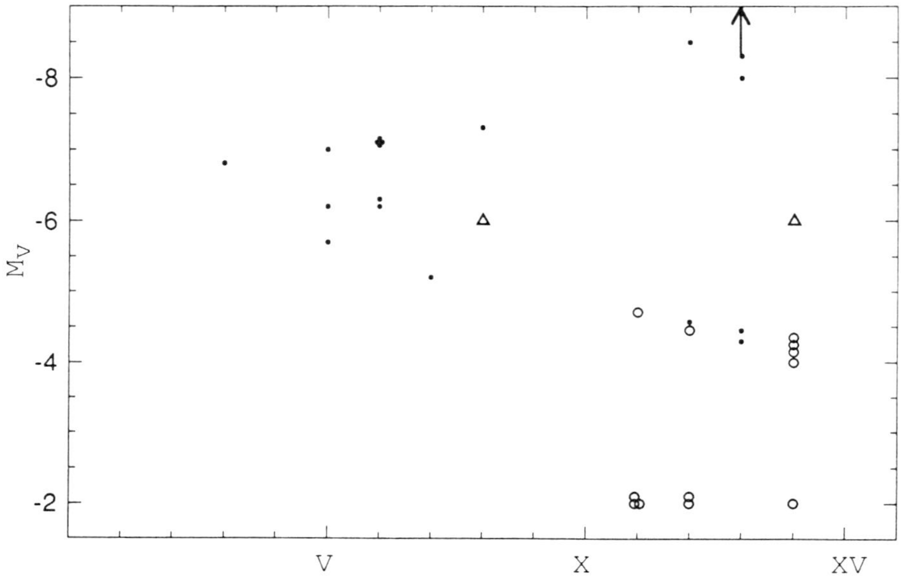

Figure 2. An HR diagram for Maury *c* and *ac* divisions with her Groups as abscissae, absolute magnitudes corresponding to MK classes as ordinates. Dots for Division *c*, open circles for *ac*.
The two triangles are for the *ac* stars 3 Pup, SB, amp. 0.24R, A2Iabe; and T Mon, 5.58-6.62V, F7Iab-K1Iab+A0V. The arrow is for the only star with luminosity zero, ρ Cas uncalibrated but brighter than -8V.

day furnish a clue to the difference of physical conditions in stars of divisions *a* and *c*" (Maury 1897, pp.124 and 125). It was eight years later that Hertzsprung (1905) discovered that distinction. Division *a* represents average line widths; division *b* the spectra showing abnormally wide lines: hence Maury interpolated between *a* and *c*, not between *b* and *c*. In a sampling of 20 stars that Maury had assigned to division *b*, 14 are now known to be spectroscopic binaries, 5 others have variable radial velocity, and only one (β Vir, A3V) is neither a known SB nor a component of a visual double or multiple system.

Maury's *c*-characteristic and Cannon's *Remarks* on sharpness of lines were assigned only to stars of early spectral classes. However, Hertzsprung (1907) pointed out that in her *Remarks* Maury had noted that some of the later type stars showed the line 4077.9 Å stronger than normal. He found that these stars showed smaller proper motions than the rest, indicating higher luminosity. For example, the line was stronger in Capella than in either 70 Ophiuchi or 61 Cygni, whose currently accepted visual absolute magnitudes are, respectively, -0.6, +5.6 and +7.6 (for 61 Cyg A). Hertzsprung thus appears to have been the first to call attention to this Sr+ line as an absolute magnitude criterion. Maury herself (1897 p.39) had cautiously stated, "Frequently also the lines 4227.0 (except in α Cassiopeiae), 4215.7, and the compound line 4076.8 to 4077.9 seem stronger,

Table 2. Stars with Moderately Sharp Lines

Name	Maury	HD	MK	M_v	Var	PL*
3 Pup	VIII*ac*	A2p	A2Iabe	-6:		
22 And	XI*ac*	F0	F2II	-2.0		
ν Her	XI*ac*	F0	F2II	-2.0	ν Her	
α Lep	XI*ac*	F0	F0Ib	-4.7	2128	
π Sgr	XI*ac*	F2	F2II	-2.0	11769	
υ UMa	XI*ac*	Fo	F2IV	+1.9	υ UMa	
ν Per	XII*ac*	F5	F5II	-2.0	1261	
α Per	XII*ac*	F5	F5Ib	-4.5	1125	
41 Cyg	XII*ac*	F5p	F5II	-2.0	13111	
α UMi	XIII*ac*	F8	F7:Ib-II	-3.2:	α UMi	-3.0
η Aql	XIV*ac*	G0p	F6Ib	-4.4	η Aql	-3.7
α Aqr	XIV*ac*	G0	G2Ib	-4.0		
10 Cam	XIV*ac*	G0p	G1Ib-IIa	-3:		
11 Pup	XIV*ac*	F8	F7II	-2.0		
δ Cep	XIV*ac*	G0	F5Ib-G2Ib	-4.2:	δ Cep	-3.4
T Mon	XIV*ac*	G5p	F7Iab-K1Iab	-6:	T Mon	-5.5
ζ Gem	XIV*ac*	G0p	F7-G3Ib	-4.2:	ζ Gem	-4.2

*Period-Luminosity Relation for Cepheids.

but this difference may be due to accidental photographic conditions. The degree of absorption indicated also varies in photographs of different density, so that while there seems to be sufficient ground for believing that the distinction is a real one, there is more or less liability to error in assigning individual stars to one or the other division." Of the six stars for which Maury cited a strong 4077.9 line, all have subsequently been classified G0 to G5 with luminosity classes Ib to III. Cannon cited 22 stars with prominent 4077.9. But of these 13 have turned out to be peculiar A stars; the 9 that are not have currently been assigned luminosity classes I or II (Table 3). These are excellent confirmations of both Maury's and Cannon's observational acuity and Hertzsprung's early inference that the strontium line constitutes a luminosity criterion. It was ten years later that the 4077.9 line was rediscovered and adopted at Mount Wilson Observatory and elsewhere as one of the more important criteria for the determination of spectroscopic absolute magnitudes and parallaxes. In the pioneering work of Adams and Kohlschütter in 1914 this particular criterion was not mentioned; its first use was by Adams and Joy in 1917.

In summary, it seems that if Pickering had heeded Hertzsprung, the MK system of two dimensional classification should have developed a third of a century before the magnificent Morgan-Keenan-Kellman Atlas appeared in 1943. But Hertzsprung was not yet famous when he published his revealing paper, published, moreover, in a journal on photography little known to astronomers. If, in addition to the luminosity criteria *a*, *b*, and *c* there had been added an additional symbol for the strong strontium lines indicative of high luminosity of late F, G and K type stars, and the letter *e* to call attention to emission lines, a fairly good first approximation to the current MK system could have been

Table 3. MK Classes for Stars for which Maury or Cannon Noted Strong λ 4077.9 (Sr+)

Maury (1897)			Cannon (1901)		
Name	Maury	MK	Name	Cannon	MK
ζ Cap	XIV	G4Ib	β Dor	F5	F6Ia
μ Per	XIV*a*	G0Ib	A Car	G5	G6II
β Lep	XIV*a*	G5II	δ Vol	F5	F6II
η Peg	XIV*a*	GII-III	ξ Pup	G	G6Iab-Ib
α Aur	XIV*a*	G5III	G Car	F5	F9II
ϵ Leo	XIVP	G1II	ξ Vel	G	G2-3Ib
			υ^2 Cen	F5	F6II
			ν^1 Sgr	G5	K2I+B9V
			ζ Cap	Gp	G4Ib:Ba2

established by 1908. By 1922, at the first meeting of the IAU, luminosity codes c, g, and d were proposed as prefixes to the Draper classes for sharp line spectra, normal giants, and dwarfs, and the suffix e for emission in mainly absorption spectra (Adams 1922).These belated recommendations were passed on the basis of the work then being carried out at Mount Wilson.

In initiating his system of classification, Pickering had stressed that it should be strictly empirical, in no way influenced by theoretical considerations. But the ever inquisitive Maury inevitably speculated on likely causes for differences in line quality, and the relative strengths of the λ 4077.9 line among stars whose major features were alike. Cannon, on the other hand, while noting the very same differences in her *Remarks*, carried out purely empirical observations and refrained from speculating on possible causes. Had Cannon given subsidiary letters for the peculiarities described in her remarks, the extensive ten volume HD catalogue would have provided a skeletal two-dimensional classification more conviently useful for astrophycists and students of galactic structure. Moreover, such codes could have been added in the tables much faster than the wordy remarks at the end of each volume.

In her examination of the spectra of the northern stars Maury made a number of other noteworthy observations and interpretations. With foresight she had commented that the primary spectral sequence probably represented an evolutionary sequence; and that the "prevalence of the Orion type [early type stars] in great nebulous regions, as in Orion and the Pleiades, indicates very emphatically that stars of this type are in an early stage of development"(Maury 1897, p. 11).

2. Spectroscopic Binaries

Spectroscopic binaries became another of Maury's strong interests. The first spectroscopic binary, Mizar, was discovered in 1889 by Pickering (1890b), the second, β Aurigae, by Maury the same year (Pickering 1890c; Shapley 1922).

Maury determined their periods as 20.5 and 3.96 days, respectively. Agnes Clerke (1902) remarked, "the new class of 'spectroscopic binaries' could never have been visually disclosed." On the basis of a parallax determined by Pritchard (1891), $0''.065$ (the current value is smaller, $0''.044$), she estimated the distance, which implied a maximum separation of the components of β Aurigae of only $0''.005$, adding, "To make this evanescent interval perceptible, a telescope eighty feet in aperture would be required."

At Shapley's request (Shapley 1922) Maury examined the nearly 600 available spectrum plates of β Aurigae and ascertained that a gaseous envelope of calcium surrounded the star, as implied by Saha's theory (Saha 1921). A similar result could not be established for Mizar. Maury then determined the orbits of the spectroscopic binaries μ^1 Sco and V Puppis (Maury 1920), ζ Centauri (1921), and β Lupi (1925a). In comparing the line quality in the spectra of ζ Cen , μ^1 Sco and V Pup she pointed out that in accordance with theory, the line widening is due to axial rotation synchronizing with revolution. Her remarks on υ Sagittarii in the same paper (Maury 1921) led Ludendorff (1924) to derive enormous masses for the components, conclusions she refuted in a subsequent paper (Maury 1925b).

The complex variable spectra of the eclipsing, as well as SB star, β Lyrae especially intrigued her the rest of her life (Pickering 1891; Maury 1924, 1926, 1933, 1935). She had 307 Harvard objective prism spectrograms taken 1886-1901 and 1921-1931, all but 50 taken with two prisms with dispersions ranging from 35 Å/mm to 45 Å/mm between H_β and H_ϵ. In each spectrogram she estimated the relative intensities of up to 15 absorption and 12 emission lines of hydrogen and helium, the H line of calcium, and the 4481 line of Mg^+. Although Rossiter in 1923 (Rossiter 1924; Henroteau 1928) had already derived an orbit, her spectrograms covered a larger interval, 1265 revolutions; hence she determined radial velocities, derived a new period, and correlated her observations with the orbital phases (Maury 1933, p. 209f). She suggested that the binary consisted of two elongated ellipsoidal stars (cB9 and B2e) surrounded by a common atmosphere – a picture not greatly different from that of Myers (1898; Struve 1958). While later astronomers (Pilans 1934; Struve 1934, 1941) tend to stress that Maury was wrong in her conclusions, F.J.M. Stratton (1934) of Cambridge, England, gave a more just opinion: "Miss Maury herself would be the first to admit that much remains to be done on this star, but the labor of love that she has so long devoted to β Lyrae will be of the greatest service to those who carry on her work to later stages." Astrophysicist Cecilia Payne Gaposchkin said of Maury, "She had a flair for picking out tremendous problems. If you look at the literature for β Lyrae you will see that nobody has ever solved the problem" (Mack, 1977).

Unlike Annie J. Cannon (1863-1941, Wellesley B.S 1884, M.A. 1907, at Harvard from 1896) who received numerous awards and honors, the intellectually superior Antonia Maury (1866-1952, Vassar A.B. 1887, one of the most talented of the students of Maria Mitchell, America's first woman astronomer; at Harvard intermittently from 1888) received just one, ironically, the Annie J. Cannon Award, in 1943.

Acknowledgments. I am grateful to John Lee who meticulously prepared my manuscript for photo-offset.

References

Adams, W.S. 1922, Report of IAU Comm. 29, *IAU Trans.* **1**, 98.

Adams, W.S. and Joy, A.H. 1917, ApJ, **46**, 313; *Mt. Wilson Contr.* No.142.

Adams, W.S. and Kohlschütter, A. 1914, ApJ, **40**, 385.

Asimov, I. 1982, *Biographical Encyclopedia of Science and Technology*, New York, items 647 and 719.

Cannon, A. J., 1901, *Ann. Harvard Coll. Obs.* **28**, Part II, 131.

Cannon, A. J., 1918-24, *Ann. Harvard Coll. Obs.* **90-99**; 1925-36 *Supp.* **100**.

Clerke, A. M. 1902, *A Popular History of Astronomy in the Nineteenth Century*, 4th Ed., 388.

Corbally, C.J. and Garrison, R.F. 1984, *The MK Process and Stellar Classification*, Toronto, 277.

Henroteau, F.C. 1928, *Handbuch d. Astrophysik*, Berlin, **6**, 432.

Hertzsprung, E. 1905, 1907, "Zur Strahlung der Sterne," *Zeitschrift für wissenschaftliche Photographie*, **3**, 423, and **5**, 86.

Hertzsprung, E. 1908, *Ast. Nach.*, **179**, 373.

Janssen, P.J.C. 1868, *Comptes Rendus*, **67**, 838.

Jones, B.Z. and Boyd, L.G. 1971, *The Harvard College Observatory*, Cambridge, MA, p.240.

Ludendorff, H. 1924, *Sitzungsberichte d. Preussischen Akad. d. Wissenschaften*, **9**, 67.

Mack, P. 1977, private interview with Payne–Gaposchkin for B.A. Thesis, Harvard.

Maury, A. 1897, *Ann. Harvard Coll. Obs.*, **28**, Part I.

Maury, A. 1920, *Ann. Harvard Coll. Obs.*, **84**, No. 6, 157.

Maury, A. 1921, *Popular Astronomy*, **29**, 636.

Maury, A. 1924, *Popular Astronomy*, **32**, 559.

Maury, A. 1925a, *Bull. Harvard Coll. Obs.*, No. 820, 3.

Maury, A. 1925b, *Bull. Harvard Coll. Obs.*, No. 824, 4.

Maury, A. 1926, *Popular Astronomy*, **34**, 625.

Maury, A. 1933, *Ann. Harvard Coll.Obs.*, **84**, No. 8, 207.

Maury, A. 1935, *Bull. Harvard Coll. Obs.*, No. 899.

Morgan, W.W., Keenan, P.C., and Kellman, E. 1943, *An Atlas of Stellar Spectra*, Chicago.

Myers, G.W. 1898, ApJ, **7**, 1.

Pickering, E.C. 1890a, *Ann. Harvard Coll. Obs.*, **27**.

Pickering, E.C. 1890b, *Observatory*, **13**, 80.

Pickering, E.C. 1890c, *MNRAS*, **50**, 296.

Pickering, E.C. 1891, *Observatory*, **14**, 341.

Pilans, H.M. 1934, ApJ, **80**, 51.

Pritchard, C. 1891, *Ast. Nach.* **127**, 201; *MNRAS*, 51, 310.

Ramsay, W. 1895, *Nature*, **52**, 7.
Rossiter, R.A. 1924, ApJ, **60**, 15, footnote.
Runge, C. and Paschen, F. 1896, ApJ, **3**, 4.
Saha, M.N. 1921, *Nature*, **107**, 488.
Shapley, H, 1922, *Bull. Harvard Coll. Obs*, No. 767, 1.
Stratton, F.J.M. 1934, *Observatory*, **57**, 165.
Struve, O. 1934, *Observatory*, **57**, 265.
Struve, O. 1941, ApJ, **93**, 164.
Struve, O. 1958 *PASP*, **70**, 5.
Vogel, H.C. 1895, ApJ, **2**, 333.

Discussion

Weaver: How did Miss Maury distinguish rotational broadening from just another class of stars?

Hoffleit: I believe she noted the profiles of the lines – deep sharp vs. shallow wide. She wanted to interpret the cause of the broadening, and rotation (and for SBs, revolution) seemed logical.

Liebert: Why did Miss Maury decide to reorder Pickering's O and B stars before the type A?

Hoffleit: She noted that the "Orion lines", later identified as helium, decreased in intensity while hydrogen increased to A0. If Pickering's order had been maintained, the order of increasing complexity of the appearance of the spectra would have had to have two branches B and F as F did not seem to follow directly from B, but from A. She may also have noted a progression of color, but that I do not recall.

MacConnell: Do you know how Hertzsprung obtained his proper motions?

Hoffleit: I do not know. Presumably the first epoch observations were visual-meridian circle. Second epoch might have been either visual or photographic. When I get back to Yale, I'll try to check.

McCarthy: What was the date of Miss Maury's death?

Hoffleit: Thank you for asking: I had not mentioned the date of Miss Maury's death. Her dates were 1866-1952.

The MK Process at 50 Years
ASP Conference Series, Vol. 60, 1994
C.J. Corbally, R.O. Gray, and R.F. Garrison (eds.)

Angelo Secchi and the Discovery of Carbon Stars

M. F. McCarthy, S.J.

Specola Vaticana

Abstract. Fifteen years ago we were commemorating the centenary of the death of Fr. Angelo Secchi (1818-1878). At that time the MK system was 35 years young. One new dimension had been firmly established and was added to the original temperature criterium of the Secchi and the Henry Draper systems: this was the luminosity parameter. Another parameter in the work of Keenan and Morgan concerned stellar abundances which had long been foreshadowed and was now confirmed for certain groups of relatively cool stars within the MK system. This foreshadowing of stellar abundances reaches back in its primitive form to Secchi's discovery in 1868 of the stars called by him Type IV, where he found evidence for much carbon and which he contrasted with the stars of his Type III, later called the M stars. We will explore some of the details which confirmed this discovery in 1868, one hundred and twenty five years ago and just seventy five years before the MKK atlas appeared.

1. Secchi's discovery of the carbon stars

The year 1868 was a banner year for Father Angelo Secchi. In that year, after being disappointed at not being able to go to India to observe the great eclipse of 18 August 1868, he stayed at home in Rome and reported to the French Academie des Sciences in a rather tentative vein his suspicions that he had detected a new spectral type (Secchi 1868a). This would be a clear distinction among the red stars he had been observing from his observatory of the *Collegio Romano* located atop the Church of San Ignazio near the *Piazza Venezia* in downtown Rome. In January Secchi (1868a) announced his discovery as follows:

> Stars which do not belong to the three established types are very rare. I have examined without success many hundreds of faint stars. I have just come across one very extraordinary star which is listed in Lalande's catalogue (RA = 4hr 54m 10s and Dec =+ 0°59′). Its spectrum is very peculiar. The red region is divided into two bands by a very broad dark line. The golden yellow is reduced to a very clear and very sharp line. After a broad dark band comes a broad green-yellow band and, after another dark interval, a zone of blue... Although I have not examined the whole sky I believe that one will find very few of these stars and that they will belong to the family of red stars and of variable stars.

In August (1868b), Secchi announced, in confirmation of his earlier suspicion, the definitive discovery of the Type IV stars:

> There is a complete contrast between the third and this fourth type... The two spectra are not merely modifications of the same type, they are evidently due to completely different substances. The light of these stars is so faint that it is not possible to use a slit spectroscope and thus we cannot identify precisely the sources of the lines and bands. We can say however that there is a marked analogy with the reversed spectrum of carbon.

These stars we know today as the carbon stars. Fr. Secchi listed, together with this announcement of the new type IV, some seventeen (17) stars and promised a detailed description which should follow. These stars had all been listed as "red" stars in a catalogue assembled by the Danish astronomer, Dr. Schjellerup (1866) just two years before Secchi's discovery of the new spectral type. The list of contributors to Schjellerup's listing is really a "Who's Who" for mid-19th century astronomy. Among the contributors we note the names of Argelander, d'Arrest, Bessell, John Herschel, Lamont, Lalande, Meyer and Piazzi-Smith.

Table 1 (Secchi 1868b) contains Secchi's presentation of his Type IV stars from Schjellerup's list with equatorial coordinates for the epoch 1870, then Fr. Secchi's estimate of the (visual) stellar magnitude and finally his notes.

Table 1. Secchi's list of Type IV stars and their positions for the epoch 1870.

No. du catal. de Schjellerup	Ascension droite	Declin.	Grandeur.	Notes.
41	4.36,2	+67,54	6 1/2	Belle.
43	4.42,8	+28,16	8	
51	4.58,1	+ 0,59	6	
78	6.26,9	+38,33	6 1/2	Belle.
89	7.11,5	-11,43	7 1/2	
124	9.44,6	-22,22	6 1/2	
128	10. 5,8	-34,38	7	Douteuse.
132	10.30,7	-12,39	6	Belle.
136	10.44,8	-20,30	6 1/2	
152	12.38,5	+46,13	6	Magnifique.
159	13.19,3	-11,59	5,8	
163	13.47,3	+41, 2	7	
229	19.26,5	+76,17	6 1/2	
238	20. 8,6	-21,45	6	
249	21.25,8	+50,58	9	
252	21.38,6	+37,13	8 1/2	
273	23.39,2	+ 2,42	6	Belle.

Secchi thus succeeded in distinguishing the stars of his new type IV among the red stars. These latter stars of Type Three, such as Antares and Betelguese show luminous bands alternating with dark ones and include most of the stars we now call M type stars. Secchi's type IV stars displayed three luminous bands in the blue, green, and red regions separated by dark intervals; these dark intervals were later named for Swan who had listed them in 1860, though their identification with carbon (C) was not confirmed until 1930. What distinguishes the stars of the type III from type IV is not merely the width of the zones, but more especially that, in the spectra of the Type IV stars, the light in the zones falls off abruptly on the edge of shorter wavelength, while it fades off gradually to darkness on the edge of longer wavelength. In spectra of type III, the distribution of light is reversed. Secchi's Type IV includes the R and N stars of the later Henry Draper Classification which are known as the C stars in the modern MK system.

In November of Secchi's "anno mirabile" his promised detailed description of these spectra was published (Secchi,1868c). This appeared in two numbers of *Memoria Sugli Spettri Prismatici Delle Stelle Fisse.* Both numbers were extracted from the Acts of the Italian Society of the Quaranta (XL) in their Series III, Tome I, Part I. Secchi combined these into his booklet and this is the basis for our discussion of his listings. For many of the stars listed, he presents, in addition to the positional and magnitude data, also a spectral curve drawn for the star; this shows how he observed the variation of intensity with wavelength. We note that in his spectral drawings Secchi will place the violet at the right and the red end at the left. In this same year (Nov. 1968) he published his list in *Astronomiche Nachrichten.*

2. Secchi's classification compared with modern data

We shall compare here the classifications for Secchi's seventeen carbon stars with the work of later spectral classifiers during the next 125 years.

In Table 2 we present the current classifications for these stars. In successive columns are listed the Schjellerup Number, the HD Number and Spectral Class, the number in the Stephenson catalogue (1989) and finally, the modern spectral types listed by Shane (1928) and Sanford (1944), by Keenan and Morgan (1941) and by Yamashita (1972). Of the seventeen stars classified as carbon stars (Type IV) by Secchi and listed in Tables 1 and 2, some thirteen are listed as variable stars in the Kukarkin and Parenago catalogue. We note that after the announcements by Schjellerup and by Secchi, most of these stars came very soon under observation for variability. Secchi suspected variations for several of his type IV stars.

Somewhat surprisingly, only six of Secchi's carbon stars are found in the Bright Star Catalogue (Hoffleit 1982). This may reflect Fr. Secchi's acute visual powers plus the fact that so many of his stars were variable with very different ranges in magnitude and so were sometimes very faint. For the carbon stars in Table 2, which are not known now as variables and which are not listed in Stephenson's catalogue as carbon stars, we have the following notes from Secchi's own commentary cited above (Secchi 1868c):

> "Schjellerup 159. The color is yellow tending toward red but without absolutely black bands. (Memoria II,p.47) [This star is HD 116870 K2.]

Table 2. Secchi carbon stars in later lists.

Schjellerup	HD	Sp(HD)	Variable Name	Stephenson	Spectral Types Sh\Sa[a]	Yam	(K-M)
41	30243	Nb	ST Cam	769	N5 (Sh)	C 5,4	C 5,4
43	30755	Nb	TT Tau	794	N3 (Sh)	C 7,4	
51	32736	Nb	W Ori	853	N5 (Sh)	C 5,4	C 5,3
78	46687	Na	UU Aur	1316	N3 (Sh)	C 6,4	C 5,3
89	54361	Na	W CMa	1565	N (Sa)	C 6,3	
124	85405	Np	Y Hya	2641	N3 (Sh)	C 4,5	C 4,5
128	88539	Na	AB Ant	2713	N (Sa)	C 6,3	
132	92055	Nb	U Hya	2803	N2 (Sa)	C 6,3	C 7,3
136			V Hya	2887	N (Sa)	C 6,5	C 7,5
152	110914	Nb	Y CVn	3283	N3 (Sh)	C 5,5 J	C 5,4
159	116870	K2	68 Vir				
163	121197	K5					
229	183556	Nb	UX Dra	4302	N0 (Sh)	C 7,3	C 7,3
238	192737	Nb	RT Cap	4774	N (Sa)	C 6,4	
249			#1[b]		[c]		
252	206750	N	RV Cyg	5425	N5 (Sh)	C 6,4	C 6,4e [d]
273	223075	Na	TX Psc	5928	N0 (Sh)	C 7,2	C 6,2

[a]Shane (Sh) or Sanford (Sa)
[b]#1 in NGC 7086 (Hoag *et al.* 1961)
[c]#1 has B-V=1.89; V=10.2
[d]Bidelman (1962)

Schjellerup 163. This star merits further study. (Memoria II,p.48) [HD 121197 K0]"

Recently Keenan (1993b) pointed out the identification of two other stars in Table 2 which appeared to lack other designations besides the Schjellerup number. These are stars Schjellerup Nos. 249 and 252.

The star Schjellerup 249 is discussed in his second listing (1874) and quotes Secchi's note that this is a "ruddy star, 11th magnitude in the midst of a beautiful cluster"; Keenan adds that this is the cluster NGC 7086 and that Schjellerup 249 must be the star called #1 by Hoag *et al.*(1961). It is located near the center of the cluster and is very red with $(B-V) = +1.89$, quite consistent with Type IV red stars.

Schjellerup 252 is probably identical, according to Keenan (1993b), with BD $+37^\circ$ 4407 (HD 206570) which is the variable star RV Cygni and listed as star No. 5425 in Stephenson's second catalogue. It was classified by Shane (1928), by Yamashita (1972) and was noted to be in emission by Bidelman (1962). I am grateful to Prof. Keenan for this information.

3. Secchi's notes on features of carbon stars

We can, I believe, give very "good marks" to Fr. Secchi regarding his discovery list for the carbon stars (Tables 1 and 2). His final notes (Secchi 1868d) list his uncertainties in assigning a classification for stars numbered 159, 163, 249 and 273 by Schjellerup. Now we know that Angelo Secchi was a devoted subject and a close friend of Pope Pius IX, in whose term of office papal infallibilty was discussed and defined. Fr. Secchi never thought of himself as "infallible" especially in scientific matters. It may be noteworthy in this regard that in *Astronomische Nachrichten* No. 1737, Secchi lists "three stars of M. Wolff" which are now the prototypes of the Wolf-Rayet stars; he noted that he found "nothing extraordinary in them" and attributed the faint luminous lines in these stars to "scintillation." *Sic itur ad astra.* Even our good Secchi could nod!

All of the stars in Table 2 certainly deserve modern observations for their spectra and especially for their infrared colors . I shall put them on our "little list" for early observations with the new Lennon telescope. For these stars we shall employ the recent work of Keenan (1993a) which gives the revised MK classification of carbon stars.

Next we present Secchi's own notes (1868c), from his Memoria II, p.47, on his favorite specimen of spectra of Type IV: Schjellerup No. 152, which he called "superba." We give the entry from Secchi's 1868 catalogue for this star. The account includes, after the positional data, the apparent visual magnitude and then a description of the spectral features as he observed them.

> "Schjellerup No. 152 RA = 12hr $38^{m}_{.}5$ Dec = $+46°13'$. Mag. 6th. Magnifico oggetto di 4° tipo, e veramente singulare per la sua vivacita. E composto di tre zone assai vive, e larghe. Una gialla, l'altra verde, e la terza blue. Sono tutte vivaci assai tagliente verso il violetto, e sfumate da lato del rosso. Il ciglio delle zone e rinforzato da righe vive all'estremo. Nel giallo queste righe sono come due bellissime fili d'oro. Cosi pure due righe rinforzate sono nel verde e nel blue, benche questo non sia tanto vivo. Vi paiono anche delle strie minori. ....Di piu, un carattere di questo 4° tipo diverso da 3° e che nel 4° le curve della luce piegano al violetto cosi nel 3° le curve piegano al rosso. oltre l'esser la metà di larghezza; .... Sicche i due tipi sono essenzialmente diversi. Vi e qualche analogia tra questo spettro e quello delle comete. Provato collo spettrometro a due prismi, ma la luce manca affatto. E la stella 4287 B. A. C."

> Translation: "Schjellerup No. 152.......A splendid object of type IV which is truly outstanding for its vividness; the spectrum is composed of three zones which are strong and broad. One zone is yellow, another green and the third is blue. These are all vivid with a cutoff on the violet side and a fading tapering from the red side. The ridge of each zone is strengthened by vivid lines at the edge. In the yellow these lines are like two most beautiful threads of gold. So too there are strong features in the green and in the blue, though this last is not so prominent. We also see some lesser streaks....In addition we note a characteristic which differentiates this type IV from type III: in type IV the light curve inclines towards the cut off at the violet side, whereas in type III the curves incline more steeply towards the red; besides they are only one half as broad. Thus the two types are essentially different. There is also some similarity between this spectrum (Schjellerup No. 152) and that of one of the comets. I

tried to observe this star with the two prism spectrometer but the light was too faint.This star is No. 4287 in the B. A. C."

Fr. Secchi noted that not all of the stars he put into type IV showed exactly the same features; some of the variety in spectral features among the carbon stars can be seen in the tracings of spectral energy he drew for these type IV stars. It is important when examining these tracings to recall that Secchi was observing these stars and their spectral features visually. He had used photography in his observations (Brück 1978) of solar eclipses but not in stellar spectroscopy. The photographic process was still in its very beginnings and Secchi was so busy with his visual observing program of the spectra of faint stars that he did not take the time to stop for experiments with this new process.

In his observations Secchi found that these carbon stars are red to very red in color in spite of the outstanding role played by the green continuum in their spectra. This phenomenon inspired the steady outreach by spectroscopists to study the carbon stars in the near infrared beginning with J.J. Nassau and his coworkers at Case Western Reserve (Nassau 1958). They used the CN bands at 7945 Å, 8125 Å and 8320 Å in their surveys of the cool red giant carbon stars.

Others studied the blue green region to detect the warmer carbon stars. Among these were the researches of Shane (1920,1928) and Sanford (1944) who used the large California instruments. With the Burrell Schmidt camera and objective prism in Cleveland, Sanduleak and Philip (1977) studied the blue green region along the plane of the Milky Way. Later, Azzopardi (1985) and his colleagues at ESO explored the distribution of warmer C stars in the center of the galaxy and in the nearby galaxies. Somewhat earlier the AURA group with the Blancos and McCarthy (1978) compared, in the near infrared, the numbers and surface densities of the cooler carbon stars in the Galactic Center and in the Magellanic Clouds and also contrasted these with the distribution of M type stars in the same fields.

Fr. Secchi was always attempting to compare the observed spectral features in carbon stars with the laboratory spectra of the known elements. He noted that the spectra of type IV, more than those of the other types, showed a pronounced resemblance to gas, especially to carbon, but in a "reversed sense" and concluded that the bands of type IV are produced by the absorption of a carbon compound. To verify this he passed a spark from platinum terminals through a benzine vapor (Secchi 1869). In England, Huggins (1868) was also using the lab spectra of carbon and was comparing the carbon features in comets and in type IV stars.

As Secchi wrote just 10 years after his announcement of the discovery of carbon stars and near the end of his life: "When we come to the end, we find that it is only the beginning." (Secchi 1878)

References

Azzopardi, M., LeQueux, J., Rebeirot, E., 1985, A&A, **145**, L4

Bidelman, W., 1962, ApJS, **1**, 202

Blanco, B.M., Blanco, V.M., and McCarthy, M.F., 1978, Nature, **271**, 638

Brück, H. A., 1978, Angelo Secchi, in Spectral Classification of the Future, IAU Coll 47, 14
Hoag A. *et al.*, 1961, USNO Publ XVII, 468
Hoffleit, D., 1982, Yale Cat of Bright Stars, New Haven Conn 4th Revised edition
Huggins, W. 1868, Astronomiche Nachrichten, No. 1704, 381
Keenan, P. C., 1993a, PASP, **105**, 905
Keenan,P.C., 1993b, (Private Communication)
Keenan, P.C., and Morgan, W. W., 1941, ApJ, **94**, 501
Nassau, J.J., 1958, in Stellar Populations, ed. D. O'Connell, p.183
Sanduleak, N., and Philip, A.G.D., 1977 Publ Warner and Swasey Obs, 2, No 5
Sanford, R., 1944, ApJ, **111**, 262
Schjellerup, H.C.F.C., 1866, Astronomische Nachrichten No 1591, 97
Schjellerup, H.C.F.C.,1874, VJS AG **9**, 252
Secchi, A., 1868a, C.R. Acad Sci Paris, **66**, 125
Secchi, A., 1868b, C.R. Acad Sci Paris, **67**, 373
Secchi, A., 1868c, Memorie II sugli Spettri Prismatici delle Stelle Fisse, Roma, p. 11
Secchi, A.,1868d, Astronomiche Nachrichten, No. 1737
Secchi, A., C.R. Acad Sci Paris, **68**, 1086
Secchi, A., 1878, Le Soleil, Gauthier Villars, Paris, **483**, 465
Shane, C.D., 1920, Lick Obs Bull 79
Shane, C.D., 1928, Lick Obs Bull 396
Stephenson, C. B., 1989, Gen Cat of Cool Carbon Stars, 2nd Ed, Publ Warner Swasey Obs 3, No 2.
Yamashita, Y.,1972, Tokyo Obs Ann (2) **13**, 1

Discussion

Weaver: How did Secchi observe lines with an objective prism?

McCarthy: Father Secchi observed lines visually with his objective prism; he mentions lines we now attribute to Hydrogen and the Sodium line and the Magnesium feature. I recall that he reports observing the Na feature as double but cannot now find this citation.

Secchi used 4 spectroscopes in the course of his investigations. His first consisted of a small spectroscope applied to the Merz refractor; in his first researches with this combination Secchi detected sodium in Betelguese. His second spectroscope was the one made by Hoffman which he used in his work on spectra between 1863 and 1866. He found that this instrument gave much more light and operated without a slit. He found that when comparison spectra were required, the Hoffman instrument could be employed with a slit placed exactly in the focus of the cylindrical lens. Because this instrument absorbed a considerable amount of light, he began to look for a different arrangement. This search led to Secchi's third spectroscope. It was a small simplified instrument fitted with a special occular. By changing the distance of the ocular the linear dispersion could be changed. Light intensity was conserved; when attached to the Merz refractor (25 cm aperture) he could reach spectra of 7th and 8th magnitude stars. The fourth instrument was what is known today as the objective prism or slitless spectroscope.Earlier prisms such as those used by Fraunhofer had prisms of 60 degrees and 38 degrees; Secchi's improvement consisted of using a prism of 12 degrees whose chief feature was the dispersion which he obtained. In fact, the dispersion proved too great for the observation of faint stars. A further feature of this fourth spectroscopic apparatus used by Secchi was the addition of a small auxiliary "finding" telescope, so mounted that the same star would be found in the field of the large telescope as was seen in the center of the field of the search telescope.

Garrison: What was the resolution of Fr. Secchi's spectroscopes?

McCarthy: I do not have any numbers to offer to you concerning resolution. The fact that Secchi did all his work visually may explain why he percieved the bands in late type spectra as "kinds of rows of columns". His keen powers of observation are demonstrated in his drawings of lines and bands in spectra of the different types. The instruments which Secchi used included the Chauchoix equatorial refractor with an aperture of 15 cm and a focal length of 224 cm and so a focal ratio of f/15. The other instrument was the Merz refractor of 25 cm aperture and to this he attached his Merz objective prism; this was the first circular prism ever constructed. He could also attach his spectroscopes to these instruments. The Merz instruments were destroyed by the fire in 1958 which consumed so large a part of the Monte Mario Observatory in Rome, now the property of the University of Rome. During the 1978 IAU Colloquium No. 47 held in Vatican City to commemorate the centennial of Fr. Secchi's death our colleagues of Monte Mario hosted a reception and demonstration of the remaining Secchi instruments which were preserved carefully until that disastrous fire in which the Director of the Monte Mario Observatory, Prof. G. Armellini, was fatally burned.

Abt: Was Secchi's prism in front of the telescope objective or after the focus?

McCarthy: Secchi's objective prism, like Fraunhofer's before him, was placed in front of the telescope objective lens. As noted above, Secchi was the first to employ a circular prism and this was his "last" in the series of spectroscopic instruments used by him; the other three, as pointed out above. were slit spectroscopes whose dispersing elements were located after (behind) the telescope objective and near the focal plane and the observer's eyepiece.

Corbally: In those wonderfully drawn spectra of Secchi, the spectra are shown as widened. Did Secchi use a widened spectrum, e.g with a barrel lens in the objective prism mode, as well as the spectrograph?

McCarthy: In 1866 Secchi did use an extended cylindrical (barrel) lens very near to the eyepiece, then put the direct vision prism train about 12 cms nearer the lens. With an eyepiece, powers ranging from 400× to 800× were measurable over the spectrum. I mentioned earlier how the spectra of late type stars appeared to Secchi as "fluted" columns or "rows of columns of colored light."

MacConnell: Was Fr. Secchi the first to use an objective prism?

McCarthy: No, it was Joseph Fraunhofer and not Fr. Secchi who was the first to use the objective prism. In 1817, Fraunhofer adapted the instrument he used to study the sun and began to study the much fainter light of the fixed stars; with his final prism, reduced from 60 degrees to 38 degrees, he surveyed several first magnitude stars and found the spectra differed widely among themselves (Cf. Fraunhofer (1823) Edinburgh Phil. Journal, Vol 9, and (1827) Edinburgh Journal of Science, Vol. 7). Secchi acknowledged his debt to Fraunhofer when he wrote in Le Soleil (1877): "In 1855, with a theodolite fitted up with a prism, we repeated Fraunhofer's observations; a little prism adapted to the great Merz Refractor furnished us with such a beautiful spectrum of Sirius that it made us eager to continue such researches; but other occupations kept us busy."

The MK Process at 50 Years
ASP Conference Series, Vol. 60, 1994
C.J. Corbally, R.O. Gray, and R.F. Garrison (eds.)

Newton's Rainbow: From a Hole in the Wall to a Window on the Universe[1]

Janet Rountree

Science Applications International Corporation, 5151 East Broadway, Tucson, AZ 85711

The science of astronomical spectroscopy, and therefore spectral classification, can be said to have begun with the experiments performed by Isaac Newton in 1666. Newton observed the continuous spectrum of the Sun, but did not detect the absorption lines because he used too wide an aperture. The solar absorption lines were first studied in detail by Josef von Fraunhofer, whose notation for the most prominent features is still in use today. Spectral classification as such was first pursued by Father Angelo Secchi in the 1860's. Secchi distinguished four classes among the most common stars; we would now identify these classes with early-type stars, solar-type stars, M stars, and carbon stars.

In the late 19th Century, the Henry Draper catalogue project was initiated at the Harvard College Observatory. The classification work was performed by a number of remarkable female astronomers, including Williamina Fleming, Antonia Maury, and Annie Cannon. Cannon alone classified over 225,000 spectra, and finalized the one-dimensional sequence that is still in use: OBAFGKMRNS.

It soon became apparent that the Henry Draper (or Harvard) system was essentially a temperature sequence, but that a second dimension was required to account for the differing luminosities of stars having the same spectral type. Thus was the Morgan-Keenan (MK) system born. But the essential difference between the MK and the Harvard classifications is not a question of dimensionality. The basic principles of the MK system, which distinguish it from its predecessors, are these:

1. the system is defined by its standard stars;
2. both standard and program stars must be observed in exactly the same way;
3. only features observed in the spectrum can be used in the classification.

Since its beginnings in the 1930's, the MK system has been used not only to classify thousands of normal stars, but also to isolate groups of astrophysically interesting objects such as Ap and Am stars. The visible arms of our galaxy were traced for the first time when aggregates of early-type stars were identified through MK classification. Suitable calibrations make it possible to go from spectral classification to astrophysical parameters such as mass, radius, age, luminosity, temperature, and distance.

[1]Summary of a public lecture at the Leo Rich Theatre, Tucson, during the Workshop

Current efforts at extending the MK system include classification in the ultraviolet and infrared spectral regions, and studies of extragalactic stellar populations. Further details are available in Rountree (1991).

References

Rountree, J. 1991, "Spectral Classification," in *The Astronomy and Astrophysics Encyclopedia*, ed. S. Maran (New York: Van Nostrand Reinhold).

PART VI.

EXTENSIONS OF THE MK SYSTEM

The MK Process at 50 Years
ASP Conference Series, Vol. 60, 1994
C.J. Corbally, R.O. Gray, and R.F. Garrison (eds.)

The MK Process in Action Today

C. J. Corbally

Vatican Observatory Group, University of Arizona, Tucson, AZ 85721

Abstract. Some of the fruitful ways in which the MK Process has been applied to spectral classification beyond the MK System are reviewed. Two particular directions are emphasized: the use of third and multiple dimensions; the extensions to wavelength regions other than from the Ca K line to Hβ. The two philosophies behind these extensions to the MK System are discussed.

1. The MK legacy – the MK Process

Many papers this week, as well as our workshop's title, reflect the conviction that the morphological classification of stellar spectra is one of the most powerful tools we have in astronomy for physical insight. This conviction has been strengthened now that we have heard something about the history of stellar classification.

We are also aware that what we owe to Dr. Morgan is not just the MK System of 1943 but its embodiment of an archetypal methodology for classification, the MK Process. Morgan defined this process to the workshop of 1983 that eventually bore the phrase "The MK Process ..." in its proceeding's title. "The MK Process is a procedure, a methodology, and operator, that makes possible the arrangement (classification) of stellar spectra into prepared structures that are defined by the spectra of arrays of individual standard stars."

Now, Morgan introduced this concept while reflecting on the incompleteness of the MK System itself.

> "... The MK System furnishes a satisfactory frame of reference for stars of Population I in the solar neighborhood; but when we consider the variety of stellar spectra encountered in our galaxy and in other stellar systems we are unable to classify many stars satisfactorily from the MK standards alone."
>
> "What we now require is a more general frame of reference - or frames of reference - that will accept with precision a greater variety of "specimens" (stellar spectra). To bring these into existence, we propose to supplement the MK System with additional systems that will be constructed according to what we shall label the MK Process." (Morgan 1984)

That need to supplement the MK System was instanced in Dr. Keenan's complementary paper, in the same 1983 workshop, when he outlined his pioneering work, the Revised MK Classification for the cooler stars (Keenan 1984). In this revision, abundance indices had been added to the temperature types

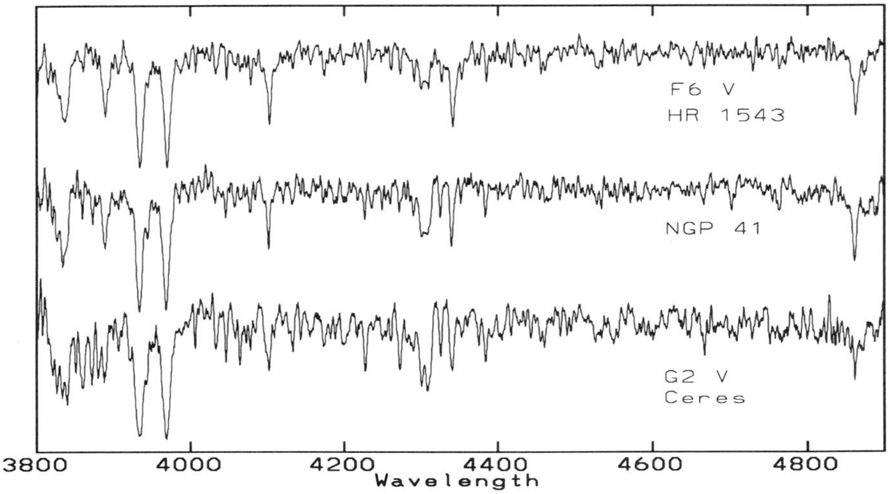

Figure 1. Within the MK System of standards, the weak-lined spectrum of NGP 41 matches a G2 V hydrogen-line strength and mainly an F6 V metal-line strength, though its G band is of a G0 V strength. A classification is possible, if awkward.

and luminosity classes when certain features in the spectra of stars indicated substantial differences from Population I stars. We have heard about the continuation of this work, of its success, and of its associated problems on the first day of the present workshop.

What I want to review in this paper is some more of the fruitful ways in which the MK Process has "gone forth and multiplied", i.e., has been applied beyond the MK System to stellar spectra. In this I shall be elaborating on the second and third roles or directions which Richard Gray (1994) outlined that spectral classification is taking: that of third and multiple dimensions, and that of extension to other wavelength regions. I shall concentrate on those examples that, to me at least, most clearly illustrate the power and principles of the MK Process. My review will start somewhat anecdotally and with the blue-green region of the spectrum, but it will broaden to the other wavelength regions.

2. A third dimension for weak-lined spectra

During a project to look carefully at the spectra of early-G dwarf stars in the Galactic Thick Disk, Bob Garrison and I (Garrison and Corbally 1993) found ourselves, like many classifiers before (Roman 1954; Bond 1970; Bidelman and MacConnell 1973; Houk and Cowley 1975; Abt 1986; ....) having to cope with weak-lined spectra. Our solution was to refine slightly but essentially to apply the method of these workers; that is, to use the grid of MK standard stars to define a temperature type from the absolute strength of the hydrogen lines and abundance independent line ratios (e.g., Fe I λ4250/Cr), a luminosity also from those line ratios relatively unaffected by abundance (e.g., Sr II/Fe I; λ4376/λ4383), and a measure of weakness from matching the metal lines to an MK standard earlier than that of the temperature type (see Figure 1. This works; and I even produced a calibration of the "measure of weakness" with [Fe/H] (Corbally 1987) that is a fairly convincing linear relationship.

However, it works under protest (and you have certainly heard Garrison's protests at this workshop). First, the strong metal lines tend to match with an MK standard that is a little later than the standard which matches the weaker lines. In other words, the overall morphology of the weak-lined star does not fit well with the matched standard and indeed is normally closest to a weakened version of the standard that gives the temperature type – as you might expect. To emphasize this mismatch in morphology, Garrison and I (Garrison and Corbally 1993) have opted to use a classification notation that is closer to Keenan's ("Fe-2") than to Houk's ("G5wF0") – not that I should want her to change now.

The second reason why the method only works "under protest" is that we find the G-band strengths to be rather variable with a given metal-weakness, and like Abt (1986), took to giving a G-band or CH index for the metal-weak spectra. However, the behavior of these G-band strengths is rather different for the early-G than for the F-type stars; the former show more scatter with respect to their metal deficiency. This difference in behavior may be because we are dealing with slightly cooler stars whose iron lines are more prominent in the G band. However, the difference does not allow a smooth sequence between the F- and the G-type stars.

You will recall that Gray (1994) mentioned his own extension of the MK System to the metal-weak, F-type, dwarf stars, in which extension he set up a grid of metal-weak standards. (His Table Ia (Gray 1989) is reproduced here in Table 1.) This extension is limited to F-stars, but it is most promising, for the

Table 1. Gray's (1989) primary metal-weak standard stars

	Metallicity Class		
T.T.	0(MK)	-1	-2
F0	V: HD 23585		
F2	V: HD 113139		V: HD 112934
F3	V: HD 26015	V: HD 9378	
F5	V: HD 27524		V: HD 32065
F6	V: HD 30652		V: HD 172914
			IV: HD 24424
F8	V: HD 27808	V: HD 91324	V: HD 188815
F9	V: HD 27383	V: HD 223400	V: HD 25704
			III: HD 17072
G0	V: HD 14802		V: HD 63077
G1	V: HD 27836		V: HD 199288
G2	V: Jupiter IV		V: HD 56274

two problems outlined above disappear. First, morphologies now match cleanly with a single extension-standard for both the strong and the weak lines in the metal-weak spectra. Second, Gray found that "in the vast majority of stars in the observational sample, the behavior of the G band was remarkably consistent with the assigned temperature and metallicity classes and that, indeed, the strength and morphology of the G band constituted very sensitive indicators of the metallicity and temperature classes of the star." I should expect that remarkable consistency to continue for the metal-weak G-type dwarfs. So, this genuine extension of the MK System by application of the MK Process is clearly the way to proceed for the weak-lined stars. All it takes is more careful work in extending the parallel grids of metal-weak standards and then in looking to the metal-rich spectra.

3. Extensions to other wavelength regions

3.1. Into the ultraviolet

The third direction in which the MK Process of classification is going is into wavelength regions other than the Ca II K to Hβ range. I start with the UV, but with no prejudice against the IR region. Initial classification work was done on low-dispersion IUE spectra by several workers: Heck, Jaschek, Jaschek, Cucchiaro, Egret (see Jaschek and Jaschek 1984). Since *The MK Process* book appeared ten years ago, we have been given two major atlases of ultraviolet high-dispersion spectra in classification sequence, that of Walborn (Walborn *et al.*, 1985) for the O stars and of Rountree (Rountree and Sonneborn 1993) for the B stars. (The historically inclined will be intrigued by the exchange between these two on standards and notation that is recorded in *Spectral Classification of the Future*, (McCarthy *et al.*, 1979).) Both atlases are featured in papers of this workshop (Rountree and Sonneborn 1994; Walborn *et al.* 1994) and are in print, so I can afford to make general comments only. While they both apply the MK Process to the "specimens" of UV spectra of similar resolution, they are interesting in taking a different approach to what constitutes those specimens. Walborn and collaborators include the wind lines as well as the photospheric lines as they order the spectra into sequences (and they continue this procedure for the B type spectra in a new atlas). Rountree and Sonneborn deliberately exclude the wind lines, even though these are the strongest lines in their spectra. These latter authors thought that "wind-line strength" might become a third parameter, apart from temperature and luminosity, in their spectral sequencing. Interestingly, the wind-lines for B stars turn out to be considerably "standardized", but always there are the exceptions, which generally proved to be the Be or shell stars in the optical region.

Gray and I have a proposal on record to classify the A-, F-, and G-type stars in the UV, and the onset of strong chromospheric activity in these types may make the use of only photospheric lines essential for their classification. The point here is not *blindly* to use every spectral feature when establishing a new extended MK classification system but to isolate those features which might best lead to useful third or further dimensions in the classification.

A further point comes from the early impression of some workers with UV spectra that the MK and the UV spectral classes only correlated roughly. As the classification became refined through Walborn and Rountree, so the correlation was found to be generally rather close. That kind of refinement is essential to bring out true differences between a star's spectrum in the UV and in the MK region, and so to allow these differences to point towards genuinely interesting phenomena in a star's atmosphere as various layers are considered. With this "confrontation" in mind I admit to personal interest in the as yet preliminary work by Faragianna and Gerbaldi (Faraggiana *et al.*, 1990) on the UV criteria and classification of λ Bootis stars. For instance, I shall look to see whether the 1600 Å dip turns out to be a resonance effect. It is also intriguing that some horizontal branch stars show similar features.

3.2. Into the infrared

The MK Process as applied to the IR will be well illustrated by the two papers following this and I need just mention them.

The work of Boeshaar and Davidge (1994) shows the power of the 2 μm IR region to investigate the very cool stars, seemingly with the Holy Grail of brown dwarfs to inspire them. One certainly feels the excitement among those exploring the spectral morphology in this wavelength region. I shall look forward to hearing these authors compare and contrast what they find with what appears at slightly shorter wavelengths, e.g., the 0.6-1.5 μm region classifications of Kirkpatrick *et al.* (1993).

The paper by Torres-Dodgen (1994), with its accompanying poster display, is based on her atlas of spectra in the $\lambda\lambda$5800-8900 Å region. Interestingly, this atlas shows that it is a prejudice, at least in the near IR, to think that this region is only useful for classification studies of the very cool stars. At least as early as the B stars, we shall be able to confront classifications in the near-IR with those in the MK region and in the UV region.

4. The two philosophies

I could continue by describing extensions which use a different spectral resolution than the traditional 2 Å of the MK (e.g., Meyers-Rice 1994) and the classification of special classes of objects, such as white dwarfs (Liebert and Sion 1994), but this would prolong the presentation and besides these are well treated elsewhere in this workshop. In finishing, I want you to be clear how, in these examples of extensions to the MK System, we have seen two philosophies of the MK Process in action. The approach of Keenan and Gray in the traditional MK blue-green spectral region has been to add additional dimensions closely tied to the present MK System. In the UV and (Near)IR, however, people have made independent systems, ones parallel to the MK System and constructed along the lines of Morgan's "MK Process". There should be no conflict between these approaches for, as Gray wrote in 1989, "Clearly, both philosophies are valid."

How are they *both* valid?

Philosophy #1. Keenan (1987), working in the traditional MK region, found that "the great majority of the so-called 'peculiar' stars fall into groups which are not sharply bounded but actually shade into the normal stars of the original MK System." Gray, too, found that the metal-weak population graded continuously into the normal Population I stars and so that the third dimension of metallicity class established grids of standards that were *very parallel* to the

MK System. Both came to their common approach of adding dimensions to the MK System, an approach that was dictated by the spectral specimens.

Philosophy #2. The success, however, of classification in regions beyond the traditional MK region (i.e. UV and [Near]IR) comes when these extensions, at least in the first instance, are kept completely autonomous from the MK System, i.e., the specimens are allowed to find their own order. This allows the (possibly different) classification from each wavelength region to reflect the different information about the layers in a stellar atmosphere, which may be only weakly coupled. When the classifications are compared, then the insight can begin. We have seen the same potential realized when the "classifications" from photometry are compared with MK classifications. This reminds me to add the caution that, just as for photometric classifications, so for the classification done by extensions to the MK System, we must find and use notation that clearly expresses the system on which the classification has been made. Such clear distinguishing notation will then invite fruitful confrontation between the systems.

I find it exciting that the current work on extensions to the MK System, whether by additional dimensions or by independent systems, ensures that there will be plenty of opportunity for future astrophysical insight as the new systems produced by the MK Process are more fully explored. The prospects are excellent for stellar classification: all we need are the workers and the telescope time. The former are represented here at this workshop; the latter can come through such projects as the Vatican Advanced Technology Telescope, whose dedication we have just celebrated.

References

Abt, H. A. 1986, ApJ, **309**, 260

Bidelman, W. P., and MacConnell, D. J. 1973, AJ, **78**, 687

Boeshaar, P. C., and Davidge, T. J. 1994, this volume

Bond, H. E. 1970, ApJS, **22**, 117

Corbally, C. J. 1987, AJ, **94**, 161

Faragianna, R., Gerbaldi, M., and Böhm, C. 1990, A&A, **235**, 311

Garrison, R. F., and Corbally, C. J. 1993, AJ, **106**, 2301

Gray, R. O. 1989, AJ, **98**, 1049

Gray, R. O. 1994,, this volume

Houk, N., and Cowley, A. P. 1975, *University of Michigan Catalogue of Two-Dimensional Spectral Types for the HD Stars*, Vol. 1 (Department of Astronomy, University of Michigan, Ann Arbor)

Jaschek, M., and Jaschek, C. 1984, in *The MK Process and Stellar Classification*, ed. R. F. Garrison (David Dunlap Observatory, Toronto), p. 290

Keenan, P. C. 1984, in *The MK Process and Stellar Classification*, ed. R. F. Garrison (David Dunlap Observatory, Toronto), p. 29

Keenan, P. C. 1987, PASP, **99**, 713

Kirkpatrick, J. D., Kelly, D. M., Rieke, G. H., Liebert, J., Allard, F., and Wehrse, R. 1993, ApJ, **402**, 643

Liebert, J., and Sion, E. M. 1994, this volume

McCarthy, M. F., Philip, A. G. D., and Coyne, G. V., 1979, eds., *Spectral Classification of the Future*, Vatican Observatory Publ., p.346

Meyers-Rice, B. A. 1994, this volume

Morgan, W. W. 1984 in *The MK Process and Stellar Classification*, ed. R. F. Garrison (David Dunlap Observatory, Toronto), p. 18

Roman, N. G. 1954, AJ, **59**, 307

Rountree, J., and Sonneborn, G. 1993, *Spectral Classification with the International Ultraviolet Explorer: An Atlas of B-Type Spectra*, NASA RP 1312

Rountree, J., and Sonneborn, G. 1994, this volume

Torres-Dodgen, A. V. 1994, this volume

Walborn, N. R., Nichols-Bohlin, J., and Panek, R. J. 1985, *International Ultraviolet Explorer Atlas of O-Type Spectra from 1200 to 1900 Å*, NASA RP 1155

Walborn, N. R., Parker, J. Wm., and Nichols, J. S. 1994, this volume

Discussion

McCarthy: Why in your discussion of SGP 106 did you use a K star rather than a G star when intercomparing it with the F star?

Corbally: Because the absolute strength of the hydrogen lines best matched a K0 star. I would have preferred to use the Fe/Cr ratios for the spectral type determination, but these lines were far too weak to use.

Parthasarathy: The classification of metal weak stars that you described, does it include metal weak giants and CN and CH stars?

Corbally: That is correct. However, since we were working under the "Hertzsprung Gap" in the HR diagram, we had few giant stars – so as not to make it difficult for ourselves.

Keenan: An alternative approach to weak-line stars is to use higher resolution. Then one can employ the same line ratios as in normal stars to estimate luminosity and type. Then, absolute line strengths are needed only in assigning metal strengths. But that is a luxury that you cannot afford in working on faint stars!

Lloyd Evans: I am worried by the procedure of estimating temperature class of metal weak stars by absolute strengths of hydrogen lines when the continuum is blanketed to different degrees by metal lines. Are your estimates backed up by independent (e.g., spectrophotometric) temperature criteria?

Corbally: In my 1987 study, I compared my spectroscopic "measure of weakness" against [Fe/H] calculated from Strömgren photometry for 37 weak-line stars. The linear relationship with the expected scatter that I found gave me reassurance that, despite continuum blanketing effects, I was correctly determining metallicity, though with an effort. The good match of my temperature classifications with those predicted by Erik Olsen from the Strömgren photometry also confirmed their temperatures.

Gray: Yes, Lloyd Evans brings up an important point. For, in my work on the classification of the metal-weak F-type stars, I used the morphology of the metallic-line spectrum as the primary temperature criterion, especially pairs and patterns of lines involving both high excitation and low excitation lines. I used the hydrogen lines only with much lower weight in the temperature classification. There are a number of reasons for this, one of which is the fact that the wings of the hydrogen lines are very severely blended with metallic lines in the metal-strong spectra, and so their appearance is quite different in the metal-weak stars. It *is* very important to look first to morphology.

Houk: I worry that introducing these complexities (i.e., new systems) for various types of stars may discourage people from classifying and especially from doing surveys at all. It is difficult enough to get people to take enough standards on the original MK system without setting up whole parallel grids. I'd say most people should use the original MK system until it becomes clear that it is not sufficient; then bring such stars to the attention of the "experts" on that particular type of star.

Corbally: I agree, but *partly* only. For it is by much experience that you yourself find classifying the weak-line stars with respect to the MK system easy enough. The less experienced will find it more straightforward to set the weak-line star into its real context of a parallel grid of weak-line standards. However, yes, the needs of surveys may not warrant the sophistication of additional parallel standards. This can be left to the time of *precision* classification.

Roman: There is no problem with the very weak-line stars such as the subdwarfs. You see immediately that they are weak-line. However, the marginally weak-line stars present a problem because the Hγ, Fe, G-band ratios are used for spectral type. It is difficult photographically to see small differences in H-line strength. If it is the hydrogen lines which define the temperature, marginally weak-line stars will be classified too early systematically.

Corbally: Both for photographic and digital spectra of F- and G-type stars, I find generally that I cannot distinguish a line-weakening of less than 2 subtypes in difference between the hydrogen and metal-line strengths. So, yes, the stars marginally weak by one subclass will be classified slightly early.

Olsen: I would like to add a comment to Roman's remarks. Ten years ago, as I mentioned Tuesday afternoon, Hans Nielsen in the MK Process volume compared *uvby*-β photometry with Nancy Houk's catalogues. For most of the F stars with [Fe/H] > -0.6, Nancy Houk does *not* detect the slight metallicity weakening. Nielsen showed that, in averages over about 170 stars, there is a difference of nearly one subtype between the temperature type *predicted* by *uvby*-β photometry and the temperature type actually assigned to the stars by Nancy Houk. This is neatly illustrating the effect just described by Nancy Roman.

Garrison: The data set used by Corbally and me for the halo G dwarfs was a difficult one to obtain. We went down to 16th mag and that meant we had to compromise on signal-to-noise, so we worked at S/N of 35-50 – not ideal. We tried very hard to get a resolution of a few angstroms so we could use the Fe/Cr lines at 4250-60 Å as abundance-independent temperature criteria.

When classifying weak-line stars, it is important to have enough resolution to be able to use abundance-independent features. If not, then you can't hope to be able to discriminate small differences.

Rountree: The history of spectral classification in the ultraviolet actually goes back before IUE. In the 1970's, ultraviolet spectra from the OAO-2 and OAO-3 (Copernicus) satellites attracted classifiers like Bidelman, Savage and Code. But these workers had two problems – first, they immediately set on the Sc IV and C IV lines which, as you point out, are the most prominent features in the UV spectrum – so they were really classifying the stellar wind. Secondly, they really had no spectra available at the appropriate resolution for MK classifications, which is several times 10^3. I believe that the classification of digital spectra in general – not only in the UV – has become possible partly because we now have this computing power to resample the spectra to this desired resolution.

Walborn: 1. I'd like to mention two important current wavelength extensions in progress for the OB stars. One concerns the 900- 1200 Å regions, not accessible to IUE or HST, which contains numerous features of interest including the very high ionization O VI resonance lines. Copernicus observed a few bright objects in this region, which have been discussed by Snow, Morton, and Jenkins, but which R. Bohlin and I are currently preparing for display in a classification atlas format. Just during the past few days, the first O3 and Magellanic Cloud stars should have been obtained by Kudritzki with ORFEUS. I have also proposed for a larger sample of the same for the second flight of the Hopkins Ultraviolet Telescope on Astro-2 late next year. Secondly, Conti is initiating a survey of the 2 μm region, which contains He I, He II, and a few metallic lines, and promises the possibility of classifying very young embedded objects not observable at shorter wavelengths.

2. The relationship of UV wind features to optical spectral types among the OB stars is a question to be investigated. That was done for the O stars, with the result that remarkably detailed correlations exist for the great majority of the stars, when known interacting binaries and rapid rotators are excluded. This result has major astrophysical implications for the origin of the winds, implying they are governed by the fundamental stellar parameters, consistent with the dominance of the radiation field in the O stars. That is no longer true in the lower-luminosity B stars, as shown in the Rountree-Sonneborn Atlas. We are now investigating the systematics of the high-luminosity B-star wind features to complete the OB group, which has not been done before at the information level of IUE high-resolution data. Preliminary indications, as shown in our poster here, are that the good correlations found for the O stars also apply to the B stars of high luminosity.

The MK Process at 50 Years
ASP Conference Series, Vol. 60, 1994
C.J. Corbally, R.O. Gray, and R.F. Garrison (eds.)

The Spectroscopic Properties of M dwarfs in the Two Micron Region

P. C. Boeshaar

Drew University, Madison, NJ 07940

T. J. Davidge[1]

Dominion Astrophysical Observatory, Victoria, BC Canada V8X 4M6

Abstract. Conspicuous changes in the strengths of both atomic and molecular features are readily apparent in the 1.5 - 2 μm spectra of M dwarfs. These differences provide a clearer distinction between the latest M-dwarf subtypes than the spectra shortward of 1 μm and may offer clues in how to distinguish brown dwarf candidates spectroscopically.

1. INTRODUCTION

Unlike the situation for other main sequence stars, the MK system has never been clearly defined for the M dwarfs. In the original papers delineating standard stars by Morgan and Keenan (1943, 1973 and references therein), there are a noticeable dearth of M dwarfs. In particular, the two stars which have been used to anchor the endpoints of the system, HD95735 (M2V) and Barnard's Star (M5V), are in themselves spectroscopically peculiar. The main reason for these difficulties lies in the fact that original MK classification was done in the photographic blue spectral region, whereas M dwarfs have spectral peaks closer to 1 μm and are intrinsically faint. The situation is further complicated since field M dwarfs represent a mixture of kinematic populations, hence metallicities, and there are no clusters close enough so that the lower main sequence can be studied with sufficient resolution spectroscopically in the blue.

Perhaps the most extensive data set for M-dwarf blue region spectral classifications is that of the Mt. Wilson system as compiled by Joy and Abt (1974), though the quality of the data is far from consistent or optimal. Morgan (1938) and Kuiper (1938, 1942) were the first to attempt to extend the M-dwarf classification to the red by utilizing the strength of the $\gamma'(0,0)$ band of TiO; however, they were unaware that this feature is contaminated by CaOH at types later than M3. As a result, their spectral types are systematically too late when compared to those obtained from other TiO bands. Recent reclassifications by Keenan and McNeil (1976, 1989) include mostly early M dwarfs; however they do include a few later types. These are based on the classification scheme of Boeshaar (1976) which analyzes the relative behavior of visual region TiO, VO,

[1]Gemini Canada Project Office, and University of British Columbia

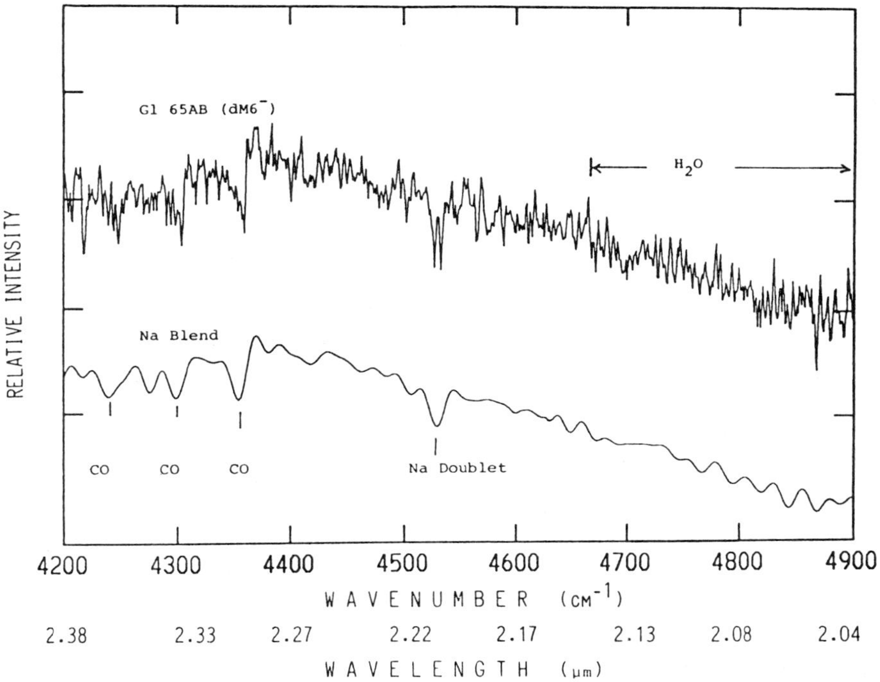

Figure 1. (A) K-band spectrum of the late M-dwarf binary Gl 65AB (UV Ceti plus companion) at a resolution of $\approx$ 1 cm^{-1}. (B) Spectrum of GL 65AB at a resolution of $\approx$ 10 cm^{-1}.

CaOH, CaH, and MgH bands in addition to several atomic features. Utilizing image-intensified photographic spectra, these later types terminate at type M6.5, but were extended to type dM8 by Turnshek *et al* (1985) using a single-array red-sensitive Reticon.

The development of sensitive infrared detectors has allowed spectroscopic observations of the intrinsically faintest M dwarfs and the introduction of several different classification schemes (Bessell 1991; Kirkpatrick *et al* 1991) based primarily upon the strength of TiO and VO bands near their peak flux. Accurate classification in this wavelength regime has been difficult for the coolest M Dwarfs as the molecular oxide bands increase in strength only slightly with decreasing temperature. Partly for this reason, Kirkpatrick *et al* (1993) have extended their wavelength coverage to 1.5 μm in order to incorporate several other relatively deep, sharply delineated spectral features, such as the FeH absorption at 9910 Å, into their classification.

This renewed attempt to understand the atmospheric parameters of the intrinsically faintest stars in our galaxy has been spurred by two recent events: vastly improved model atmosphere codes (Allard, 1990) for the M dwarfs and the non-discovery of any bona fide brown dwarfs to date. As a consequence there exists great interest in determining just how the coolest disk M dwarfs may differ spectroscopically from brown dwarfs, if indeed they do. The wavelength region longward of 1.6 μm is the logical place to obtain spectra of suspected brown dwarf candidates since this region contains many features sensitive to small changes in temperature among the coolest M dwarfs as well as the slight changes in surface gravity that are theoretically predicted to distinguish M dwarfs from brown dwarfs.

2. OBSERVATIONS

As part of a program to examine the behavior of the strong infrared molecular features in suspected brown-dwarf candidates, intermediate-resolution spectra (mostly at $\approx$ 10 cm^{-1} or $\approx$ 40 Å) were obtained for a sample of M1-M8 dwarfs, M subdwarfs, and M4-M9 giants. The data were obtained with the Canada-France-Hawaii Fourier-Transform Spectrometer in the 1.5 to 2.3 μm wavelength interval at a S/N of 50–100. This resolution has been found sufficiently good to allow ratios of individual molecular bands and atomic line blends to be studied empirically, even though some of the intrinsically faintest stars were observed at $\approx$ 20 cm^{-1} or $\approx$ 80 Å resolution. The 2 μm spectra in Figure 1 provide an indication of how well atomic features remain identifiable as the resolution is reduced by a factor of 5-10 from approximately 4–8 Å to 40–50 Å. A complete description of these observations and the data reduction techniques has been described previously by Davidge and Boeshaar (1991, 1993).

3. DISCUSSION

Recent high-resolution observations have established that many features in the 2 μm spectra of cool stars are sensitive to temperature and surface gravity. Hall and Kleinmann (1986) have made detailed comparisons between the 2 μm spectra of high-luminosity M stars and dwarfs between types M0 and M6.5. The most noticeable differences between luminosity classes II and V lies in the behavior of the CO and H_2O absorption bands, the latter of which is also insensitive to changes in [Fe/H]. The first overtone bands of CO continue to strengthen with decreasing temperature in all M stars, with the absorption much stronger in more luminous stars. Changes in the behavior of the second overtone CO bands located in the H-band filter are much less dramatic in the M dwarfs. In contrast, the H_2O absorption is stronger in both the H and K band passes in less luminous M stars, continuing to strengthen to type dM8. The relative strengths of the H_2O/CO bands coupled with the increasing equivalent width of the Na doublet blends can provide an indication of small changes in log g, as noted in the spectrum of the marginal subdwarf, Gl 754.1b, in Figure 2.

Conspicuous differences between the dwarfs themselves due to decreasing temperature are noted in the strengths of the CO and H_2O bands in addition to the behavior of several line blends such as the Na I doublets at 4280 cm^{-1} and 4530 cm^{-1} plus Ca I triplet at 4420 cm^{-1}. This behavior is illustrated in the dwarf temperature sequence in Figure 2. For the latest dM7-9 dwarfs, the increasing absorption of the first overtone CO bands is in marked contrast to the much more subtle differences that are seen shortward of 8500 Å as described by Bessell (1991). We have found that the strength of the first overtone CO

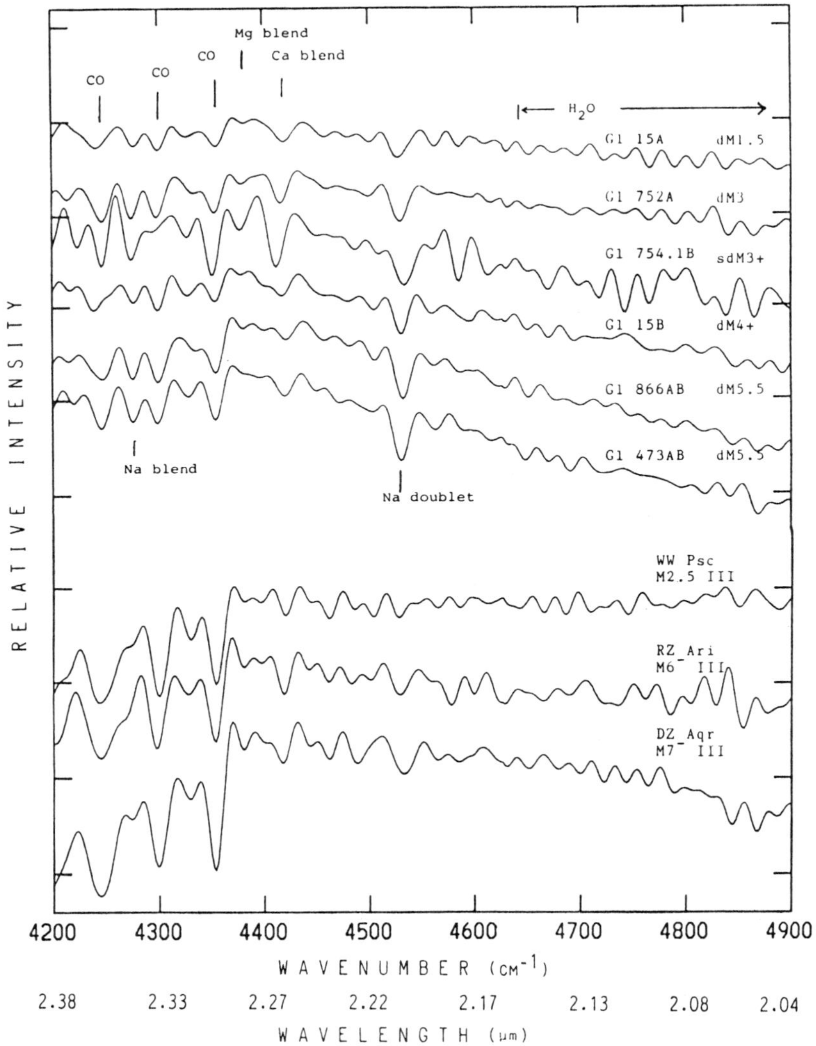

Figure 2. K-band spectra of M stars at a resolution of $\approx 10\ cm^{-1}$ arranged by luminosity class in order of decreasing temperature. Noticeable differences between the spectra of M dwarfs due to small changes in surface gravity/opacity can be seen by examining Gl 754.1b (marginal sdM3+).

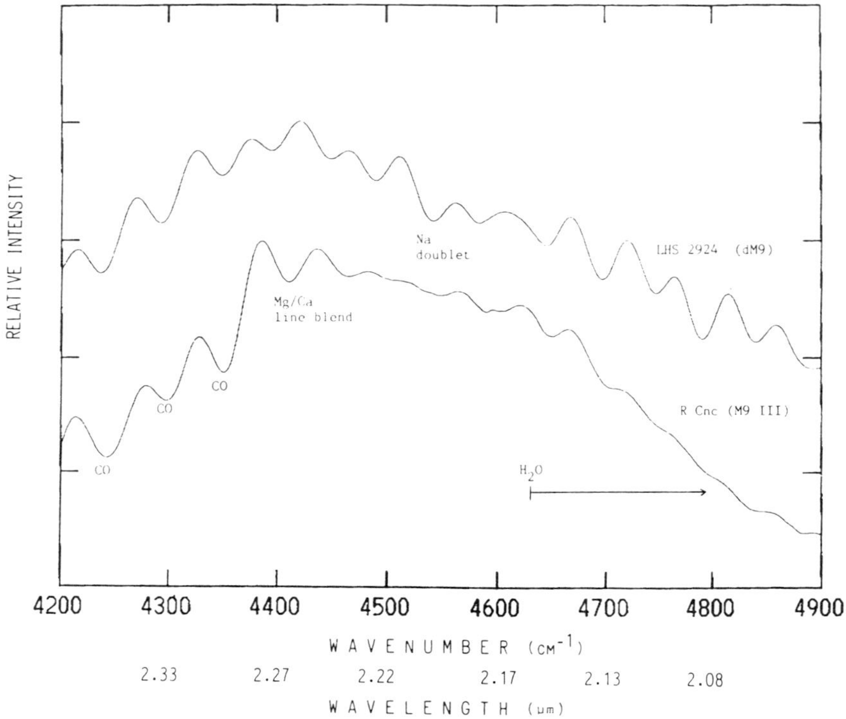

Figure 3. K-band spectra at a resolution of $\approx$ 20 cm^{-1} of an LPV M9 III and the M9 dwarf LHS 2924.

bands in M dwarfs is correlated with the spectral type derived shortward of 1 μm by Kirkpatrick *et al* (1991) utilizing the TiO and VO band strengths in the 0.69 - 0.9 μm and with I-K color. This does not appear to be the case for H_2O absorption which appears weaker in one of the latest M dwarfs, LHS 2924, a dM9. Furthermore, a number of strong unidentified features are present in the spectra of LHS 2924 and GL569B (dM8.5) which do not appear in the spectrum of the M9 giant R Cnc, as can been seen in Figure 3. For a detailed discussion of these differences see Davidge and Boeshaar (1993).

The overall appearance of the fluxed 1-2.4 μm spectra of several dM8-9 stars presented by Tinney *et al* (1993) and Jones (1993) seems to support the observation that noticeable differences exist between the infrared spectra of the latest M dwarfs with similar 0.6 - 0.9 μm spectral types, further suggesting the presence of unidentified molecules. We have found that some of the molecular transitions which occur in the 2 μm region may provide important constraints on the temperature scale for M dwarfs as well as be indicative of small luminosity differences between M dwarfs and their subluminous counterparts; however, the presence of unidentified molecular features complicates the interpretation at the present time.

References

Allard, F. 1990, *Ph.D. thesis*, University of Heidelberg

Bessell, M. S. 1991, AJ, **101**, 662

Boeshaar, P.C. 1976, *Ph.D. thesis*, Ohio State University

Davidge, T. J. and Boeshaar, P. C. 1991, AJ, **102**, 267

Davidge, T. J. and Boeshaar, P. C. 1993, ApJ, **403**, 47

Hall, D. N. B. and Kleinmann S. G. 1986, ApJS, **62**, 501

Jones, H., Longmore, A., Jameson, R. and Mountain, M. 1993, *Eighth Cambridge Cool Star Conference*, PASP Conference Series, in press

Joy, A. H. and Abt, H. A. 1974, ApJS, **28**, 1

Keenan, P. C. and McNeil, R. C. 1976, *An Atlas of Stellar Spectra of the Cooler Stars: Types G,K,M,S, & C* (Columbus: Ohio State University Press)

Keenan, P. C. and McNeil, R. C. 1989, ApJS, **71**, 245

Kirkpatrick, J. D., Henry T. J. and McCarthy, D. W. 1991, ApJS, **77**, 417

Kirkpatrick, J. D., Kelly, D. M., Rieke, G. R., Liebert, J., Allard F., and Wehrse, R. 1993, ApJ, **402**, 643

Kuiper, G. P. 1938, ApJ, **87**, 592

Kuiper, G. P. 1942 ApJ, **95**, 201

Morgan, W. W. 1938, ApJ, **87**, 589

Morgan, W. W.. Keenan, P. C., and Kellman, E. 1943, *An Atlas of Stellar Spectra* (Chicago; University of Chicago Press)

Morgan, W. W. and Keenan, P. C. 1973, ARA&A, **11**, 29

Tinney, C. G., Mould, J. R., and Reid, I. N. 1993, AJ, **105**, 1045

Turnshek, D. E., Turnshek, D. A., Craine, E. R., and Boeshaar, P. C. 1985, *An Atlas of the Digital Spectra of Cool Stars* (Tucson: Western Research Co.)

Discussion

Liebert: First a comment: There is now a real temperature scale for M dwarfs based on model atmosphere fits to spectrophotometry, rather than a blackbody fit. This is in the Kirkpatrick *et al.* 1992 paper. And the Te, L values *actually agree* with the ZAMS predicted by interior models.

My question concerns the fact that Mike Bessell and Bob Wing favor a different spectral classification system for M dwarfs than yours (ours). Since this is the topic of the conference, would you explain the differences and why we favor ours?

Boeshaar: An excellent discussion of the differences between the M-dwarf classification system used by myself and Kirkpatrick vs. that used by Wing and Bessell can be found in Appendix A of Kirkpatrick *et al.* (1991). Basically, we agree very well with Wing's types to dM6. The disagreement occurs primarily for the later types. Wing's narrow band photometric system calibrated the depths of the 7000-10,000 Å molecular bands using M III standard stars of Keenan. These standards themselves had been classified originally in the blue/visual spectral region. Bessell has adopted Wing's system and assigns his spectroscopic temperature classes using the absolute depths of the red/near infrared TiO and VO bands relative to a blackbody curve drawn through the highest "continuum" points in this spectral region, as can be clearly seen in Allard's 1990 model atmospheres, making absolute band strengths unreliable. As Morgan and Keenan pointed out in their 1973 *Annual Reviews* paper on spectral classification, using many temperature sensitive molecular band and atomic line ratios is a better way to classify these stars. This is exactly what Kirkpatrick and I have attempted to do. One final point worth mentioning is that field M dwarfs encompass a broad range of populations, hence of metallicity. Any variations in the abundance will also affect the absolute TiO and VO band strengths.

Craine: Given that we see many types of variability (brightness and color) in M dwarfs do you have enough data to comment on stability of these 2 μm spectra in detail? Does this issue pose a quantifiable problem in establishing MK type standard anchors?

Boeshaar: Many M dwarfs do indeed have large spots and are active flare stars. My impression is that the increased flux due to flare activity primarily affects the continuum shortward of 1 μm. At least in the case of cataclysmic variables, the dM companion is often easily classified by identifying the molecular oxide bands longward of 7000 Å. At present, I see the recognizable differences between the 2 μm spectra of M dwarfs with the same spectral type when classified in the 5000-9000 Å region as posing a greater problem in establishing a good set of temperature standards.

Warren: I noticed in one of your series of dwarf M spectra that the VO band is stronger at M7 and M9 than it is at M8 and M8.5. As you mentioned using the VO band for classification, what do you do when your sequence in the IR does not correspond to the original MK sequence?

Boeshaar: The problem is that no original MK standard stars later than about M5 V were obtained in the photographic blue region and the VO bands appear in the visual and near infrared spectral regions. The 5736 Å VO band has been used by Keenan to classify cool M giants but the latest M dwarfs were just too faint to obtain spectra photographically. I have used the 5736 Å and 7400 Å bands of VO to classify dwarfs to type dM8 (VB10), while the spectra of Kirkpatrick *et al.* (1991) that you refer to, contain additional infrared VO features which allowed them to classify to spectral type dM9+. The TiO bands appear to saturate while the depth of the VO bands in red/near IR appear to change only marginally with decreasing temperature in the coolest M dwarfs. As a result, one must scrutinize the behavior of all molecular bands present carefully since there are no real continuum points for comparison. Quick glances at one band may be deceiving.

The MK Process at 50 Years
ASP Conference Series, Vol. 60, 1994
C.J. Corbally, R.O. Gray, and R.F. Garrison (eds.)

Near-Infrared Spectra of A-Type Stars

Ana V. Torres-Dodgen

Monterey Institute for Research in Astronomy, 900 Major Sherman Lane, Monterey, CA 93940

Abstract. As part of a program to develop a near-infrared (NIR) classification system for all stars, we show 15.5 Å resolution spectra of A0 to A9 stars, luminosity classes V to I, in the wavelength range 5750-8950 Å. The stars follow well defined sequences in temperature and luminosity, but cannot be classified as well as in the blue part of the spectrum. We confirm that luminosity classes I and II are easily distinguished from each other, and from III, IV and V stars. Solely on the basis of line ratios and line strengths, luminosity classes III, IV and V are indistinguishable. Temperature classes can be determined to ±1.5-2 MK types.

1. Introduction

Taking advantage of the high sensitivity of modern detectors in the red part of the spectrum, Torres-Dodgen and Weaver (1993, hereafter TW) present NIR spectra of O to M stars, luminosity classes V, III and I, and show that "normal stars" follow well defined morphological sequences in temperature and luminosity at low resolution. An example of their work is shown in Figure 1 for G2 to K6 giant stars. References prior to 1993 are cited in TW.

Other advantages of using the NIR to classify stars are that cool stars emit most of their light in this wavelength range, and that heavily reddened OB and supergiant stars are more easily detected than in the blue. Some applications of a NIR classification system are: 1) Star counts, e.g. to better determine the ratio of blue to red supergiants, or to find more binary systems consisting of a blue and a red star with very different visual magnitudes; 2) stellar population synthesis; 3) initial mass function and luminosity function; 4) stellar variability; and 5) automatic classification.

2. Observations

In order to see how well the A-type stars can be classified in the NIR, 43 A-type stars were observed, covering most MK types. All of them are either MK standards or are well classified in the MK system (Gray and Garrison 1987, 1989a, 1989b).

The observations were taken with the MIRA 36-inch telescope, a spectrograph and an unintensified 512 pixel Reticon. The wavelength range is ~5750-8950 Å and the resolution 15.5 Å. The usual corrections were applied for dark

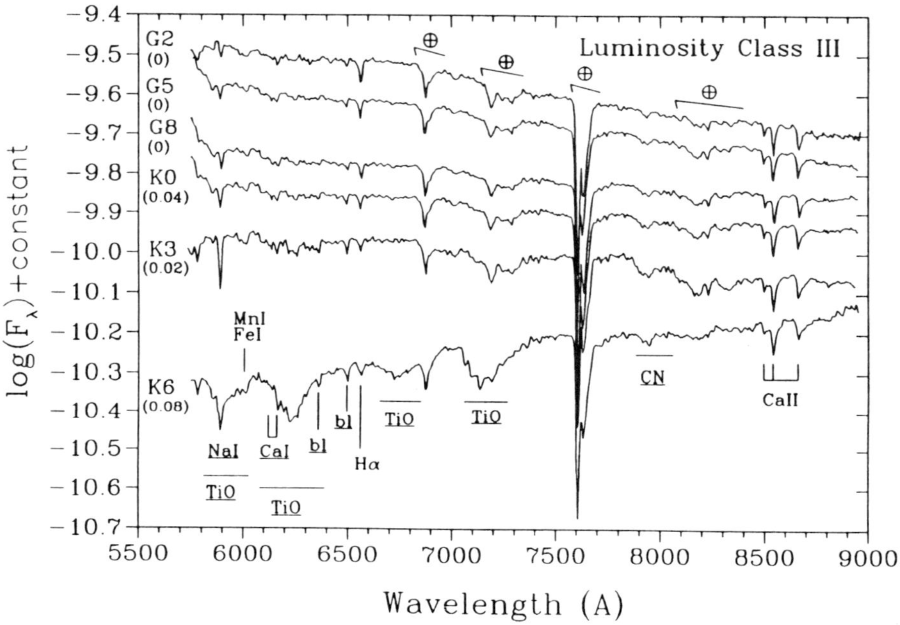

Figure 1.
A sample of G2 to K6 giant stars (from TW) showing the spectral changes with temperature (underlined features). The spectra have been displaced vertically by an arbitrary constant. Color excesses are given in parentheses under the spectral type. Telluric lines are marked with ⊕, and metallic blends are indicated with "bl".

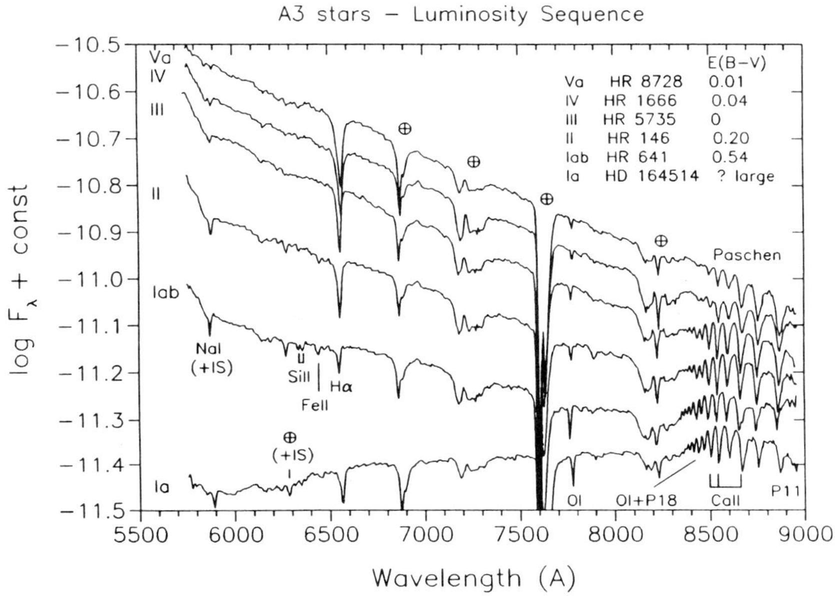

Figure 2. Sequence of A3 stars at different luminosity classes. The main spectral features are indicated, including telluric (⊕) and interstellar (IS) lines. The A3 III star looks like a luminosity class II in the NIR.

noise, pixel-to-pixel variations, atmospheric extinction and flux calibration. The error in the fluxes is less than 15%, except in the Paschen area, where they can be worse. A detailed explanation of the observations and the errors can be found in TW.

3. The Spectra

Examples of the spectra are shown in Figures 2 and 3 for A3 and A7 stars at various luminosities. The main stellar features and the telluric and interstellar bands are marked. The Si II lines $\lambda\lambda$6347, 6371 are conspicuous in supergiants. The strength of Hα decreases in II and I stars; it can be variable and even absent in Ia stars (Figure 6a). As is well known, the O I lines $\lambda\lambda$7774 and 8446 (the latter blended with P18) increase in strength in II and I stars; they appear to be equally strong in III, IV and V stars (Figure 6b). The Paschen jump starts to appear at luminosity II, and is very pronounced in the supergiants. The Paschen lines are narrower in stars of luminosities II and I: P22 can be easily resolved, while only lines up to P19 can be resolved in V, IV and III stars. The $\lambda\lambda$8498, 8542, 8662 lines of the Ca II triplet are blended with P16, P15, and P13, respectively.

Two examples of the temperature sequence are shown in Figures 4 and 5 for A III and A Ia stars, respectively. The Fe I λ6495 line is absent in the early

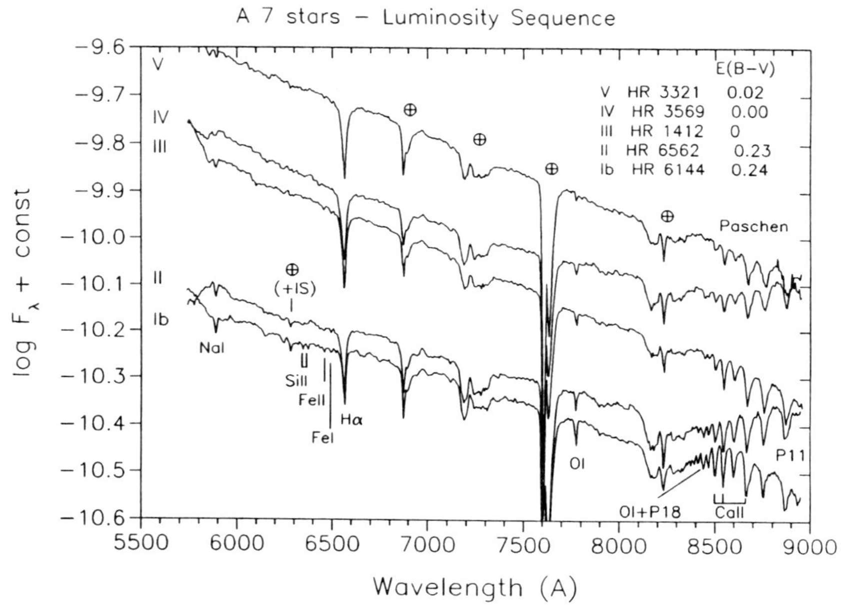

Figure 3. Same as Figure 2 but for A7 stars. Continuum fluxes in the Paschen area are not well calibrated.

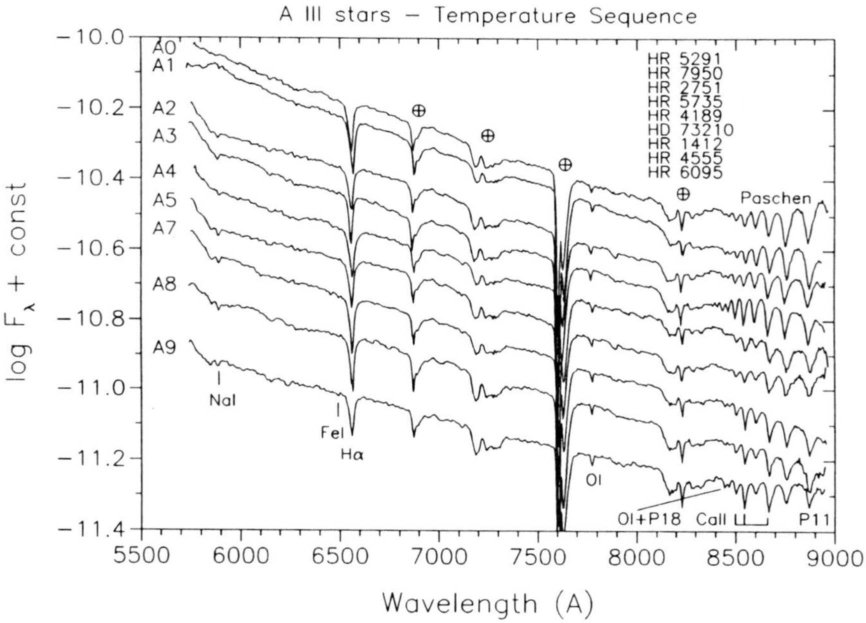

Figure 4. Temperature sequence for A giant stars. The A3 III star HR 5735 appears to be of luminosity class II in the NIR. All stars have $E(B-V) \leq 0.05$.

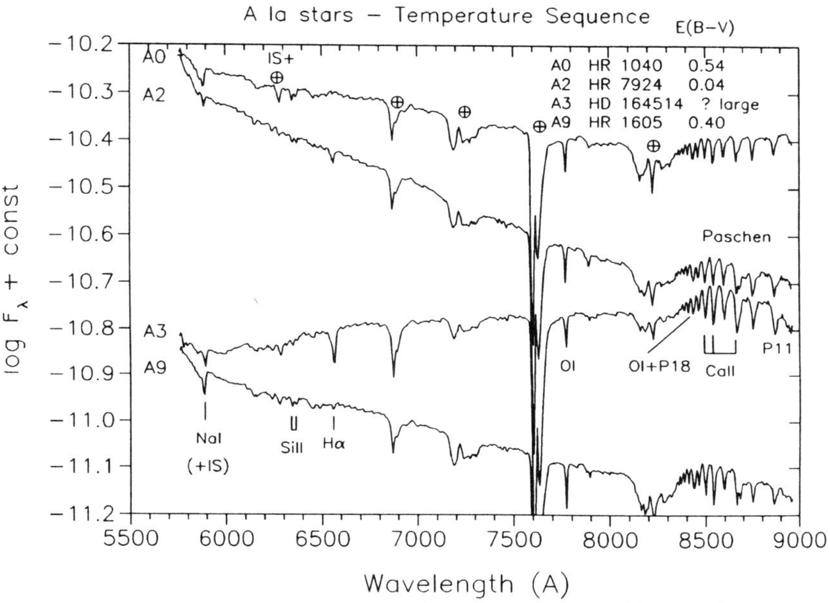

Figure 5. Temperature sequence for A supergiants. Notice the strong Paschen jump. Changes in the continuum slopes are due to interstellar reddening.

types, appears around types A3-A4, and increases strength with later types. This is analogous to the behavior of the metallic lines and blends in the blue part of the spectrum, which are used as a criterion for temperature classification in the MK system. The Paschen lines dominate over the Ca II triplet in the early types, whereas the opposite is true in the later types. Figure 7a shows this behavior clearly; it continues to hold true for F-type stars (Mantegazza 1991). In addition, the P13 line in supergiants has another blend in the long wavelength side which further increases its strength (Figure 5). Therefore, supergiants are also separated from other luminosities in Figure 7a. The strength of P12 is also a useful temperature classification criterion, as can be seen in Figure 7b.

The A3 III star HR 5735 (Figures 2 and 4) appears to be of luminosity class II in our three spectra, which were taken in 1993 April and June. *The Bright Star Catalogue* (Hoffleit and Jaschek 1982) mentions that this is a shell star with variable hydrogen and Ca II, but it appears normal in Gray and Garrison's (1989b) spectra. A "II" has been placed next to it in Figures 6 and 7, although the symbol for an A III star (triangle) has been used in the plots.

HR 8518 is a spectrophotometric standard. Slettebak (1954) classified it as an A0 V star, Parenago (1958) discovered it was a λ Bootis star, and Gray and Garrison (1987) reclassified it as a B9.5 III star. It is plotted in Figures 6 and 7 under A0 V stars and is marked with a "B". The line strengths clearly indicate that this star does not belong with the A-type stars.

Most stars were observed more than once. The symbols in Figures 6 and 7 are the average of all observations for a given star, and the error bars are the

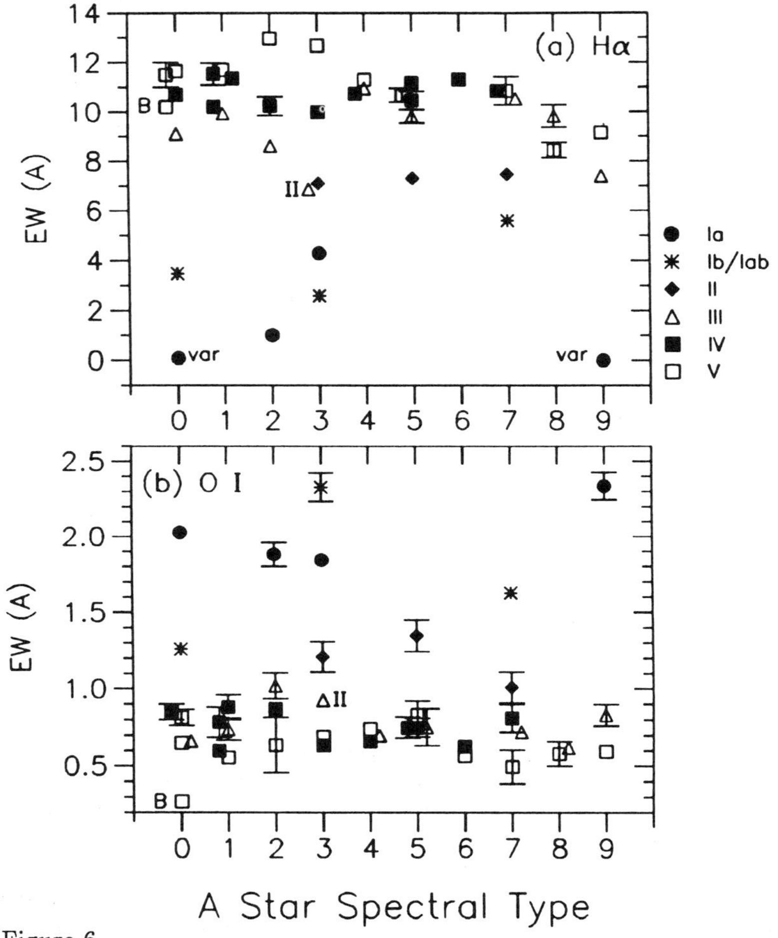

Figure 6.
Equivalent widths in Å for the program A-stars. (a) Hα, (b) O I λ7774. Error bars are the standard deviation of the mean for multiple observations of the same star. The "B" indicates the B9.5 III star HR 8518 (see text), and the "II" points to the A3 III star that looks like an A3 II in our spectra. Luminosity classes I and II are easily separated from III, IV and V in these plots.

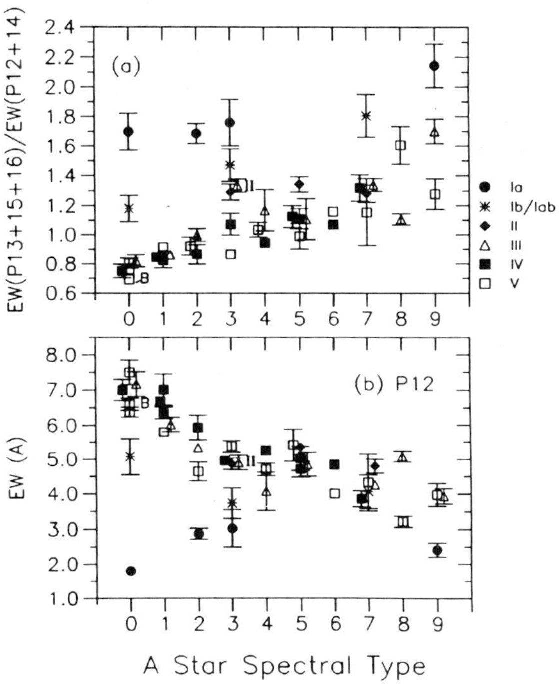

Figure 7.
(a) Ratio of the equivalent widths of the three Paschen lines blended with calcium and two unblended Pachen lines. (b) Equivalent width (in Å) of the Paschen 12 λ8750 line. The symbols are the same as in Figure 6. Supergiants are separated from other luminosity classes. Temperature types can be derived from these two graphs to an accuracy of ±1.5-2.

standard deviations of the mean. The effect of interstellar reddening in the NIR can be seen easily in Figures 2 and 5.

4. Conclusions

1. A-type stars are more easily distinguished in the MK system (2 Å resolution) than in the NIR at 15 Å resolution. Nonetheless, the NIR region at low resolution is very helpful in determining approximate types of faint, highly reddened A-type stars.
2. We confirm that luminosities I and II are easily distinguished from each other, and from luminosities III, IV, and V.
3. There is no single line or line ratio in the NIR that distinguishes luminosities III, IV, V. *But* for A5 and later types, the Fe I λ6495 line may be helpful for this purpose.
4. Spectral types (i.e. temperature) can be determined to an accuracy of ±1.5-2 MK types using the P12 line strength and the ratio of equivalent widths

(P13+P15+P16)/(P12+P14). The lines in the numerator are blended with the Ca II lines.
5. Using the observations presented here, statistical and neural network multivariate analysis can classify A-stars to about 0.5 luminosity classes and about 1.0 temperature class (Weaver 1994).

Acknowledgments. This work was supported by NSF grant no. AST-8914577.

References

Gray, R. O., and Garrison, R. F. 1987, ApJS, **65**, 581

Gray, R. O., and Garrison, R. F. 1989a, ApJS, **69**, 301

Gray, R. O., and Garrison, R. F. 1987, ApJS, **70**, 623

Hoffleit, D., and Jaschek, C. 1982, *The Bright Star Catalogue*, (New Haven, Yale University Observatory)

Mantegazza, L. 1991, A&AS, **88**, 255

Parenago, P. P. 1958, Soviet Astron. AJ, **2**, 151

Slettebak, A. 1954, ApJ, **119**, 146

Torres-Dodgen, A. V., and Weaver, Wm. B. (TW) 1993, PASP, **105**, 693

Weaver, Wm. B. 1994, this volume

Discussion

Gray: Are you planning to apply this system to higher resolution and a smaller spectral region? Perhaps this would help to improve the discrimination especially among the V-III stars.

Torres-Dodgen: Not in the near future. I agree it would help, especially in the region between 5800 Å and Hα.

Garrison: If you use poor resolution so that you are not able to discriminate types of the same level as with traditional MK classifications, why go to the near-IR for B stars, when one can go fainter with better resolution in the blue? I understand why in the case of M stars, but don't see any advantages for the OB-type stars – at least not enough to justify all the work.

Torres-Dodgen: A near-IR classification will probably not work well for O-stars. The advantage for B and A stars is that you can observe highly reddened stars ($E(B\text{-}V) > 1.0$) easier in the near-IR than in the blue and still get a good idea of their spectral type.

MacConnell: I am pleased to see that your atlas was recently published in the *PASP*. If you want to start applying it to classification of program stars with implications for galactic structure, I suggest you observe the heavily-reddened, non-banded stars in Bruce Stephenson's recent list in the *AJ*. These are good candidates for distant A and F supergiants. These stars are very faint in the "classical" classification region, due to the interstellar reddening, so observing them in the near-IR is the most efficient way to classify them.

The MK Process at 50 Years
ASP Conference Series, Vol. 60, 1994
C.J. Corbally, R.O. Gray, and R.F. Garrison (eds.)

Post-Asymptotic-Giant-Branch Supergiants

M. Parthasarathy

Indian Institute of Astrophysics, Bangalore 560034, India

Abstract. From an analysis of IRAS PSC data a new class of stars was detected. These stars have circumstellar dust shells with far infrared colors similar to planetary nebulae and most of them show A,F,G and K supergiant-like spectra in the optical region. On the MK classification-dispersion spectra some of these stars have spectral characteristics similar to those of normal supergiants. However they show very unusual characteristics at other wavelengths. The MK classification-dispersion spectra of metal-poor post-AGB supergiants indicate their peculiar chemical composition. The observed characteristics of the spectra of these stars in the UV, optical, infrared and millimeter wave region, their high galactic latitude and/or high velocity, and their chemical composition suggest that these stars have evolved from low-mass stars. These results indicate that low-mass stars during their evolution from the tip of the AGB to the left (post-AGB stage) in the HR diagram can mimic the spectra of supergiants. The sequence of post-AGB supergiants was found to extend from K type to B type, which indicates the evolutionary sequence of these objects to be from the tip of the AGB towards the left in the HR diagram into the region of planetary nebulae.

Some of the post-AGB A and F supergiants are very metal-poor. However the abundances of C,N,O,S and Zn are not reduced by the same factor. The depletion of refractory elements and nearly normal abundances of volatile elements in the photospheres of post-AGB stars are similar to those observed in the λ Boo stars. The processes that produce photospheres with depleted refractory elements in λ Boo stars and post-AGB A-F supergiants may be the same. Formation of the dust close to the star (during the pre-main-sequence phase in the case of λ Boo stars, and during AGB or post-AGB mass loss phase in the case of post-AGB A-F supergiants) and subsequent gas and dust separation and dust driven mass loss (driving out mostly the dust) may have taken place, resulting in the depletion of refractory elements in the photospheres of these stars.

1. Introduction

From an analysis of the IRAS point source catalogue a new class of stars was detected. These stars have circumstellar dust shells with far-infrared colors and flux distribution and the dust-shell parameters similar to the dust shells of planetary nebulae. Most of these stars show A,F,G, and K supergiant-like spectra in the optical region (Parthasarathy and Pottasch 1986; Lamers *et al.*

1986). Parthasarathy and Pottasch (1986) suggested that the dust shells around these stars are the result of severe mass loss during their AGB stage of evolution and that these stars are evolved, low-mass stars and are now in the post-AGB stage of evolution.

Recent studies in the optical, infrared and millimeter wave region indicate that the following types of stars are most likely post-AGB supergiants: (i) High-galactic latitude supergiants, (ii) IRAS sources with far-IR colors similar to planetary nebulae and identified with B,A,F,G,K supergiants, (iii) RV Tauri and UU Her stars, (iv) A-F supergiants with high Strömgren c_1 indices, (v) very metal-poor supergiants, (vi) UV-bright stars in globular clusters.

On the MK classification-dispersion spectra, some of the post-AGB supergiants appear nearly normal, some show underabundance of metals, some may show enhanced carbon features, and some show emission lines due to [NII], [SII]. Most of them show H-alpha emission. However most of the post-AGB supergiants show unusual characteristics at other wavelengths and seem to form a class of stars not associated with the massive young supergiants.

2. High-Galactic-Latitude Supergiants

The presence of F-type supergiants at high-galactic-latitude was first noticed by Bidelman (1951). The problem of how such luminous F supergiants, if young, could be so far from the galactic plane led Abt (1960) to study their chemical composition. The fact that some bright high-galactic-latitude supergiants have planetary-nebula colors in the far-infrared was first pointed out by Parthasarathy and Pottasch (1986). Some of the high-galactic-latitude F supergiants which were found to show far-infrared colors similar to planetary nebulae are: HD 161796 (F3Ib), IRAS 18095+2704 (F3Ib), HD 187885 (F3I), HD 101584 (F0Iape) etc. These stars show CO molecular envelopes, and some show OH-maser emission. The observed characteristics of these stars in different wavelength regions suggests that they are in a post-AGB stage of evolution (Parthasarathy 1993a).

Garrison and Lopez-Cruz (1993) reviewed the MK spectra of F-type supergiants at high galactic latitudes. Lopez-Cruz and Garrison (1993) have classified on the MK system 56 high-galactic-latitude supergiants and their list includes some post-AGB supergiants. Bidelman (1993) suggests that a possible fruitful way to distinguish post-AGB supergiants from normal high-luminosity stars is to look at the stars of high Strömgren c_1 indices, in view of the fact that many post-AGB stars are known to show high c_1 indices. Based on this criteria Bidelman (1993) has given a list of MK types and c_1 indices for several A and F supergiants which includes a considerable number of known and possible post-AGB stars.

3. Metal-Poor Post-AGB Supergiants

Several post-AGB supergiants were found to be metal-poor. Some of these stars are also high-galactic-latitude and/or high-velocity stars. Some of them are IRAS sources with far-infrared colors similar to planetary nebulae (examples: HD 56126 (F5I), HD 44179 (B8I)). However some of them show warm dust shells (examples: HR 4049 (A0Ibpec), HD 52961 (FI)). Morgan (1984) and Abt (1984) discussed the peculiar nature of HR 4049. Morgan (1984) remarked that "there is no MK box that can accept the spectrum of HR 4049 satisfactorily, the line weakening is pronounced." Abt and Morgan give the spectral type of HR 4049 to be A0Ib-IIp. Abt comments that relative to the spectrum of the standard star η Leo the strong lines of HeI, FeII and MgII are absent from the spectrum of HR 4049. Waelkens *et al.* (1991) find the presence of several carbon and oxygen lines usually not seen in the spectra of late B and early A supergiants. Waelkens *et al.* (1991) find the spectrum of HD 52961 to be similar to that of HR 4049. The MgII 4481 is not detectable at MK classification-dispersion, however there is an indication of the presence of a G band. Recently Abt (1993) found HD 105262 (A0pIb) (galactic latitude = 72.5°) to show a spectrum similar to that of HR 4049. Abt (1984) has included HR 4049 in the list of λ Boo stars. We know now of several metal-poor, post-AGB stars that have photospheric abundance patterns similar to that of λ Boo stars (i.e. depletion of refractory elements).

4. IRAS Sources with Far-IR Colors Similar to PN

In recent years several IRAS sources with far-infrared colors similar to planetary nebulae have been identified with A,F,G,K supergiant-like stars (Pottasch and Parthasarathy 1988; Hrivnak *et al.* 1988, 1989; van der Veen *et al.*1989; Slijkhuis 1992; Garcia-Lario *et al.* 1990; Manchado *et al.* 1989; Parthasarathy 1993a). Some of these stars are fainter than 10th magnitude and are not in the HD catalogue. MK classification-dispersion spectra and high-resolution spectra have been obtained by several above mentioned authors and are classified as A,F,G,K supergiants. The presence of circumstellar dust shells with far-infrared colors similar to planetary nebulae, and CO molecular envelopes with characteristics similar to that observed in evolved stars suggests that these are post-AGB supergiants.

5. The Characteristics of Post-AGB Supergiants

The main observed characteristics of post-AGB supergiants are: (i) Most of them are far-infrared (IRAS) sources with far-infrared colors similar to planetary nebulae. (ii) Their spatial distribution is quite far from the galactic plane – possibly indicative of an older stellar population. (iii) These stars are not associated with star forming regions, OB associations and star clusters. (iv) Their spectral types range from K1 to B1 with many F–G supergiants, which is the temperature range expected as the low and intermediate mass stars evolve from the AGB to PN. (v) These stars are undergoing or have very recently undergone a very strong mass-loss episode deduced from the presence of strong far-IR emission indicating recently ejected dust, H-alpha emission indicating mass loss, and CO millimeter-wave emission indicating high mass-loss rates. (vi) Several post-AGB supergiants show unusual 3 μm and 21 μm emission features. (vii) Some post-AGB supergiants show nebular emission lines of [NII] and [SII]. (viii) Several post-AGB supergiants show small-amplitude-light and radial-velocity variations of relatively long period. (ix) Some of these stars are high-galactic-latitude and/or high-velocity stars and some are metal poor.

6. The Chemical Composition

The high galactic latitudes, high velocities, low metallicities, circumstellar dust shells and molecular envelopes indicate that these are evolved low-mass stars and can become supergiant-like during their post-AGB evolutionary stage. From an analysis of high-resolution spectra, low gravities were also derived for most of them (Luck *et al.*1990). The C,N,O abundance ratios also indicate that these are evolved low-mass stars (Luck *et al.* 1990; Parthasarathy *et al.* 1992). However, explaining the spectra and abundances of extremely metal-poor, post-AGB supergiants such as HR 4049, HD 52961, HD 44179, etc. (Waelkens *et al.* 1991; 1992) is still a problem. MK classification-dispersion spectra indicate the peculiar chemical composition of these stars. Abt (1984) has put HR 4049 in the λ Boo type. The λ Boo stars and metal-poor, post-AGB supergiants have similar abundance patterns (i.e depletion of refractory elements Fe, Ca, Mg, Si, Ti, Al, etc., and normal abundance of volatile elements C, N, O, S, Zn). The photospheric abundances in these stars are similar to the gas-phase abundances of the interstellar medium. The presence of circumstellar dust shells lends strong support to the idea that the depleted refractory elements are locked up in circumstellar dust grains (Bond 1991; Van Winckel *et al.* 1992; Parthasarathy *et al.* 1992). In the metal-poor, post-AGB supergiant HD 52961 Van Winckel *et al.* (1992) find that Zn is not depleted. In the gas phase abundances of the interstellar medium the element Zn is not depleted. These observations support the idea that the depleted refractory elements are locked up in the circumstellar dust grains.

The processes that produce photospheres with depleted refractory elements in λ Boo stars and post-AGB A-F supergiants may be the same. Mass loss, circumstellar dust shells, cold outflows and extended atmospheres are understood to be common during the pre-main-sequence stage and also during the AGB and post-AGB stages of evolution. The formation of dust close to the star (during the pre-main-sequence stage in the case of λ Boo stars, and during AGB and/or post-AGB mass loss phase in the case of post-AGB A-F supergiants) can deplete the refractory elements from the surface. The subsequent gas and dust separation and dust driven mass loss, driving out mostly the dust can explain the photospheric abundances of these stars.

7. Hot Post-AGB Stars

Recent work indicates that some of the high-galactic-latitude, low-gravity, blue stars may be actually low-mass, hot, post-AGB stars (McCausland *et al.* 1992; Parthasarathy 1993b), e.g., LB 3193, LB 3116, LS IV -12 111, LS II + 34 26, etc. MK classification-dispersion spectra show that these are B-type supergiants (Kilkenny and Lydon 1986; Turner and Drilling 1984). LS IV -12 111 and LS II +34 26 are IRAS sources with far-infrared colors simliar to planetary nebulae. LS IV -12 111 is a high-galactic-latitude and high-velocity star. The presence of a circumstellar dust shell with far-IR colors similar to planetary nebulae and the presence of nebular emission lines in the MK classification-dispersion spectrum and abundances (McCausland *et al.* 1992) suggests that it is a hot post-AGB star.

Turner and Drilling (1984) classified the spectrum of LS II +34 26. They find the MK spectral type of LS II 34 26 to be B1.5Ia-Iabe. They have concluded that it is a massive, Population-I B supergiant located at a distance of 17.8 kpc. Parthasarathy (1993b) found it to be an IRAS source (IRAS 20462+3416 = LS II +34 26) with far-infrared colors similar to planetary nebulae. Parthasarathy (1993b) suggested that LS II +34 26 is a low-mass post-AGB supergiant and not a massive, Population-I B star located near the outer edge of the Galaxy. The radial veocity of LS II +34 26 (V_{lsr} = -75 km per sec) indicates that it is a high-velocity star.

8. Luminosities

The MK–M_v luminosity calibrations (Corbally and Garrison 1984) yield high luminosities (M_v in the range -6 to -8) for most of the post-AGB supergiants. For example for LS II +34 26 Turner and Drilling (1984) find M_v = -6.35 which means M_{bol} = -8.0 and the distance d= 17.8 kpc. Similarly for LB 3193 and LB 3116 Kilkenny and Lydon (1986) find high luminosities and large distances (25kpc) if MK–M_v calibration is used. Similarly, for the high-galactic-latitude, A-F supergiants we find large distances above the galactic plane if we use MK–M_v calibrations. Maybe we should not apply the MK–M_v calibration for the post-AGB supergiants. Maybe they are not supergiants in the real sense; their atmospheres are temporarily in an extended state and they are mimicking the spectra of normal supergiants. Maybe because of rapid and severe mass loss on the AGB they are left with a low-mass carbon and oxygen core with an extended outer envelope.

9. The Evolution

The MK classification in fact gives the position of the star in the HR diagram. The sequence of post-AGB supergiants was found to extend from K1 to B1 which indicates the evolution sequence of these stars from the tip of the AGB towards the left in the HR diagram into the region of young planetary nebulae (Parthasarathy 1993b). The duration of the transition phase (from the tip of the AGB to young planetary nebula stage) is short as is evident from the variations in the spectrum of SAO 244567 (Hen 1357) (Parthasarathy *et al.* 1993). SAO 244567 (CPD -59 6926 = Hen 1357 = IRAS 17119-5926) was classified as a B or A H-alpha emission star (Henize 1976, Wackerling 1970). It is found to be an IRAS source with far-infrared colors similar to planetary nebulae (Parthasarathy and Pottasch 1989). Hill *et al.* (1974) and Kilkenny and Hill (1975) give a spectral type of B3e and find V=10.75, $B-V$=-0.02 and $U-B$=-0.89. The spectrum obtained by Kilkenny in 1971 and the *UBV* colors suggest that SAO 244567 was a B3 giant or supergiant in 1971. However the IUE ultraviolet spectra obtained recently and the recent optical spectrum show remarkable variations and show that SAO 244567 has evolved into a planetary nebula within the last 20 years (Parthasarathy *et al.* 1993).

Recently Parthasarathy, Garcia-Lario and de Cordoba analyzed the high resolution sectrum of LS II +34 26 (B1.5Ia-Iabe) and found nebular emission lines of [NII] and [SII]. The HeI lines show P-Cygni profiles indicating mass loss.

The SiII lines were also found in emission. The spectrum obtained by Turner and Drilling (1984) did not show nebular emission lines, and the HeI and SiII lines were in absorption. These results indicate that the spectrum of LS II +34 26 has changed within the last 10 years. Parthasarathy (1993b) found LS II +34 26 to be an IRAS source with far-infrared colors similar to planetary nebulae and interpreted that LS II +34 26 is a high velocity post-AGB supergiant. The observed recent spectral characteristics of LS II +34 26 suggests that it is rapidly evolving into a planetary nebula similar to SAO 244567.

10. Conclusions

Several high-galactic-latitude and or high-velocity B,A,F,G supergiants were found to be IRAS sources with far-IR colors similar to planetary nebulae. Several IRAS sources with far-IR colors similar to planetary nebulae were found to have optical counterparts showing B,A,F,G,K supergiant-like spectra. The presence of warm or cold dust shells, CO molecular envelopes, OH-maser emission, unusual 3 μm and 21 μm emission, large mass loss rates derived from CO millimeter profiles, high galactic latitudes and/or high velocities, low metallicities, and C,N,O abundance ratios indicate that these stars have evolved from low mass stars. They experienced severe mass loss on the AGB and now in the post-AGB-stage evolution. The sequence of post-AGB supergiants is found to extend from K1 to B1 which suggests an evolutionary sequence of these stars towards the left in the HR diagram into the region of planetary nebulae. The observed spectral characteristics from the UV to the far-IR suggest that low-mass stars, after experiencing severe mass loss on the AGB and during their evolution from the tip of the AGB towards the left in the HR diagram, can mimic the spectra of supergiants. These stars have a white-dwarf-like, carbon-oxygen core and a very extended envelope. The evolution of these stars towards higher temperatures and also the post-AGB mass loss will cause the spectrum to vary from K,G,F,A supergiant-like spectrum to B supergiant-like spectrum to planetary nebula spectrum.

Acknowledgments. I am very much thankful to Dr. Chris Corbally for his kind encouragement and support which enabled me to participate in this workshop.

References

Abt, H. A., 1960, ApJ, **131**, 99

——, 1984, in *The MK Process*, edited by R. F. Garrison, p. 340

——, 1993, PASP, in press

Bidelman W. P., 1951, ApJ, **113**, 304

——, 1993, in *Luminous High Latitude Stars*, ASP Conference Series 45, edited by D. D.Sasselov

Bond, H. E., 1991, IAU Symposium 145, 341

Corbally, C. J., and Garrison R. F., 1984, *The MK Process*, ed. R. F. Garrison, p. 277

Garcia-Lario, P., Manchado, A., Pottasch, S.R., Suso, J., Olling, R., 1990, A&AS, **82**, 497

Garrison, R. F., Lopez-Cruz, 1993, in *Luminous High Latitude Stars*, ASP Conf. Ser., 45, ed. D. D Sasselov

Henize, K. G., 1976, ApJS, **30**, 491

Hill, P.W., Kilkenny, D., van Breda, I.G., 1974, MNRAS, **168**, 451

Hrivnak, B. J., Kwok, S., Volk, K. M., 1988, ApJ, **331**, 832

———, 1989, ApJ, **346**, 265

Kilkenny, D., Hill, P. W., 1975, MNRAS, **172**, 649

Kilkenny, D., Lydon, J., 1986, MNRAS, **218**, 279

Lamers, H.J.G.L.M., Waters, L.B.F.M., Garmany, C.P., Perez, M.R., Waelkens, C., 1986, A&A, **154**, L20

Lopez-Cruz, Garrison, R. F., 1993, in *Luminous High Latitude Stars*, ASP Cofer-ence Series 45, ed. D. D. Sasselov

Luck, R. E., Bond, H. E., and Lambert, D. L., 1990, ApJ, **357**, 188

Manchado, A., Garcia-Lario, P., Pottasch, S.R., 1989, A&A, **218**, 67

McCausland, R.J.H., Conlon, E.S., Dufton, P.L., Keenan, F.P., 1992, ApJ, **394**, 298

Morgan, W. W., 1984, in *The MK Process*, ed. R. F. Garrison, p. 18

Parthasarathy M., 1993a in *Luminous High Latitude Stars*, ASP Conf. Ser., 45, ed. D. D.Sasselov, p. 173

———, 1993b, ApJ, **414**, L109

Parthasarathy, M., Garcia-Lario, P., Pottasch, S.R., Manchado, A., Clavel, J., de Martino, D., Van de Steene, G.C.M., 1993, A&A, **267**, L19

Parthasarathy, M., Garcia-Lario, P., and Pottasch, S.R., 1992, A&A, **264**, 159

Parthasarathy, M. and Pottasch, S. R., 1986, A&A, **154**, L16

———, 1989, A&A, **225**, 521

Pottasch, S. R. and Parthasarathy, M., 1988, A&A, **192**, 182

Slijkhuis, S., 1992, Thesis, University of Amsterdam

Turner, D. G. and Drilling, J. S. 1984, PASP, **96**,292

Van der Veen, W.E.C.J., Habing, H.J., Geballe, T.R., 1989, A&A, **226**, 108

Van Winckel, H., Mathis, J.S., Waelkens, C., 1992, Nature, **356**, 500

Wackerling, L. R., 1970, MNRAS, **73**, 153

Waelkens, C., Van Winckel, H., Bogaert, E., Trams, N.R., 1991, A&A, **251**, 495

Waelkens, C., Van Winckel, H., Trams, N.R., Waters, L.B.F.M., 1992, A&A, **256**, L15

Discussion

Walborn: The grain-depletion hypothesis for metal-deficient post-AGB stars was first suggested by Howard Bond and recently confirmed by Henry Lamers and others through the presence of strong lines of zinc, which is undepleted in the interstellar medium. The identification of the latter in the post-AGB spectra had initially been rejected due to the known metal deficiency, which provides a nice example of why descriptions of the observations should not be constrained by presumed interpretations.

What would be the physical origin of an analogous phenomenon in the λ Boo stars?

Parthasarathy: Yes, Howard Bond was the first one to suggest that in metal-poor, post-AGB stars the depleted refractory elements are locked up in circumstellar dust grains. The evidence for normal abundance of zinc in metal-poor, post-AGB stars was first found by van Winckel, Mathis, and Waelkens (*Nature*, **356**, 500, 1992). You will find an account on chemical composition of post-AGB stars in the printed version of my paper. Lack of time did not permit me to go into details in my talk.

The answer to your question is: maybe the processes that produce photospheres with depleted refractory elements in λ Boo stars and post-AGB A-F supergiants may be the same. Mass loss, cold outflows and extended atmospheres are understood to be common during the pre-main-sequence stage of evolution and also during the AGB and post-AGB stage of evolution. The formation of dust close to the star during the pre-main-sequence stage of λ Boo stars and during the AGB and/or post-AGB A-F supergiants can deplete the refractory elements from the surface. The subsequent gas and dust separation and dust-driven mass loss (driving out mostly the dust) can explain the photospheric abundance of λ Boo stars and post-AGB A-F supergiants.

Abt: A good confirmation of your ideas is the newly-discovered star called HD 105262 that has a A0 Ib p spectrum like HR 4049. But it has a significant proper motion (0.06 ″/yr). If it had the luminosity of a Ib supergiant, the space motion would be three times the escape velocity from the galaxy. So it must be a low-mass giant with an extended atmosphere that mimics a supergiant.

Parthasarathy: Yes, these stars have supergiant-like spectra. However, they may not be supergiants in the real sense. Maybe we should not use the MK luminosity class and M_v calibration to estimate absolute luminosities and distances of these stars.

Garrison: I have enough objections about your presentation of the data that I am not at all convinced that all high-latitude supergiants are post-AGB.

My main objection, in general, is that you have used selective data and have included many different kinds of objects in the same interpretation, using extreme cases to infer an interpretation for stars that appear normal. In some cases, the *only* characteristic is the IRAS data showing circumstellar dust. However, that is not sufficient, because if the star is massive, there are several possible mass-loss phases – e.g., it could have been a very massive Of or WR star.

It seems to me that you are throwing away some possibly interesting interpretations by assuming that, because these stars are at high latitude, they therefore cannot be normal supergiants, and by only looking at results which support that interpretation. I have specific objections, but there isn't time to elaborate. I will only say that 89 Her and HD 161796 remain as anomalies and should not be lumped together with stars like HR 4049 and many others for which there is much better evidence of their post-AGB nature.

The detailed analyses for these two stars in particular show only very marginal results, which depend too much on the particular choices for temperature, $\log g$, microturbulence, etc. Your interpretation is based mainly on their locations. If they were located in the plane, they would not be called "post-AGB." I'm not saying that these aren't post-AGB; I am saying that we must keep an open mind until we have more convincing evidence. As yet I'm not convinced by any of your arguments for these two stars.

Parthasarathy: I am not saying that all the high-galactic-latitude supergiants are post-AGB stars. I considered only those high-galactic-latitude supergiants which have some of the following characteristics: (1) the far-infrared color similar to planetary nebulae; (2) expanding molecular envelopes similar to evolved stars as revealed by CO millimeter-wave observations; (3) OH-maser emission; (4) extremely metal-poor, supergiant-like spectrum; (5) C/N/O abundance ratios; (6) high velocity; (7) unusual 3 μm and 21 μm emission features. We have found several of the above-mentioned characteristics in several high-galactic-latitude supergiants. We have found about 50 supergiant-like stars brighter than 10th magnitude with some of the above-mentioned characteristics indicating that they are evolved low-mass stars.

The evolution in the transition region (i.e., the region between the tip of the AGB and the region of young planetary nebulae in the HR diagram) is very rapid, and stars between 1 $M_\odot$ and 8 $M_\odot$ pass through this. Therefore one can find objects with different characteristics in this region.

I am not sure of the nature of 89 Her. I think HD 161796 is a post-AGB supergiant based on the observed characteristics of HD 161796 in the far-infrared, CO millimeter-wave region, OH-maser emission, C/N/O abundance ratios, etc.

The MK Process at 50 Years
ASP Conference Series, Vol. 60, 1994
C.J. Corbally, R.O. Gray, and R.F. Garrison (eds.)

A Calibrated System for Low Resolution Spectral Classification

B. A. Meyers-Rice and E. T. Young

Steward Observatory, University of Arizona, Tucson, AZ 85721

Abstract. We have spectra for 364 stars, spanning the wavelength range $\lambda\lambda$ 3690–6960 Å at approximately 10 Å resolution. These stars were selected from our multiband database of 4900 stars in the young galactic cluster NGC 2264. We describe the methods we used to estimate the MK spectral types of these stars and the accuracy of these determinations.

1. Introduction

The cluster NGC 2264 has long been known to contain a population of young stars. In 1954 Herbig discovered a number of stars with Hα emission which were later shown by Walker (1956) to be members of the pre-main-sequence population in the cluster. Walker's color-magnitude diagram demonstrated that all the cluster stars later than spectral class A0 lie above the main sequence, and he estimated the cluster's age to be approximately 5×10^6 years. More recent studies (Warner, Strom, and Strom 1977, Adams, Strom, and Strom 1983, and others) further established the youthful nature of the stellar population. Many of these stars exhibit large infrared excesses indicative of some form of interstellar medium. This material could be diffuse cluster gas and dust, or might be concentrated around the stars in disks or shells. Research efforts have also concentrated on the giant molecular cloud (GMC) behind the cluster (Blitz 1977, Margulis and Lada 1986, Margulis, Lada and Young 1989, and others). These projects identified many molecular outflows and pre-main-sequence IRAS sources, confirming the presence of active star formation in the region. This diverse assembly of pre-main-sequence phenomena makes N2264 and the associated GMC an important laboratory for understanding star formation and the nature of interactions between young stars and the natal interstellar medium.

In our own work, we have amassed V, R, I, J, H and K photometry for 4900 sources in the direction of NGC 2264. Of these sources, 1645 have unsaturated detections at V, R, and I, and 591 have complete coverage in all six bands. While many of these have $J-H$, $H-K$ excesses, photometry alone does not provide all the necessary information to determine unambiguously the nature of these objects. To understand these sources more fully we obtained optical spectra of 364 stars in our survey that had good photometric coverage, i.e. at least in four bands, and especially sources that exhibited significant infrared excesses. With these spectra in hand, we required a method of spectral classification—a method that could be applied to all the stars as consistently as possible, and that would expediently classify them to an accuracy of a few spectral subclasses. These data will help to answer many questions about the stars in our sample. For example, how many of the stars exhibit emission lines, and how many of these fall into the typical spectral types associated with T Tauri stars? Are the stars with infrared excesses pre-main-sequence objects, or just reddened background stars? How much extinction are the stars experiencing? How many of the stars are not cluster members?

2. Spectral Data

Our sample of 364 stars was culled from sources in our survey with complete or nearly complete photometric coverage. The stars were observed at the Steward Observatory 2.3m telescope, using the MX spectrometer designed and built by John Hill. The MX is a fiber-optic-coupled multiplex spectrometer capable of simultaneously observing thirty-two sources. Observations of this sort require accurate ($\sim$ 2 arcsec) coordinate information for the program stars, a requirement easily met by the stars in our survey. The first stars observed were the brightest (m_V=10–15), then we progressed to fainter stars (m_V=15—19). The raw spectra were extracted, dispersion-corrected to a linear wavelength scale, and finally sky-subtracted. The spectra span the region 3690–6960Å at a resolution of approximately 10Å, but because of decreasing efficiency of our instrument at short wavelengths and telluric O_2 bands, the majority of our spectra are usable in the spectral region 3880–6750Å. The signal to noise ratio ranges from extremely high for the brightest stars in the sample to zero for discarded spectra of the faintest stars. All spectra were divided by an approximate continuum fit to ease their comparison with our choice of spectral standards.

2.1. Reference Spectra

For our spectral standards, we used the library of stellar spectra published by Jacoby, Hunter, Christian 1984 (hereafter JHC). These spectra were smoothed to the spectral resolution of our MX data. For our program stars, the continua were fit and removed from these spectra. Figure 1 shows a few sample spectra from this library, trimmed in wavelength range to show important features.

3. Performing the Classifications

We wished to classify our stars quickly and reasonably accurately, at least to within 3 spectral subclasses. Since we had a large number of spectra it was a priority to develop a way to classify them as automatically as possible, since this would be relatively speedy and objective.

3.1. Seeking and Quantifying Useful Spectral Features

The main-sequence spectra in the JHC library were inspected for features which varied with spectral subclass. Some of these features were simple line ratios, such as Ca K/Hϵ. Others involved broader or more complicated spectral structures such as the G band of CH, the Mg "b" line-blend near 5167–5184Å, or MgH 5210Å (a 300Å wide feature prominent in late G–early M stars, enhanced

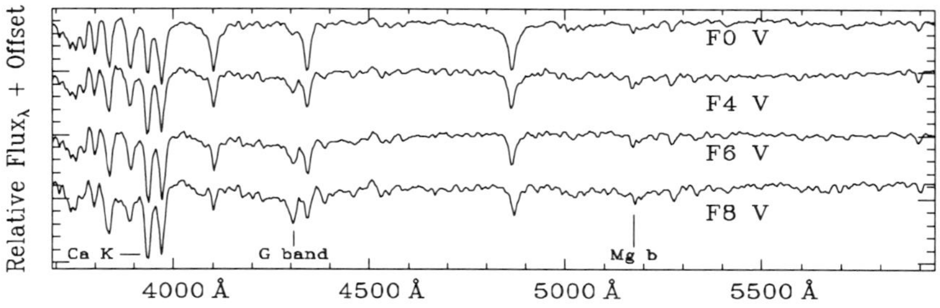

Figure 1. A temperature sequence of F stars selected from the JHC library, continuum divided and trimmed to the wavelength range 3690-5937Å. Three temperature-sensitive features are indicated.

in dwarfs). Other features were more artificial, such as a blend of lines near 4019–4084Å (for F–K stars). The temperature-sensitive features judged most useful were those that were independent of luminosity class. Table 1 shows the most important temperature-sensitive features used in this study. Some features which initially appeared promising were discarded because emission lines in many of our spectra limited their use. For example, the 4019–4084Å blend was only marginally useful because of luminosity sensitivity in Sr II 4077, and because components occasionally occurred in emission. Line ratios involving Hα were usually avoided since that line often appeared in emission, or with an emission reversal in the line core. Once a set of useful diagnostics was identified for a spectral class, a fit (almost always first order) of each feature's variation vs. subclass was computed. Figure 2 shows typical fits, in this case for the three temperature-sensitive features used in classifying F stars. Since our methods ultimately proved successful, we applied this technique not only to estimating spectral type but also to performing luminosity classifications, using features found to be luminosity dependent. Discriminants useful in making luminosity estimates are listed in table 1.

3.2. Using the Spectral Features to Classify Our Stars

We measured the relevant features in each of our spectra as we did with the stars in the JHC spectral library. Using the relations derived from the fits to the JHC measurements, we obtained a set of tentative spectral subclass and luminosity class estimates for each star. These estimates were averaged to produce the best estimate for each star's spectral subtype. In computing these averages, more weight was given to those features judged to be more reliable. As a final check, each spectrum was examined to see if the computed classification seemed reasonable. While it would be unrealistic to expect that all the spectral diagnostics would predict the same spectral type for a given star, the scatter in calculated spectral type was acceptably small (for examples, only ± 1 subtype for B stars, ± 1.5 subtype for F stars). Using as many features as possible to classify each

Table 1. Useful Spectral Features

Spectral Class	Temperature Indicators	Luminosity Indicators
O	He I 4471, 4686, 5412	
B	He I 4388, 4471, 5876	
A	Ca K	
F	Ca K, G band, Mg "b"[a]	Mg "b"[a]
G	G band, Mg "b", MgH 5210	Fe 6495, Mg "b", MgH 5210[a]
K	TiO bands, MgH 5210, 4150–4300 blend	MgH 4780, MgH 5210
M	MgH 5210, TiO, VO bands	MgH 4780, 5210, CaH 6385, Ca I 6103, 6122, 6162

[a]This feature was considered less reliable than others in this temperature/luminosity range.

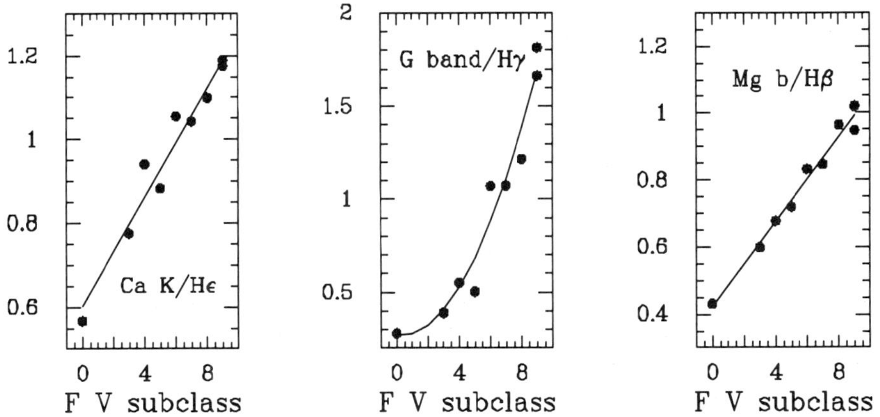

Figure 2. The behavior of three features—the G band of CH, the Ca K line, and the Mg "b" blend—are plotted as a function of spectral subclass. These fits were used in classifying our F spectra.

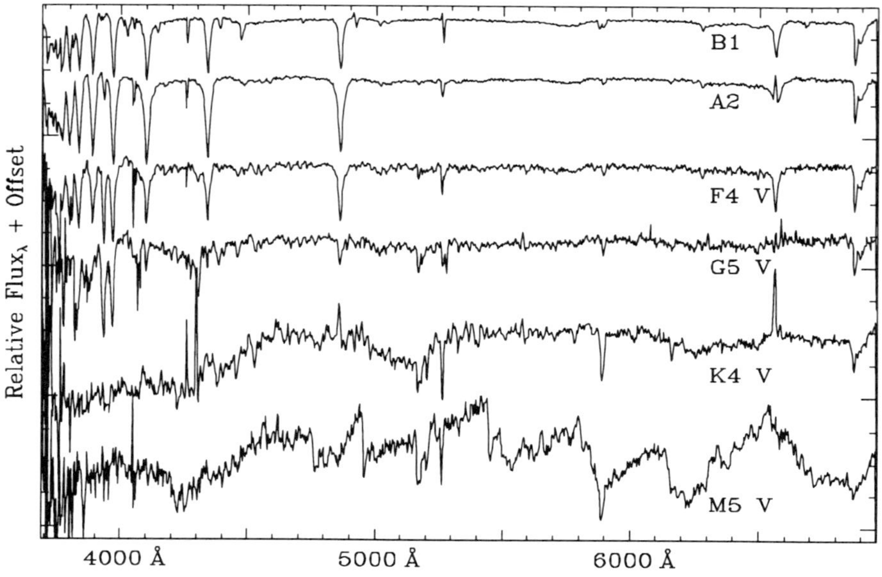

Figure 3. A few spectra from our survey. Notice the Hα emission in the A and K spectra. Cosmic rays have not been removed from these spectra.

star is essential, since the core to the philosophy of MK classification is to use all the features in a spectrogram.

4. Results

In general the methods we used successfully estimated the spectral types of our stars to the desired level of accuracy. Only occasionally would automatic estimates be discarded or modified, and these cases could always be traced to obvious spectral defects such as cosmic rays or bad pixels. Our method was completely automatic on the earlier spectral classes, but by late K and M it degraded to a procedure of manually looking for the presence of molecular absorption bands, particularly the TiO bands for the M stars. In figures 3–4 we show some sample spectra. Our survey consists of the following numbers of stars in each spectral type: 0 O stars, 14 B stars, 26 A stars, 32 F stars 36 G stars, 118 K stars, and 138 M stars. While we are most interested in the results for the lower mass stars, there are some surprises in the 26 A stars. Of these, at least five show strong Hα emission, and of the remaining stars the Hα/Hβ ratio is smaller than for JHC stars, suggesting Hα emission may be occurring.

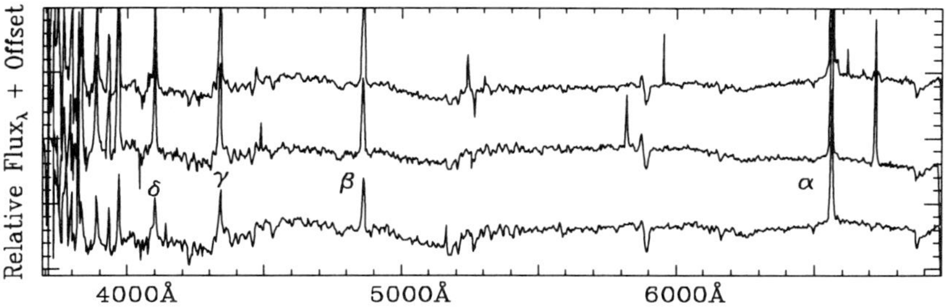

Figure 4. Three K4 V survey stars showing strong hydrogen and Ca K emission lines. We have detected emission lines in 179 of our 364 survey stars, most of which are mid-K T Tauri stars.

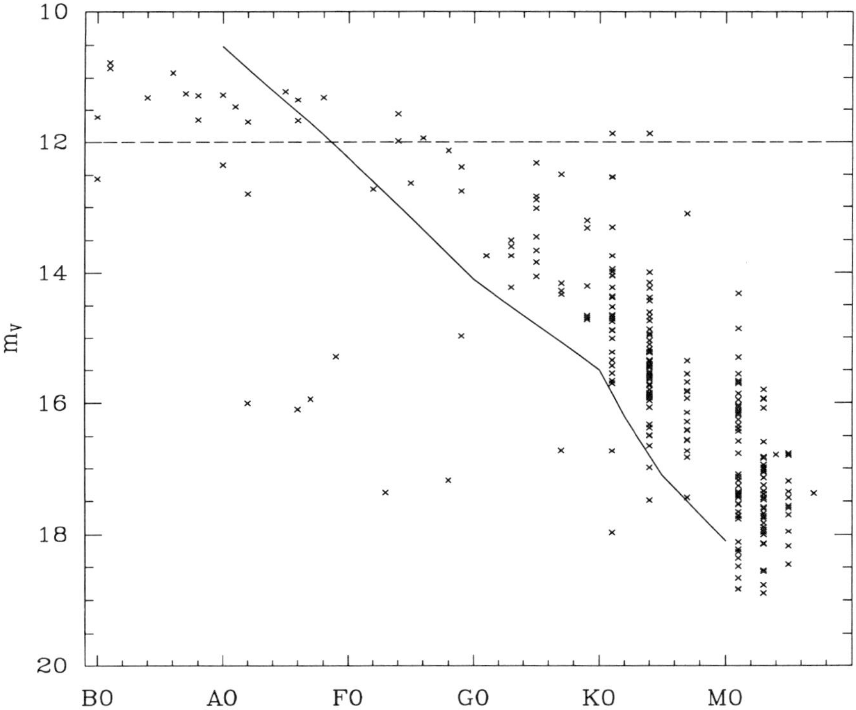

Figure 5. An H-R diagram using our spectral classifications vs. m_V. A dashed horizontal line at $m_V = 12$ indicates the point at which the brightest stars in our sample become saturated, and the fluxes are greatly underestimated. A main sequence is drawn in as a solid line, using data from Johnson (1966) and a distance modulus of 9.7. The majority of stars are clearly pre-main sequence.

5. A Final Truth Test

Six spectroscopic standards were observed on the same run as our program stars. After the methods to classify our stars had been developed, we used them to classify these standards. Our estimates of their MK types and their published spectral types differ by an average of 3 subclasses, a result entirely acceptable for our purposes. Clearly, by using a variety of spectral features simultaneously, normal stars can be sufficiently accurately classified even at low resolution. It is interesting to see that while the MK system was devised at shorter wavelengths, it still makes sense to assign MK types to spectra classified with our wavelength coverage.

Acknowledgments. We are grateful to Chris Corbally for useful explanations of the philosophy of the black art of MK classification. We also thank Pat Boeshaar for her warnings on the dangers of M star classification.

References

Adams, M. T., Strom K. M., and Strom S. E. 1983, ApJS, **53**, 893.

Blitz, L. 1977, PhD Dissertation, Columbia University.

Herbig, G. H. 1954, ApJ, **119**, 483.

Jacoby, G. H., Hunter, D. A., and Christian, C. A. 1984, ApJS, **56**, 257.

Johnson, H. L. 1966, ARA&A, **4**, 193.

Margulis, M. and Lada, C. J. 1986, ApJ, **309**, L87.

Margulis, M., Lada, C. J., and Young, E. T. 1989, ApJ, **345**, 906.

Walker, M. F. 1956, ApJS, **2**, 365.

Warner, J. W., Strom, S. E., and Strom K. M. 1977, ApJ, **213**, 427.

The MK Process at 50 Years
ASP Conference Series, Vol. 60, 1994
C.J. Corbally, R.O. Gray, and R.F. Garrison (eds.)

Spectral Classification with the IUE: An Atlas of B-Type Spectra

Janet Rountree

Science Applications International Corporation, 5151 East Broadway, Tucson, AZ 85711

George Sonneborn

Code 681, NASA/Goddard Space Flight Center, Greenbelt, MD 20771

1. The Classification System

A set of criteria for the spectral classification of B stars in the ultraviolet was published by Rountree and Sonneborn (1991). In that paper it was shown that photospheric absorption lines in the 1200-1900 Å wavelength region can be used to classify the spectra of B-type dwarfs, subgiants and giants in a two-dimensional system consistent with the optical MK system. The observational material consisted of high-resolution spectra from the International Ultraviolet Explorer (IUE) archives, suitably resampled and displayed. Insofar as possible, the standard stars on which the ultraviolet classification was based were chosen from among the MK standards.

2. The Atlas

We have recently published an *Atlas of B-Type Spectra*, based on this classification system (Rountree and Sonneborn 1993). The spectral scans in this atlas are displayed at the scale actually used to develop the system. Unlike photographic prints, they can be used directly in practical classification, provided that the user's program stars were observed with the IUE Short-Wavelength Prime (SWP) camera and processed as described in Rountree and Sonneborn (1991).

There are three parts to the Atlas. Part 1 contains sequences of spectra of standard stars; spectral-subtype (or temperature-type) sequences are included for dwarf (class V), subgiant (class IV), and giant (class III) stars. Almost all of the stars illustrated in this section are standards listed by Morgan and Keenan (1973) or by Rountree Lesh (1968). Next come luminosity sequences at seven spectral subtypes. In this set, some of the standards in classes III - V have been replaced with other stars whose spectra were judged to be equally representative of their type, so as to give as many examples as possible of normal ultraviolet spectra.

Part 2 of the Atlas illustrates the effect of rapid stellar rotation on the ultraviolet spectrum. The spectra are presented in pairs, consisting of a slowly rotating star and a rapidly rotating star, closely matched in spectral type. Although most of the rapid rotators are not standards, these illustrations should be helpful to the user who wishes to classify the spectrum of a broad-lined program star.

Finally, spectra with anomalous stellar wind lines, primarily strong absorption lines of C IV, Si IV, and/or N V, are shown in Part 3 of the Atlas. These are stars whose ultraviolet spectral types have a "w" suffix in the classification of Rountree and Sonneborn (1991). The spectrum of at least one "normal" star of identical type is presented with each anomalous spectrum to show the good match in the photospheric absorption lines, which alone are used in classification. We recall that in B stars, the stellar wind lines are not always correlated with spectral type, and are sometimes variable. Many – but not all! – of the stars that exhibit this phenomenon turn out, upon subsequent investigation, to be Be or shell stars.

Copies of the Atlas may be obtained free of charge from one of the above authors.

References

Morgan, W. W., and Keenan, P. C. 1973, ARA&A, **11**, 29

Rountree Lesh, J. 1968, ApJS, **17**, 371

Rountree, J. and Sonneborn, G. 1991, ApJ, **369**, 515

Rountree, J. and Sonneborn, G. 1993, Spectral Classification with the International Ultraviolet Explorer: An Atlas of B-Type Spectra, NASA Reference Publication 1312

Discussion

Walborn: I wish you would specify that you are talking about low luminosity B stars, because what you are saying about wind lines is simply not true of the O stars.

Rountree: True, I am not talking about the O stars. However, I *am* talking about philosophy: I am trying to classify without reference to the optical types. You can only do that for the B stars if you are careful to exclude the wind lines.

Garrison: Perhaps one of the differences between the two philosophies is one of emphasis. Nolan seems to be saying that the wind lines are generally correlated with the optical spectral type, except in 10-20% of the cases, and Janet seems to be saying that while there is some correlation, 10-20% don't fit, so we should not rely on wind lines. Note that both say $\sim$ 20%.

Even in the O stars, it seems to me that if *only* the wind lines are used, the information about the wind is being thrown away. From a practical point of view, however, there are few or no photospheric features in the far UV, so one is *forced* to use wind lines.

Rountree: I don't know exactly what percentage of B stars have "anomalous" wind lines. I think this is a potential third dimension in ultraviolet classification, but I don't have a big enough sample to define a "wind" sequence.

The MK Process at 50 Years
ASP Conference Series, Vol. 60, 1994
C.J. Corbally, R.O. Gray, and R.F. Garrison (eds.)

International Ultraviolet Explorer Atlas of B-Type Spectra from 1200 to 1900 Å

Nolan R. Walborn

Space Telescope Science Institute, 3700 San Martin Drive, Baltimore, Maryland 21218, USA

Joel Wm. Parker

Laboratory for Astronomy and Solar Physics, Mail Code 681, Goddard Space Flight Center, Greenbelt, Maryland 20771, USA

Joy S. Nichols

Science Programs, Computer Sciences Corporation, 10000-A Aerospace Road, Lanham-Seabrook, Maryland 20706, USA

We are preparing an atlas of B-type spectra consisting of short-wavelength, high-resolution data from the International Ultraviolet Explorer archive, designed to complement the widely used O-star atlas from the same source (NASA Reference Publication 1155). The intention is to complete the OB natural group, i.e., to spectral type B3 on the main sequence and B8 at Ia. A primary objective is to chart in detail the gradual disappearance of the stellar-wind features in normal spectra as a function of spectral type and luminosity class. As in the O-star atlas, which first demonstrated the strong correlation between the optical spectral types and the UV wind behavior in the majority of the stars, the primary selection will be well-defined optical classifications, preferably of standard quality, to ensure that a consistent reference frame of normal objects is derived. Some peculiar categories will also be included, particularly the BN/BC classes and superluminous objects such as P Cygni – as well as any unexpected peculiarities which emerge from the survey. Approximately 80 images have been selected from the IUE archive for the atlas, including 25 of types B0-B3 V-III, 15 B0-B5 II-Ib, 25 B0-B8 Iab-Ia+, and 15 BN/BC/Peculiar. The processing and presentation will be as similar as possible to the O-star atlas, with the data rectified and rebinned to a uniform resolution of 0.25 Å. The recently published Rountree-Sonneborn atlas (NASA RP 1312) will be complemented by the emphasis here on the more luminous stars. A preliminary sample (half-) page is shown in Figure 1, where the pronounced luminosity effect in the Si IV wind features at type B0 can be seen.

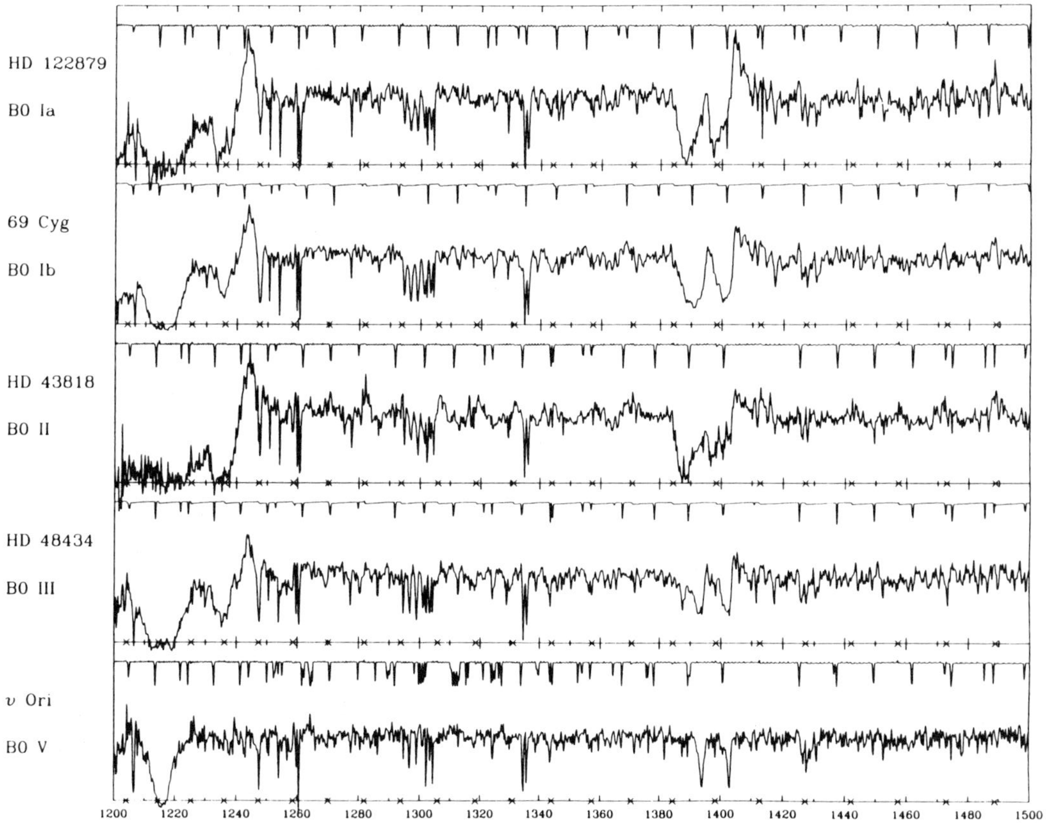

FIGURE I A luminosity sequence of IUE high-resolution data at spectral type B0. Note the luminosity effect in the Si IV λλ 1394, 1403 stellar-wind profiles.

Discussion

Roman: The increase in the strength of wind lines with temperature and luminosity is convincing, but how do you disentangle luminosity and temperature?

Walborn: Features of different ionization, e.g., Si IV vs. C IV, have distinct behaviors which in combination provide discrimination of the two dimensions. However, there are ambiguities at some types, usually along diagonal lines in the HR diagram, which may not be easily resolved from the available wind features above.

Torres-Dodgen: Could you remind us what is the resolution of Copernicus spectra?

Walborn: The majority were obtained at 0.2 Å. There are a few scans at 0.05 Å.

PART VII.

SURVEYS AND AUTOMATED CLASSIFICATION

The MK Process at 50 Years
ASP Conference Series, Vol. 60, 1994
C.J. Corbally, R.O. Gray, and R.F. Garrison (eds.)

The Michigan Survey and the Continuing Importance of Spectral Surveys

Nancy Houk

Univ. of Michigan, Ann Arbor, MI 48109

Abstract. The Michigan Spectral Survey has so far increased the number of stars with two-dimensional MK types by a factor of six and has provided a large set of data usable for statistical purposes and for galactic structure studies, as well as for providing spectral types of individual stars. A history of the project is outlined. The importance of similar photometric surveys as well as deeper spectral surveys is stressed. It is estimated that Volume 5 of the Michigan Spectral Survey ($\delta = -12°$ to $+5°$) will be published in late 1994, with the entire sky being completed in 7 volumes by $\sim$ 2004. Data from the Michigan Spectral Catalogue will be used to provide learning sets for automating spectral classification. Reliable automation is a necessity for dealing with the huge number of stars to be encountered in deeper surveys.

First, I would like to remind you that before I started the Michigan Spectral Survey, fewer than 23,000 stars had MK classifications. These had been classified by many different people, some with inadequate standards or unsuitable plates. For example, some of the southern spectral types were obtained using spectra that had been taken for radial velocity work. Also, most people were only interested in certain types of stars, and only classified those groups, so no meaningful statistics could be done with these 23,000 classified stars, and people had to fall back upon the HD catalogue for galactic structure problems, etc.

I wonder if possibly the situation might be somewhat the same now for photometry; different systems used and no large body of measurements on the same system and measuring all stars down to a certain magnitude, i.e., a photometric survey of, for example, the HD stars. It is disappointing that for the stars for which I now have good two-dimensional types, the largest source of uncertainty in getting their distance is the lack of good apparent magnitudes and colors for many of them. I hope the Hipparcos results will help here, but even a less sophisticated ground-based program would be very useful.

I was inspired by the history session yesterday to give a little of the history of the project, so we can get it straight now instead of trying to speculate 50 years from now. In 1963 or '64, shortly after I had taken and enjoyed a graduate course in spectroscopy, including spectral classification, at Warner and Swasey Observatory at Case in Cleveland, Billy Bidelman, who had just come to the University of Michigan in 1963, came down and gave a colloquium at Warner and Swasey. He mentioned how advantageous it would be if the Michigan Curtis Schmidt telescope were moved to the great southern site at Cerro Tololo, and systematic reclassification of the HD stars undertaken along with a survey of peculiar stars. (Bidelman does not remember giving this talk at all, but after my oral presentation here John Drilling told me that he was also at that colloquium and remembers it well). A few years later Orren Mohler and Bidelman were instrumental in getting the Michigan Curtis Schmidt moved to the Cerro Tololo Inter-American Observatory in Chile with the help of an NSF grant. Bidelman directed the southern blue spectral survey observing program from 1967-69, started a survey of peculiar stars, and then left for the U. of Texas at Austin. I was employed in a post-doctoral position at Case at that time, but watching progress at Michigan with interest and was dismayed when he left, thinking the HD reclassification plans might end with his departure. I visited some astronomer friends in Ann Arbor and intimated that I could probably be talked into doing the HD reclassification. After meetings of several astronomers both at Case and Michigan, Anne and Charles Cowley agreed to write a proposal to NSF along these lines if I would agree to do the classification. Meanwhile, under the direction of F. D. Miller and Jack MacConnell, Michigan continued the southern plate-taking and MacConnell and several graduate students continued the peculiar star survey. The NSF grant was obtained in December of 1970 and I started work then. Since I was young and a complete unknown at that point, I was not surprised when I got a call from W. W. Morgan to come and give a colloquium at Yerkes on the project in early February of 1971. There Morgan showed me a number of spectra to classify, which I did. I had used binocular zoom microscopes both at Case and Michigan, so I was somewhat dismayed when he handed me a small eyepiece magnifier with which to check them out, but I evidently passed the test because Morgan has been very supportive of the project ever since then.

The MK Process volume of 1984 gives most of the information about the HD reclassification project that you would want to know and many of you have heard about it from me before, so I shall not go into any details here. However, I shall show a table that will bring you up to date and tell you what you *really* want to know, judging from the questions I have been asked here! We expect Volume 5, which will cover declinations $-12°$ to $+5°$, to be out sometime in 1994. Our original plan was to cover from $-12°$ to $0°$ or at most $+1°$, but the staff I had was not equipped or experienced enough with computers to produce Vol 5, so I added another zone in declination, and am now classifying to $+5°$. Meanwhile new assistants have been found capable of going further with the production of Volume 5. Since they are new to the project, when it will come out is difficult to predict, but I hope a year from now for the hard-cover copy. As usual the magnetic tape version will be ready several months before the book is available.

Volume	Declination	No. of HD Stars	Running Total Classified	Year of Publ.
1	$-90°$ to $-53°$	36,382	36,382	1975
2	$-53°$ to $-40°$	30,400	66,782	1978
3	$-40°$ to $-26°$	30,314	97,096	1982
4	$-26°$ to $-12°$	33,301	130,397	1988
5	$-12°$ to $+5°$	$\sim 32,500$	$\sim 162,900$	1994
6 & 7	$+5°$ to $+90°$ (2 volumes)	$\sim 62,400$	225,300	~ 1999, ~ 2004

This table also clearly shows the predominance of southern stars in the HD. Only two volumes will be required for all the HD stars north of +5°. So with completion of Vol.5, only about 1/4 of the stars will remain to be classified. Some people not only ask when Volume 5 will be out, but also ask when the whole project will be completed. My rough estimate is 2004, but that of course, assumes continued funding, good health, and competent assistants.

Turning to the future, I want to emphasize the importance of other astronomers continuing to do spectral and photometric surveys. Those who have listened to the papers given earlier in the meeting must have noticed how often results from the Michigan Spectral Survey and other surveys were used.

I do not keep up with what all of you are doing as well as I should, and am forgiven for this and am even encouraged by some to keep my eyes to the microscope when it comes to a choice of that or keeping up with the literature. Consequently I am not well-informed on many current and future surveys, but I have heard that even some of those which are primarily aimed at galaxies will be scanning stars as well, and I would advise the spectral classifiers to accept, when applicable, these 'crumbs from the rich man's table,' so to speak, as well as planning surveys of our own. It would be too bad if data were lost because the people were really looking for something else at the time. It seems especially important to have a well-run adequate data center, such as already exists for some data from instruments in space. Our current data centers at Goddard and Strasbourg may be overwhelmed, so we should plan ahead. It is important that at least after a reasonable time all data be available to any who wish to use it. Without going into detail, I would just like to emphasize that we do need surveys of fainter stars. As Garrison said on the first day, to get the fainter secondary stars we need, people have to look at a lot of candidates.

The rest of this session is going to be on the automation of spectral classification, and I want it to be known that I strongly support this if it is done well. In fact, it is the only way to go for the huge numbers of fainter stars that we are going to be dealing with. People have been working on this without very much success since I first started the MK classification project back in 1971. One astronomer predicted that within 4 years of then he would be in a position to do it all automatically much faster than I. Another, after years of work found a way that worked pretty well, if a trained human checked each spectrum as it appeared, for overlaps, defects, and other problems! However, now the time seems to be ripe for making another serious try at the so-far-intractable problem. One thing that will help a lot is a large and consistent set of data for the computer to learn from. I have already in the past provided both Ted von Hippel and Kurtz and LaSala with some unused but not poor quality Schmidt plates from the Michigan survey to be scanned by Mike Irwin using the machine at U. of Cambridge, England. I am providing Ted von Hippel with over 100 of the plates actually used in the spectral classification for scanning at Cambridge. Von Hippel and his collaborators hope to eventually scan most of the plates I have classified in order to have a very good learning set for the computer. This will provide a better foundation for automation schemes of more than one type, perhaps. I wish all those who are working on this problem the best of luck and am always ready to classify special groups of stars if a comparison with machine classifications is needed. In fact, that sort of job may be what I shall be spending my time on after the Michigan Spectral survey is completed! I thank all of you as well as the National Science Foundation for all the support you have given me over the years and I especially thank Morgan and Keenan for providing the system that made it all possible.

Discussion

Abt: Perhaps I should know, but how close are you to the MK system?

Houk: I seem to be pretty close, judging from comparisons I have made with Garrison's types, etc. Although they are objective-prism spectra, the Michigan spectra seem to have slightly higher resolution than the original MK slit spectra, and they are also very widened (to 0.8 mm). We titled the volumes 'Two-Dimensional' rather than MK because at the time we were starting, MK was defined from slit spectra. We did not want to change the title in mid-stream, but I have a lot of confidence in the quality of my objective-prism plates.

Levato: How many spectra do you classify in a standard day?

Houk: I do not have a standard day, unfortunately! (Laughter) Some days are spent in administration, preparing proposals for funding, etc, and overseeing my 8 student assistants. So, some days I might classify one or two hundred stars and other days none. The average Volume of about 33,000 stars takes an average of 4.6 years to produce, and part of that period is spent actually preparing the 300-400 page volume. We do not do very much analysis. That is why I am so thankful for Jim Sowell and his student Shannon Austin helping us put this poster paper together, because I do not like to take too much time away from the classification, yet naturally we like to show some of the results of the work.

The MK Process at 50 Years
ASP Conference Series, Vol. 60, 1994
C.J. Corbally, R.O. Gray, and R.F. Garrison (eds.)

Automated Classification of Stellar Spectra: Where Are We Now?

T. von Hippel, L. J. Storrie-Lombardi,
M. C. Storrie-Lombardi
Institute of Astronomy, Madingley Rd., Cambridge CB3 OHA, U.K.

M. J. Irwin
Royal Greenwich Obs., Madingley Rd., Cambridge CB3 OEZ, U.K.

Abstract. We briefly review the work of the past decade on automated classification of stellar spectra and discuss techniques which show particular promise. Emphasis is placed on Artificial Neural Network and Principle Component Analysis based techniques, due both to our greater familiarity with these and to their rising popularity. As an example of the abilities of current techniques we report on our automated classification work based on the visual classifications of N. Houk (Michigan Spectral Catalogue, Vol. 1 - 4, 1975, 1978, 1982, 1988).

1. Introduction

The classification of stellar spectra has long been a fundamental tool for determining the physical parameters of bright stars (e.g., Cannon and Pickering 1918-24; Morgan *et al.* 1943). Earlier in this conference Osterbrock (1994), Hoffleit (1994), and McCarthy (1994) reviewed the history of stellar classification, and Mihalas (1994) showed the early achievements and the continuing challenges being met by the conflict of classification with theory. Stellar classification encompasses both the rare objects which refine our understanding of the physics of stellar evolution and stellar atmospheres, and the common objects which we employ towards understanding the general questions of Galactic structure and stellar populations. All of these successes have been built upon the classification work of a few experts – a system which worked well when the data rates where relatively low. Currently, however, very large scale projects are being initiated which will acquire many thousands of spectra per night and which should "reduce" those spectra in nearly real time. The Sloan Digital Sky Survey (Gunn and Knapp 1992), for example, plans on acquiring $\geq 10^6$ spectra over a few year period, though most of these spectra will probably not be stellar. Likewise, a 400 fiber spectrograph is currently being constructed for the Anglo-Australian telescope, making this instrument capable of providing nearly 10^4 fully reduced spectra per night.

In order for spectral classification to leap into this age of very large digital surveys automated techniques must take over much of the time-consuming work done by the human visual classifier. This has been recognized for some time and considerable effort has been put into automating spectral classification (see Kurtz 1984 for efforts of the previous decades). Some successes have been achieved, though no readily usable and robust techniques have yet emerged. Besides an increase in classification speed, an automated, objective and repeatable technique should also provide a basis for building classification "expertise" at different survey centers.

Techniques which have been developed to automate the classification of stellar spectra can be broken down into two basic categories: criterion-evaluation and pattern-recognition techniques. Criterion-evaluation techniques were tried earlier as they are computationally easier, though they are only a crude representation of the method employed by the human classifier. An early example of criterion evaluation can be found in Jones (1966) where the equivalent widths of a number of hydrogen, calcium, iron, manganese and strontium lines were correlated with spectral type, luminosity class and metallicity. A more modern example of criterion evaluation can be found in Malyuto and Shvelidze (1994, and references therein), where the relative fluxes in a large number of wavelength regions are compared, through stepwise-regression techniques, to the spectral types, luminosities and metallicities. Pattern-recognition techniques have been employed by LaSala and Kurtz (LaSala 1994, and references therein), Weaver (1994) and our group (von Hippel *et al.* 1994). The technique of LaSala and Kurtz is to find the minimum in a weighted vector distance between the program object and the library of standards. Weaver has used supervised artificial neural network (ANN) techniques (1994) for the specific case of Wolf-Rayet stars as well as for the general case of spectral classification. Our work (see also von Hippel *et al.* 1994 for a more detailed look at our techniques) also uses supervised ANN techniques for the general problem of spectral classification. We are creating an expert spectral classification system based on Houk as our expert, and using her original data.

Over the last 20 years, Houk and co-workers (Houk and Cowley 1975; Houk 1978, 1982; Houk and Smith-Moore 1988; Houk 1994) have been creating one of the greatest modern stellar classification libraries by determining the 2- and 3-dimensional spectral types of all stars in the Henry Draper Catalogue (Cannon and Pickering 1918-24, hereafter HD). The 4 catalogues published to date by Houk include over 120,000 stars covering the full range of temperatures and luminosities, and a wide range of metallicities down to the HD magnitude limit of 8.5 to 11, depending on the particular area of the sky. Errors in human classification are hard to quantify, but Houk reclassifies $\approx$ 10% of her stars without knowledge of her earlier classification, and finds an rms error of $\approx$ 0.6 spectral subtypes and $\approx$ 0.25 luminosity types.

Within the spirit of this conference we also note that we believe automating the MK classification scheme is preferable to building a crude model atmosphere spectral evaluation tool since: (1) not enough stars have their astrophysical parameters precisely defined to build the type of bulk classification system we envision; (2) classifications can be mapped into the astrophysical parameters with well-studied relations (e.g., Popper 1980, and references therein); (3) this mapping can be repeated whenever more up-to-date and accurate relations are determined; and, (4) classifications are considered inviolate, since they are based on a series of standards, and therefore should not change with time. The recent advances in plate measuring (e.g., the Automatic Plate Measuring Facility in

Cambridge; Kibblewhite *et al.* 1984, hereafter APM) and the steady advances in computer speed, are also key ingredients in our automation strategy.

Here we examine an expert system based on a back-propagation neural network. We expect neural networks and other pattern-recognition techniques to be free of many of the disadvantages of criterion-evaluation techniques. Line fitting techniques, for example, have the disadvantage of requiring knowledge of the approximate stellar type before they can determine which lines to fit, since very different features will be found at the same wavelengths for different stellar types. Additionally, with criterion evaluation, when different line indices indicate divergent results it is not readily apparent to what degree noise, overall abundances, or abundance patterns are responsible. The most straight-forward pattern-recognition technique is probably cross-correlation, but, in its simplest form, it weights the strongest lines most heavily, though they are not necessarily the features with the highest weight in classification determination. Furthermore, all the classical techniques are based on linear operations. Since we expect to find subtle non-linear relationships between the temperature, surface gravity and metallic-line indicators, a classification scheme which copes with non-linear relationships between parameters ought to offer significant advantages.

2. Data Reduction and Pre-processing

Our data reduction is essentially the conversion of Houk's objective-prism plates and her catalogue into one-dimensional digital spectra tagged with her spectral types and luminosity classes. The "raw" data are ≈ 1000 IIaO high-dispersion (108 Å mm^{-1} at $H\gamma$) objective-prism plates taken with the Michigan Curtis Schmidt telescope at Cerro Tololo, Chile (see also Houk 1994). Each plate is 20cm x 20cm in size and at a plate scale of 96.6″/mm subtends an area of $5^\circ \times 5^\circ$ on the sky. We report here the results from the scanning of 6 plates containing 575 HD stars which Houk classified and assigned a quality of 1 or 2 (on a 1 to 4 system). Of these 575 stars 65% are quality 1 and 35% are quality 2. Details of the plate material can be found in Table I, where we list the date each plate was taken, Houk's plate codes, the J2000.0 equatorial coordinates and galactic coordinates of the plate centers, and the number of non-overlapped quality 1 and 2 spectra extracted from each plate.

The APM facility in Cambridge was used to measure the plates. General details of this facility are given in Kibblewhite *et al.* (1984) and the general use of this system for measuring objective prism plates is given in Hewett *et al.* (1985). The scanning procedure used the HST guide-star catalogue as a position driver for the scanner. Digitized scans 18mm x 2mm in area were extracted around each widened spectrum at 22.8μm pixel resolution (corresponding to a 768 x 80 array). After finding and subtracting off the general "sky" background level, the digitized scans were summed orthogonally to the dispersion axis to create one-dimensional, pixel versus density spectra. Following this the extracted spectra were matched against the Houk catalogues, and only those for which Houk has determined a spectral type were kept. In future work we plan to keep all spectra with sufficient S/N for classification, but at present we are just building the expert system, and not classifying previously unclassified stars.

Table 1. Plates Scanned

date	plate	RA	Dec	l	b	num
27/ 5/68	2280	4:13.2	-1:46	194	-35	76
28/ 5/68	2303	3:28.6	1:23	182	-43	99
16/12/68	3316	20:26.9	-0:44	44	-21	96
12/ 1/69	3504	5:29.7	4:43	199	-16	136
15/ 7/69	4825	23:14.6	-4:45	73	-58	73
9/ 8/74	16506	4:35.6	2:58	193	-28	95

The next data reduction step was the differential wavelength calibration of all the spectra within each plate. An objective prism introduces a quadratic distortion of the position of a spectrum on the prism plate compared to a direct plate (in this case the HST Guide Star Catalog) thereby causing the spectrum to shift slightly (up to several pixels) from the expected position. This distortion was mapped as a function of the standard coordinates of the objects by cross-correlating each high-quality spectrum with a high S/N template star of approximately the same type. A quadratic transformation relating direct and prism coordinates was then derived and the solution applied to all stars regardless of S/N. This places all the stars on the same (arbitrary) wavelength system.

The final stage in the pre-processing was to define a stellar "continuum" so that the information in the spectra could be partitioned into line and continuum features. The continuum estimation was accomplished by first running a 51 pixel median filter (Tukey 1971) over each spectrum followed by a 25 pixel boxcar filter. All absorption (or emission) features with scale length less than ≈25 pixels are removed by this processing. This is sufficient to track over all the main absorption lines though it underestimates the continuum in areas where the spectral shape changes rapidly, especially through the major molecular features. This will have no effect on our analysis as long as the continuum fits are consistent for a given spectral type and no residual continuum slope remains. We fit the continuum so that we can remove it and thereby produce a classification scheme based on the line information alone, as that is the general method employed by human classifiers. We also examine classification based on the complete spectra. In von Hippel *et al.* (1994) we also study classification based on the continuum information alone. The purpose of studying the continuum-only method was its equivalence to classification based on photometric colors, whereas the difference between the line-only method and the full spectrum method may eventually provide an independent means of determining reddening.

As samples of the spectra used in this work, Figure I presents spectra of 6 stars of various spectral types, along with their continuum fits. From top to bottom the stars are classified by Houk as B9 V, A2 V, F3 V, G3 V, K5 III and M2 III, respectively. The horizontal axis is wavelength in Å, and the vertical axis is emulsion density in arbitrary units. The spectra have $S/N \gg 100$ and an effective resolution of ≈ 2 - 3 Å, resulting in 382 resolution elements over the useful range of 3843 - 5150 Å. Note that for the later spectral types the more limited flux in the blue increases the noise markedly shortward of 4000 Å.

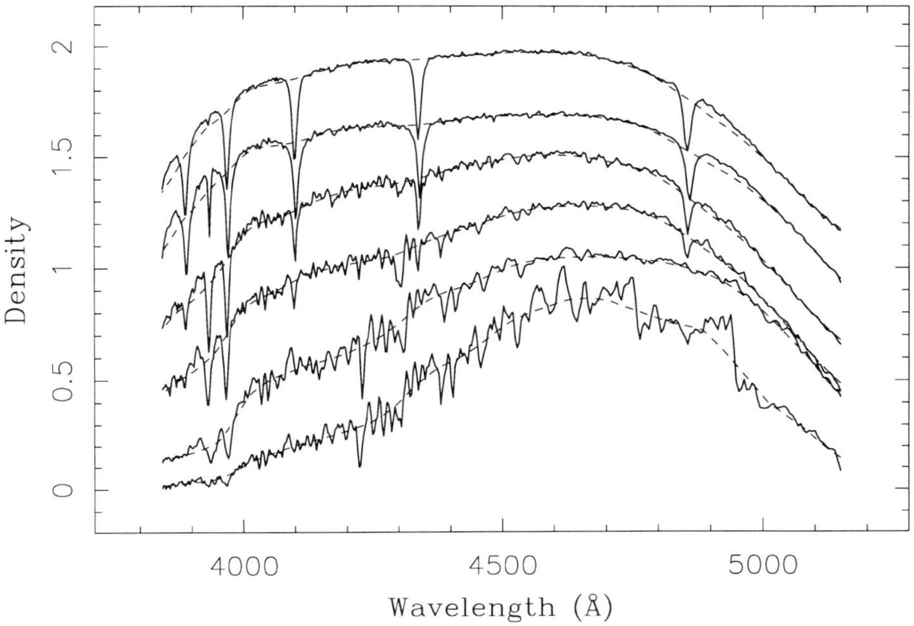

Figure 1. Sample spectra plotted with flux in arbitrary units.

3. Classification Algorithm

3.1. Background

For this study we have utilized an artificial-intelligence technique known as Artificial Neural Networks (ANN) for stellar classification. Originally derived from simplified models of human central nervous system activity (McCullogh and Pitts 1943), ANNs have found utility in signal processing, (Widrow and Winter 1988), adaptive optics (e.g., Merkle 1988), star/galaxy distinction (Odewahn *et al.* 1991), and galaxy classification (Storrie-Lombardi *et al.* 1992). Discovery of an error minimization algorithm for training multiple layers of ANN nodes (Werbos 1974; Parker 1985; Rumelhart *et al.* 1986) has produced considerable interest in a very simple gradient-descent, non-linear minimization ANN technique known as back propagation. Back-propagation ANNs consist of a series of nodes arranged in layers with inputs, summation nodes, outputs, and weighted connections between layers. The classical back-propagation algorithm requires a minimum of three layers: an input layer for bringing the data into the ANN, at least one hidden layer to affect a non-linear mapping from input to output space, and an output layer to provide classification. We have used 383 input nodes, 1 for each of the 382 spectral resolution elements and 1 as a bias node which has a constant value (see Figure II). Weights connecting the bias to the hidden layer and output nodes are modified during training exactly like all other weights. We term any layer between the input and output layers as a hidden layer. We have confined ourselves to the case of a single hidden layer and a single output node for the purposes of this study to minimize the number of free parameters in our system. The output node is used to provide the temperature classification.

In the standard back-propagation algorithm, the output is defined as a vector in the classification space with an individual node assigned for each class. In our case this would produce a 59-dimension output vector if we considered each temperature classification as a discrete class. The ANN starts with a random initial weight state and guesses at the identity of an incoming target vector by producing an output vector with the value of each output node lying somewhere between 0 and 1. For each star, it then compares its output to the desired classification vector $\mathbf{d}$ generated by the human expert. For example, we might define $\mathbf{d} = (1,0,0,0,...)$ for class B0, $\mathbf{d} = (0,1,0,0,0)$ for class B1, etc.

The comparison is done in terms of a cost function, usually of the form

$$E = \frac{1}{2}\sum_k (o_k - d_k)^2 \quad (1)$$

summed over the vector components. Because during training the ANN adjusts its weight space according to the decisions of a human expert, we term it a "supervised" network. This cost function is minimized with respect to the free parameters, the weights w_{ij}, connecting each layer. Minimization over the weights w_{ij} is done using the chain rule (gradient descent). The weights are updated *backwards* from the output layer through the hidden layer(s), and the rate of updating is controlled by the learning coefficient, η. A second coefficient, α (often called the momentum), can dampen the oscillations in minimization sometimes seen as the ANN searches through weight space for an optimal solution. Training set size determines the optimal values for η and α (Eaton and Olivier 1992).

At each node at layer s of the ANN a linear combination over the input x_i from the previous layer $s-1$ is given by $I_j^{(s)} = \sum_i w_{ij}^{(s)} \, x_i^{(s-1)}$. The node fires according to a non-linear sigmoid threshold function usually of the form $f(z) = 1/(1+\exp(-z))$ (in the interval [0,1]) or $f(z) = \tanh(z)$ (in the interval [-1,1]). The ANN functions as a non-linear operator transforming the representation of objects in the input parameter space to the classification space.

After completion of the learning process (i.e. optimizing the weights w_{ij} according to a given training set) the ANN is ready to analyze new data. As we present each input vector the ANN produces an output vector containing its classification coding. In the one output per class configuration the output of an ANN provides a Bayesian *a posteriori* probability for class membership (e.g., Richard and Lippmann 1991; Storrie-Lombardi *et al.* 1992).

3.2. ANN Design

ANN design depends on finding an optimally efficient network architecture for a particular data set. The principal factors include size of the training set, the number of input parameters, number of hidden layer nodes, and number of output nodes. We have maximized the size of the training set (to ≈ 500 objects) by training with the data from 5 of the 6 plates, and thereafter testing on the data from the unseen 6th plate. For our output layer we used 1 node since temperature classification is a continuous function rather than a discrete set of

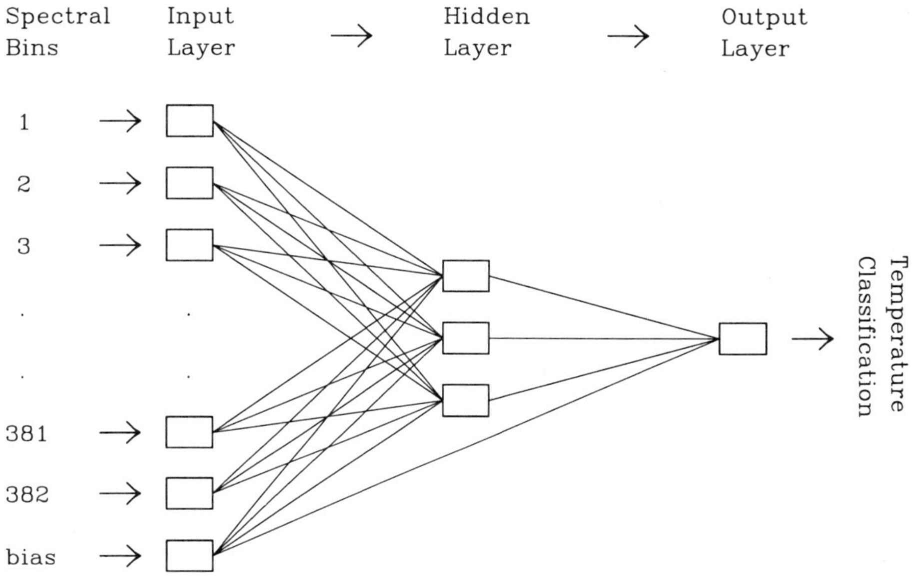

Figure 2. ANN architecture.

output classes. The different spectral type subclasses were converted to floating point numbers from 0 (O3) to 1 (M9), following the 59 subclasses recognized by Houk. This simple linearization of the Houk spectral types mimics the human ability to recognize the effects of small changes in temperature. While we could have constructed an ANN with 59 output nodes, and such an ANN would have had the added advantage of reporting probabilities of class membership, the amount of data presently available to us is not yet sufficient to warrant the greater complications of training and testing such an ANN.

For the present study we have chosen to leave our network fully connected and decrease the total number of weights by decreasing the number of nodes in the hidden and output layers. We have constructed a very simple, minimalist ANN. Back-propagation networks using a single hidden layer train more rapidly and fall into inefficient local minima less frequently than ANNs with two hidden layers (de Villiers and Barnard 1993). We utilize only a single hidden layer containing three nodes. In tests on hidden layer size we have investigated 3, 5, and 7 nodes. The 5 and 7 node cases have nearly the same random and systematic errors as the 3 node case over 1000 training cycles.

At this time we have confined our classification to the temperature domain since we do not have a significant number of stars of different luminosity class at a given temperature class. The spectra described in this series of experiments do not all reside in one luminosity class, but predominantly come from dwarfs at the earlier types and giants at the later types. At this point we have no stars at the classification extremes, either earlier than B3 or later than M4. Figure II depicts our fully connected (each node connects to every node in the next layer) back-propagation network.

3.3. ANN Training

Once the basic ANN architecture has been established, values for the learning (η) and momentum (α) coefficients must be fixed. The best values for η and α seem to be best determined by experience with the data. Poor values for these coefficients cause slow or incomplete learning. We have found that for our current data set $\alpha = 0.9$ and $\eta = 0.01$ produce stable temperature classifications.

Training the ANN can take hundreds or even thousands of iterations. The major portion of the error minimization occurs in the first few dozen iterations, followed by an exponential decrease in the rate of learning. Reasonable criteria for stopping learning include: (1) the rate of learning has effectively approached zero, and (2) a comparison of ANN performance with its training and unseen testing sets, indicates good generalization to the testing set without memorization of the training set. We have found that our networks reach a stable state and show good generalization after 1000 passes through our training samples.

4. Results

In order to maximize the number of stars used to train the ANNs we trained our ANN on the spectra from 5 plates and tested on the spectra from the remaining 1 plate. At present, with $\approx$ 500 training spectra containing 382 independent data points (resolution elements) and an ANN architecture with 383 (= 382 input nodes + 1 bias node) $\times$ 3 (hidden layer nodes) + 4 (= 3 hidden layer nodes + 1 bias node) free parameters (weights), the final ANN weights are overdetermined by a factor of $\approx$ 150.

A persistent problem at present is the large number of objects contaminated by overlapping spectra. For the present these objects were identified and removed by eye. Undoubtedly some spectra remain which are affected by overlapping objects and this may degrade the quality of the classification results.

Figure III shows the result of a typical training and testing session for our ANNs. Panel 3a is the training case and panel 3b is the testing case. The horizontal axis for both panels is the catalogue classification of Houk and the vertical axis is the ANN classification. A line of slope unity is plotted in each panel to allow the eye to assess the quality of the ANN determinations. The self-test plots are created by passing the training set back through the ANN, essentially showing how well the ANN reached the global minimum of the solution space during training. The testing plots are created by passing the previously unseen data through the trained ANN. The testing data will generally have a larger scatter than the self-test data due to the imperfect ability of the ANN to generalize. Nonetheless, in all cases with these data, the test data have only slightly more scatter than the training data, and high quality classifications are achieved.

The quality of the classifications can be seen quantitatively from the systematic and random errors of the training and testing processes. For the line-only case the systematic classification offsets for the training data was 0.19 spectral subtypes, in the sense of the Houk value - the ANN value, (e.g., 1.0 is the difference between an F6 and an F7 star). The systematic classification offset for the testing data was 0.24 spectral subtypes. The classification uncertainties (root-mean-square differences of Houk - ANN) are 1.39 and 2.15 spectral subtypes for

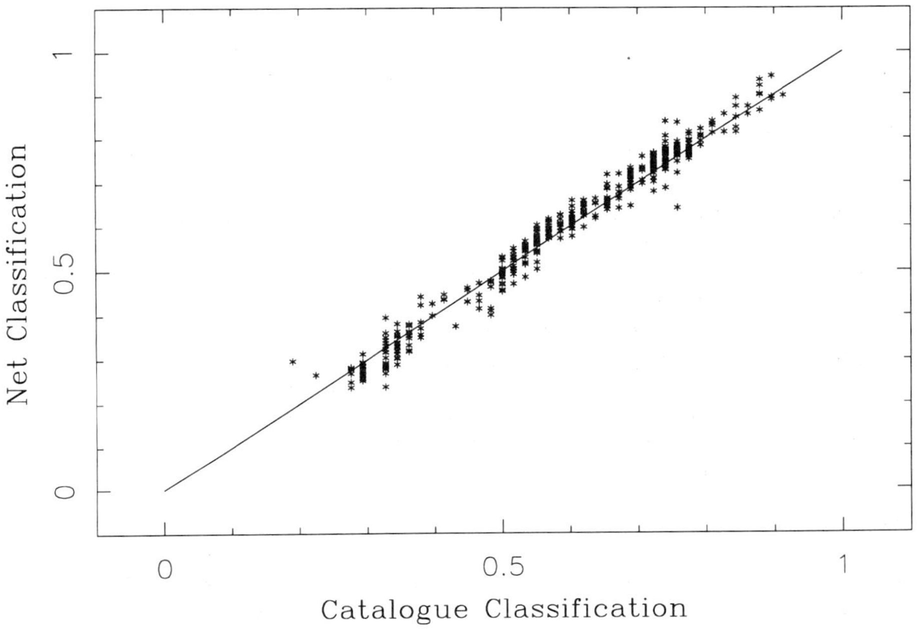

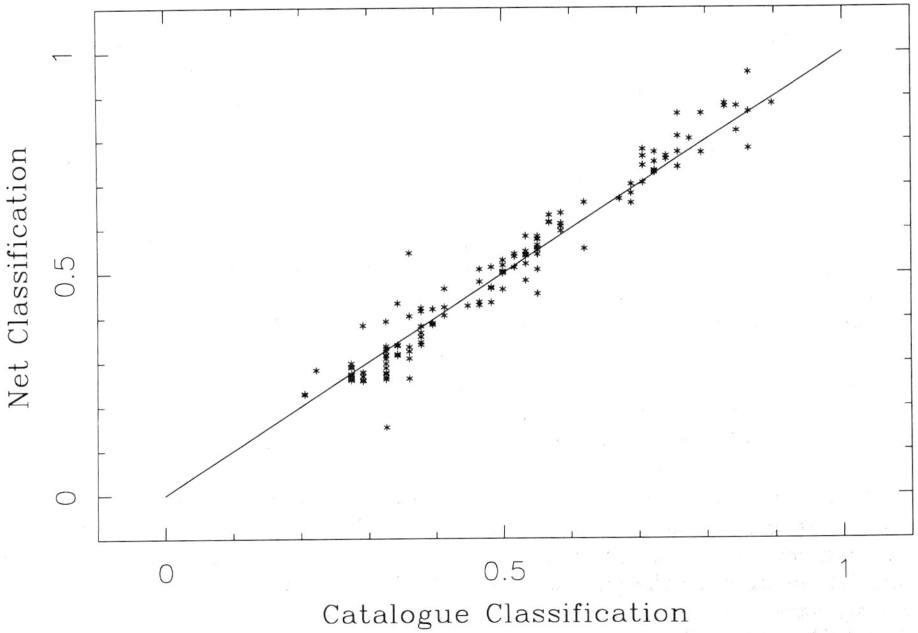

Figure 3. ANN results for typical training (Panel A, above) and testing (Panel B, below) sessions. The horizontal axes are the Houk classifications and the vertical axes are the ANN classifications. Unit slope lines are overplotted.

training and testing, respectively. Since the distribution of errors was somewhat non-normal for the line-only case we also computed the distance from the mean which contained 68% of the classification differences. Thus 68% of the training spectra are within 1.18 spectral subtypes of the mean and 68% of the testing spectra are within 1.72 spectral subtypes of the mean. The complete spectra (line + continuum) case had somewhat lower uncertainties and also low systematics. The complete spectra systematic classification offsets were 0.59 and 0.64 subtypes for training and testing, respectively. The classification rms uncertainties were 1.38 and 1.72 subtypes for training and testing, respectively. Thus our ANNs are currently able to classify stars from B3 to M4 with and rms error of $\approx$ 1.7 spectral subtypes.

Figure III and the above statistics also show that no stars are grossly misclassified. Approximately 5% of the spectra are still contaminated with overlapping stars, however, and this probably degrades the quality of the ANN results. We see systematic errors of $\approx$ 2 spectral subtypes at the blue and/or red ends for some of the training and testing sets. The systematics for the bluest stars (B3-B5) are the result of the subtle features which cause the early B stars to differ from the later B stars, and the lack of examples in this temperature region. The systematics for the reddest stars (M3-M4) are the result of the diminishing amount of distinguishing spectral information available in blue spectra of very late type stars, as well as the limited number of examples of these objects.

Our classifications for the line-only case show only a small overall systematic offset but a somewhat high classification rms uncertainty ($\sigma = 2.2$). The classification uncertainty is elevated by the $\approx$ 5% of the classifications which are furthest removed from their nominal catalogue value, leading to the non-normal error distribution. These objects had a large amount of noise somewhere in their spectra, generally in the blue. A further $\approx$ 10% of the stars had considerable noise in the blue and were still properly classified. This highlights the degree to which the ANNs were generally robust to noise.

As an alternate method of transforming spectra into classifications we used the data compression property of Principal Component Analysis (PCA) to reduce our spectra to 20 principal components. We found that the first PCA component accounted for 71% of the spectral variation, the first 2 components 84%, the first 5 components 94%, the first 10 components 97%, and the first 20 components 99%. The rest of the spectral variation seemed to be noise and the contaminating effects of overlapping spectra. Use of PCA had the advantage of removing much of the contaminating effects of overlapping spectra.

We combined the 20 PCA components for all of our stars with the Houk classifications, and trained and tested a smaller ANN. Use of the PCA components had the advantage of allowing us to build an ANN with only 21 input nodes (20 PCA inputs + bias), nearly 20 times fewer input nodes than our standard ANN. We thus were able to train these ANNs 20 times faster. This method also allowed us to experiment with PCA as a form of data compression and to compare different principal components to the classifications. We found no simpler correlations between the principal components and the classifications than previously have been recognized in the criterion evaluation work of other authors. Our ANNs with PCA inputs produced nearly identical answers to those ANNs which used the full spectral information, at least for temperature classifications. We are still actively investigating the full utility of PCA preprocessing.

5. Conclusions

We have presented initial automated stellar spectral classifications based on 575 stars scanned from 6 objective prism plates at the APM Facility. We have employed a simple back-propagation artificial neural network which is trained to duplicate the classifications of Houk (1975, 1978, 1982, 1988). We find that even with the limited data set now in hand we can determine decent quality temperature classifications for spectral subtypes B3 to M4, though we do not yet have the ability to determine luminosity classifications. We classify spectra to within 2.2 spectral subtypes based on the spectral line information alone and to within 1.7 spectral subtypes based on the full spectral information.

Future work will be based on more plate material so that we can increase our classification precision and extend our techniques to the determination of luminosity classes and, eventually, stellar abundances. We shall also explore alternative classification techniques, including unsupervised ANNs. The supervised networks presented here serve as error minimization algorithms, replicating the choices and decision patterns of their trainer. Unsupervised ANNs, however, are capable of deciding on an appropriate classification scheme without benefit of a trainer. Such unsupervised ANNs can learn to group or cluster objects into bins, usually by determining the Euclidean (or any other appropriate metric) distance between individual vectors. These ANNs operate much like classical vector analysis strategies, and they may allow us to identify previously unrecognized stellar classifications.

We believe our results indicate the potential precision, repeatability and objectivity of automated stellar spectral classification. Furthermore, only automated techniques have the speed to extend the expertise of the human classifier to the vast amount of data which should be produced by the next generation of spectral survey instruments.

Acknowledgments. It is a pleasure to thank Nancy Houk for her considerable efforts in helping us begin this program and for loaning us the plate material. We had assistance with numerical techniques from Geraint Lewis and Derek Richardson. We thank Ofer Lahav for very useful discussions.

References

Cannon, A.J., Pickering, E.C. 1918-1924, *The Henry Draper Catalogue*, Harvard Annals, **91-99**

de Villiers, J., Barnard, E. 1993, *IEEE Trans. Neural Networks*, **4**, 136

Eaton, H.A.C., Olivier, T.L. 1992, *Neural Networks*, **5**, 283

Gunn, J.E., Knapp, G.R. 1992, *Astronomical Surveys*

Hewett, P.C., Irwin, M.J., Bunclark, P.S., Bridgeland, M.T., Kibblewhite, E.J., He, X.T., Smith, M.G. 1985, MNRAS, **213**, 971

Hoffleit, D. 1994, this volume

Houk, N. 1994, this volume

Houk, N., Cowley, A.P. 1975, *University of Michigan Catalogue of Two-Dimensional Spectral Types for the HD Stars*, Vol. 1

Houk, N. 1978, *University of Michigan Catalogue of Two-Dimensional Spectral Types for the HD Stars*, Vol. 2

Houk, N. 1982, *University of Michigan Catalogue of Two-Dimensional Spectral Types for the HD Stars*, Vol. 3

Houk, N., Smith-Moore, M. 1988, *University of Michigan Catalogue of Two-Dimensional Spectral Types for the HD Stars*, Vol. 4

Jones, D.H.P. 1966, *Roy. Obs. Bull.*, **126**, 219

Kibblewhite, E.J., Bridgeland, M.T., Bunclark, P.S., Irwin, M.J. 1984, *Astronomical Microdensitometry Conference*, ed. D.A. Klinglesmith, Washington D.C.: NASA, 277

Kurtz, M.J. 1984, *The MK Process and Stellar Classification*, ed, B.F. Garrison, Toronto: David Dunlop Observatory, 136

LaSala, J. 1994, this volume

Malyuto, V., Shvelidze, T. 1994, this volume

McCarthy, M.F. 1994, this volume

McCullogh, W.S., Pitts, W.H. 1943, *Bull. Math. Biophysics*, **5**, 115

Merkle, F. 1988, *Messenger*, **52**, 5

Mihalas, D.M. 1994, this volume

Morgan, W.W., Keenan, P.C., Kellman, E. 1943, *An Atlas of Stellar Spectra*, Chicago: University of Chicago Press

Odewahn, E.B., Stockwell, R.L., Pennington, R.M., Humphreys, R.M., Zumach, W. 1991, AJ, **103**, 318

Osterbrock, D.E. 1994, this volume

Parker, D.B. 1985, *Learning-logic*, Report TR-47, Cambridge, MA: MIT Press

Popper, D.M. 1980, ARA&A, **18**, 115

Richard, M.D., Lippmann, R.P. 1991, *Neural Computing*, **3**, 461

Rumelhart, D.E., Hinton, G.E., Williams, R.J. 1986, *Parallel Distributed Processing*, eds. J.L. McClelland, D.E. Rumelhart, and the PDP Research Group, Cambridge, MA: MIT Press, Vol. 1

Storrie-Lombardi, M., Lahav, O., Sodré, L., Storrie-Lombardi, L. 1992, *MNRAS*, **259**, 8p

Tukey, J. W. 1971, *Exploratory Data Analysis*, Reading: Addison-Wesley

von Hippel, T., Storrie-Lombardi, L.J., Storrie-Lombardi, M., and Irwin, M. 1993, *MNRAS*, (submitted)

Weaver, Wm.B. 1994, this volume

Werbos, P.J. 1974, Ph.D. Thesis, Harvard University

Widrow, B., Winter, R. 1988, *Computer*, **21**, 25

Discussion

Gray: Have you gone back into your network and examined the weightings that the network has come up with after training? Is it classifying the lines, the continuum or some ill-defined combination of the two?

von Hippel: We have looked at this in a cursory way. Our tests with three nodes in the hidden layer indicate that one node is generally looking at all the major features, another is looking at Ca H&K and Hδ and the third at the continuum shape (if given that information).

Lloyd Evans: The IRAS catalogue has revealed new types of spectra with magnitudes 12-15. More must remain to be discovered but lack IR excesses to pinpoint them.

von Hippel: If we find new types of stars, we will require calibration by humans. Houk and Garrison have indicated their willingness to help us in this regard.

Lloyd Evans: This technique can only carry conviction if sample results are subjected to blind crosscheck by experienced classifiers. Automated variable star surveys – a simpler problem – have produced some very inaccurate results because there were inadequate human checks.

von Hippel: I showed you a "blind crosscheck" in the testing sample. These spectra had not been seen by the network and were classified to within $\sim$ 1.5 spectral subtypes.

Rountree: Is there a systematic effect in the differences between the neural net classification and Houk classification?

von Hippel: Yes, at early and late types there are small systematic departures between Houk's types and the neural net types. This is due to the few examples at the temperature extremes and to the small amount of information in the blue for the late types, which has also given human classifiers difficulty.

Rountree: In the architecture of your neural net, is there one input node that by-passes the hidden layer?

von Hippel: Yes, this is the "bias" node (constant input signal = 1) which gives the neural net an additional free parameter for each hidden and output node, allowing these nodes to use the most appropriate part of the sigmoid function.

Drilling: Have you devised tests which shows how well your procedure will detect peculiarities?

von Hippel: We have not dealt with peculiarities yet, but intend to try unsupervised neural networks for this purpose. Such networks may even identify new groups of peculiar stars.

La Sala: How do you remove the continuum residual? Continuum effects can contain as much power as the lines and can bias the results.

von Hippel: We remove the continuum by a combination of linear and median boxcar smoothing, then dividing this "continuum" out.

La Sala: Have you dealt with reddened spectra?

von Hippel: Not yet.

La Sala: Do you need to retrain your system for each new instrument or instrumental set-up? How many spectra are needed to train?

von Hippel: Probably. As long as the flux scale can be accurately remapped a previously trained neural network could be used. Houk's plate data, however, probably cannot be precisely flux calibrated. The number of spectra needed to retrain depends on the accuracy and the spectral type–luminosity domain which is required, but I don't yet know the actual figure.

Craine: I would like to comment that the efforts under discussion in this session have very close analogs in medicine related to comparison of various medical systems, particularly those involving, at some point, a human expert. A very useful tool has been developed from signal-detection theory, known as the Receiver Operating Characteristic (or ROC analysis). This tool is unknown in astronomy and has considerable potential power. The classical reference is: Swets, J. A. and Pickett, R. M., *Evaluation of Diagnostic Systems*, Academic Press: New York (1982). Examples of applications can be found in Craine, E. R. and Seely, G. W. *SPIE Infrared Technology*, **572**, 15 (1985) and Craine, E. R. and Cramer, B. L., *SPIE Medical Imaging*, **1444**, 389 (1991).

von Hippel: I appreciate the comparison with medicine and the references. My collaborator, Michael Storrie-Lombardi, is a medical doctor and his interests lie in this domain as well.

Corbally: Could you explain why the quality and density of the data do not have to be uniform? In contrast, for visual classification both quality and exposure (or S/N) *are* important for learning and accuracy.

von Hippel: The library of training spectra may have an uneven distribution of types and quality as the neural network generalizes from all the spectra it has seen. By density I did not mean plate density, but rather the density of types at any given location in the stellar library.

The MK Process at 50 Years
ASP Conference Series, Vol. 60, 1994
C.J. Corbally, R.O. Gray, and R.F. Garrison (eds.)

Neural-Network Classification of Normal, WN, and A-type Stars

Wm. Bruce Weaver

Monterey Institute for Research in Astronomy, 900 Major Sherman Lane, Monterey, CA 93940

Abstract. With digital spectra and a classification system in the near-infrared (NIR) where modern detectors are most efficient, major classification survey projects need only an adequate automatic classification capability. Three examples of morphological classification by neural networks are shown here. Classification of WN-stars from equivalent-width measures were performed to an accuracy of 0.6 classes. Preliminary classification of dwarf stars of all temperature classes from spectral scans in the NIR at 15 Å resolution shows that an error of *at most* two types can be expected from this approach. A more complete study of A stars in the NIR at the same resolution provided a mean absolute error of 0.4 types in temperature and 0.15 classes in luminosity.

1. Introduction

The ubiquity of computer-compatible digital spectra and the power of spectral classification for a wide variety of astronomical problems naturally raises the issue of automatic computerized classification. One of the original motivations given by Morgan and Keenan for the creation the MK classification system was the study of Galactic structure. We are approaching the point where astronomers will be able to do Galactic structure research using Kapteyn's approach – not just in selected areas but by extending Nancy Houk's work to all the stars down to 16th - 18th magnitude.

In order for this to be practical, even with modern, nearly quantum-perfect detectors, we need the lowest possible dispersion that will permit us to classify to an error of about 0.5 in spectral type (where O, B, etc., are spectral *classes* and F7 is a spectral *type*) and about 0.5 in luminosity class. In addition, such a system should be able to classify major classes of peculiar stars and detect minor peculiarities. For reasons of detector quantum efficiency and minimizing interstellar reddening, the preferred spectral region is the near infrared region where the current, silicon-based, detectors are most sensitive (see Torres-Dodgen and Weaver 1993, hereafter TW, for a more complete discussion). The missing component for this to be practical is an automatic classification technique. I will show here that neural networks are a leading contender to fill this need.

Neural networks are introduced elsewhere in this volume (von Hippel *et al.* 1994). Inspired by biological networks of nerve cells, neural networks are more like highly parallel computing memories where the operations are association and similarity rather than the serial algorithmic processors of information usually found in scientific programming of statistical or rule-based classification. Because of the high degree of parallelism and interconnectivity, the neural-network solutions are insensitive to missing or incorrect data. Their nonlinear nature makes them well suited for complex morphological classification.

All the results described here were from three-layer back-propagation networks. Neural networks can overlearn or memorize a data set; thus, the error estimates described below for networks are obtained by extracting small randomly selected test sets from the data and learning on the residual. Typically, this was repeated many times, starting with naive networks, to provide a reasonable sample of the network classification accuracy. Thus the tests always occur on data that the network had not seen during learning.

2. Classification of WN Stars

Unlike WC stars (Torres *et al.* 1986), the characteristics of WN stars are not well correlated with their classification. Conti and Massey (1989) provide spectrophotometric data for leading emission-line features for nearly all the Wolf-Rayet stars in the Galaxy and in the Large Magellanic Cloud (LMC). When one examines the correlation of these emission-line characteristics with WN classification (Table 1), the highest r^2 is only 0.38. (The Early/Late description is merely a rough classification used by Conti and Massey and ignored in this study.) These poor correlations make this problem a good test for the capabilities of neural networks and were initially explored by Weaver (1990).

Table 1. Correlation Coefficients for WN-Stars.

	Correlation with WN Class	r^2
Gal/LMC	-0.36	0.13
He II λ 4686	-0.49	0.24
FWHM (λ 4686)	-0.61	0.38
Early/Late (not used)	(0.88)	(0.77)
N III/V λλ 4604 - 4640	-0.026	0.0007
He II λ 5411	-0.46	0.22
C IV λλ 5801,5812	-0.17	0.03
He I λ 5876	0.50	0.24
N IV λλ 3478,3483	-0.44	0.19

There were 136 stars with enough data for this study. As a control, a multiple linear regression was calculated for the data set. The distribution of classification errors from the multiple linear-regression fit to the data is shown in Figure 1. The large number of unclassified stars is due to one or more missing equivalent widths for that star. Another advantage of the neural network approach is that neural networks can perform a classification with missing data.

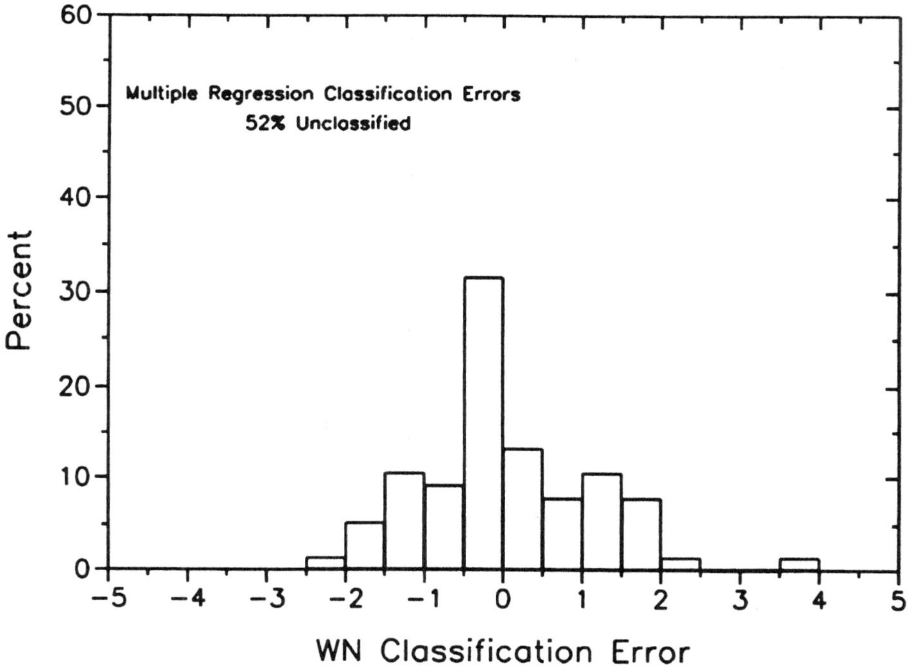

Figure 1. Histogram of Classification Errors of WN Stars by a Multiple Linear-Regression Fit.

Figure 2 shows the error distribution for 100 neural network solutions; 20 stars were used as test stars in each solution.

Figure 3 shows the distribution of errors by the neural network solutions displayed by WN class. Unlike a multiple regression fit, solutions outside the range of examples used in the learning are not found. A comparison of the results is shown in Table 2 .

Table 2. Comparison of WN-Star Classification Techniques.

Technique	Error (classes) ave abs error
Multiple Regression	1.1
Neural Network	0.62

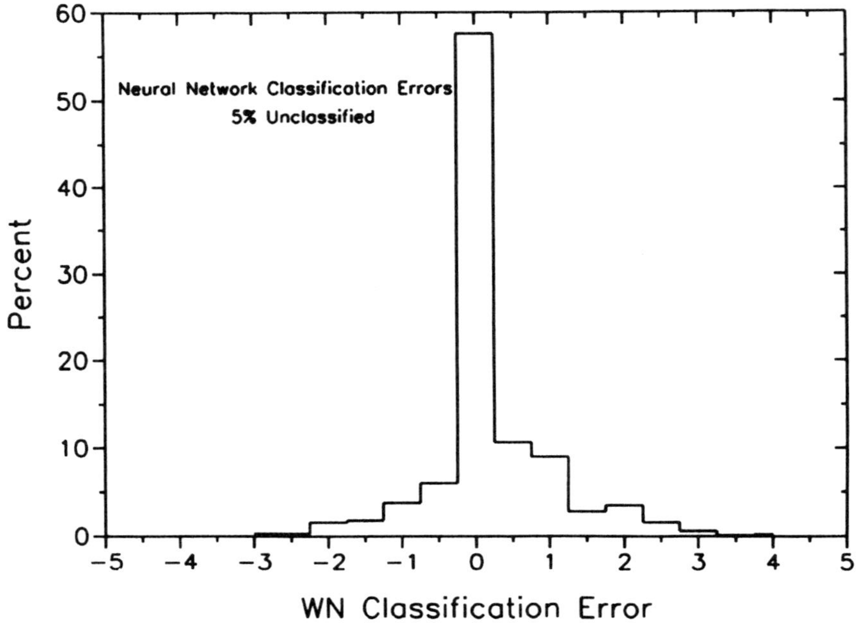

Figure 2. Histogram of Classification Errors of WN Stars by a Neural Network.

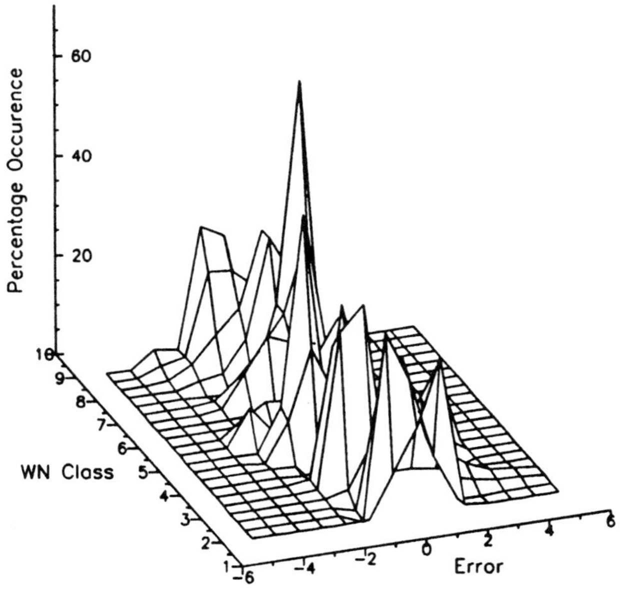

Figure 3. Distribution of Classification Errors of WN Stars by a Neural Network.

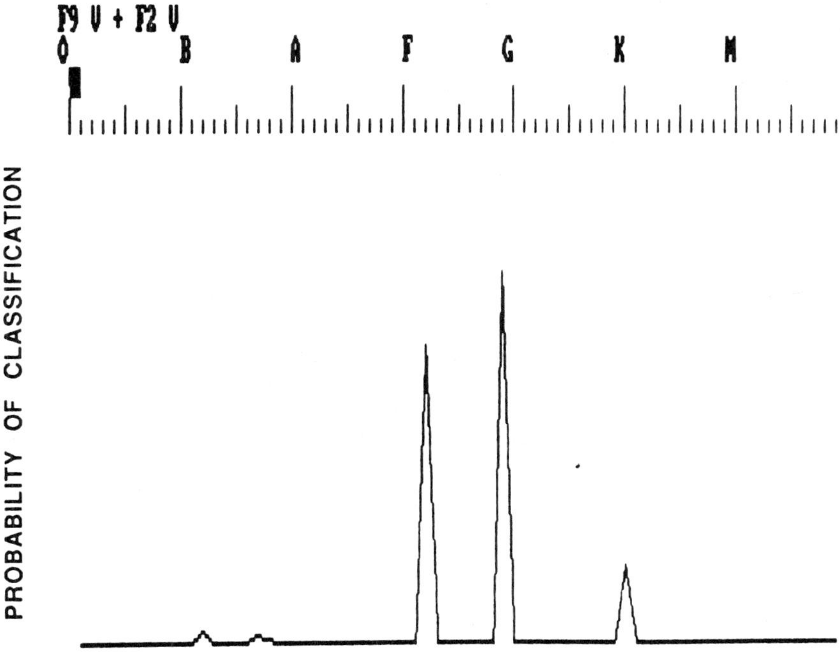

Figure 4. Discovery of a Synthetic Composite Spectrum (F9 V + F2 V) by a Neural Network

3. Classification of NIR Spectral Scans

The previous and subsequent examples of neural network classifications depend on measured equivalent widths. For large surveys, such as studies of Galactic structure, the goal is to perform the classifications with the minimum of human attention. Thus, I have taken a preliminary look at using neural networks to classify NIR spectral scans of the type shown in TW and by Torres-Dodgen (1994).

These spectra were taken with a 512-element unintensified Reticon in the wavelength range $\lambda\lambda$ 5750 - 8950 with a resolution of 15 Å. A set of spectra of 27 dwarf stars with spectral classes ranging from B2 to M6 were learned while, for each solution, one star was held out for testing. This is much too small a data set for adequate learning but it is adequate to test if precise classification from these data is a reasonable goal. The network classified to a precision of a mean absolute error of 2.2 classes. This result, similar to that obtained by von Hippel based on classification of a set of Houk's objective-prism data, should be taken as an upper limit on the error expected after a learning with a more reasonably sized data set.

In addition, a number of synthetic composites were created by simple addition of two sample spectra. The neural network, which had not been trained on composite spectra, was quite successful in determining that it was classifying combined spectra and was especially good at determining the two underlying classes when the types were within a subclass or two (e.g., Figure 4). Contrary to human classifiers, the network was less accurate in determining the underlying classes when the composite comprised widely separated classes. This performance is best described as a *discovery* of composite spectra.

4. Classification of A Stars in the NIR

Continuing the development of the NIR classification system of TW, Torres-Dodgen (1994) measured equivalent widths for a sample of A-star standards which fairly completely cover the temperature and luminosity range. Based on visual inspection of spectra and plots of the variation of luminosity and temperature with equivalent widths, a classification accuracy of ± 1.5 spectral types and about one luminosity class was estimated. These data provided an opportunity to compare these classical classification techniques with linear statistical methods and a nonlinear neural network. The correlation coefficients displayed in Table 3 show that, as is typical in the NIR, there is good correlation of these EWs with luminosity but relatively poor one with temperature.

Table 3. Correlation Coefficients for A Stars.

Abs. Line	Temp	Luminosity
H alpha	-0.057	0.86
O I	0.025	-0.84
P_{18} + O I	0.048	-0.87
P_{17}	-0.10	-0.84
P_{16} + Ca II	0.16	-0.80
P_{15} + Ca II	0.36	-0.66
P_{14}	-0.45	-0.46
P_{13} Ca II	-0.11	-0.34
P_{12}	-0.55	0.51

Equivalent-width data were available for 143 scans of A stars. The first statistical technique employed was multivariate regression. In this approach both temperature and luminosity are simultaneously treated as dependent variables. This fit gives a mean absolute error of about one type in temperature and about a half class in luminosity (Table 4). These results are comparable to or modestly better than the manual (i.e., human visual) results.

Factor analysis, similar to principal component analysis where the eigenvectors are rotated to simplify the loadings, was also attempted. Since, in the factor analysis, fewer variables are ultimately used, poorer results are expected than those found in the full multivariate regression. When three factors are included, an accuracy similar to that obtained with manual classification is achieved.

Table 4. Comparison of A-Star Classification Technique Results

Linear Statistical Techniques

	Factor Analysis (2 factors)		Factor Analysis (3 factors)		Multivariate Regression	
	Temp	Lum	Temp	Lum	Temp	Lum
Mean abs. error	2.02	0.53	1.30	0.51	0.97	0.47
Correlation coef (r)	0.44	0.88	0.79	0.88	0.89	0.90

Neural Networks

	2 hidden neurons		3 hidden neurons		8 hidden neurons	
	Temp	Lum	Temp	Lum	Temp	Lum
Mean abs. error	0.79	0.51	0.65	0.38	0.42	0.15
Correlation coef (r)	0.94	0.88	0.95	0.93	0.98	0.96

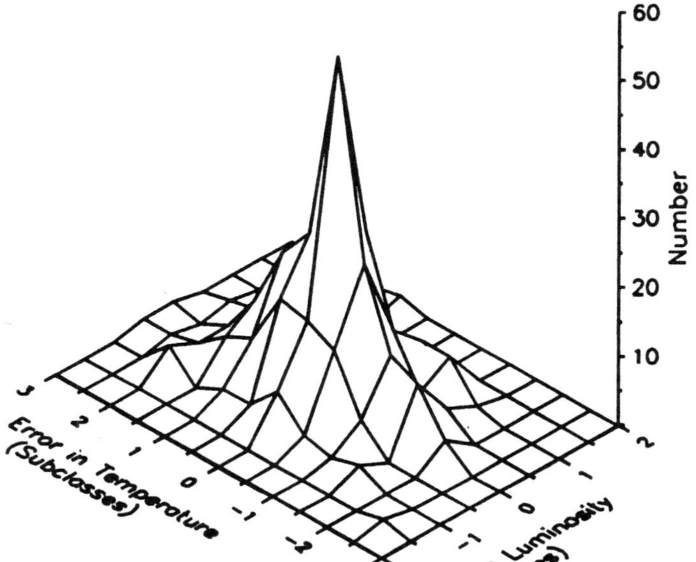

Figure 5. Distribution of Classification Errors of A-Stars by a Neural Network.

For the neural network analysis, two, three, and eight neurons were used in the hidden layer. When compared to the corresponding linear statistical techniques (two factors, three factors, and full multivariate regression, respectively), the neural network classifications show about half the error in temperature as the linear classifications. The ability to classify luminosity is fairly constant at an error of 0.5 classes until eight hidden neurons when it drops to 0.15 classes.

Unlike statistical approaches, there should be no stigma associated with increasing the number of neurons in a neural network fit. This is analogous to requiring the human classifier to use less of his brain! Since the mean absolute error is calculated from test stars excluded from the learning set, it is an accurate estimate of the ability of the network to classify. For example, in the eight-neuron case, the neural-network solution was performed 51 times, each time excluding about 10% of the stars selected at random upon which the evaluation of the network fit was performed. The distribution of errors of classification in temperature and luminosity for the eight-neuron case is shown in Figure 5.

An examination of the factor loadings for the 3-factor analysis and the neuron weights for the three-hidden-neuron network is interesting. The factor loadings reflect the Ca II and hydrogen line strengths in the classification as discussed by Torres-Dodgen. The neuron weights, however, show the primary linear effects plus weights which clearly represent the nonlinear terms which reduce the errors by a factor of two.

5. Conclusions

Neural networks are able to classify WN stars on the basis of equivalent widths as well as human classifiers and significantly better than linear statistical techniques. They can perform spectral classifications based on spectral scans of 15 Å resolution at, at least, 2 spectral type accuracy. With a reasonable learning-sample size, they are able to classify A stars to an accuracy of 0.5 spectral types and 0.2 luminosity classes.

Neural networks are capable of the classification accuracy needed to perform many of the desired automatic classification tasks. Since current networks have only a fraction of the neural interconnections found in simple invertebrates, substantial improvement may yet be possible.

Acknowledgments. This work was supported by the Friends of MIRA and the Mericos Foundation.

References

Conti, P. S. and Massey, P. 1989, ApJ, **337**, 251

Torres, A. V., Conti, P. S., and Massey, P. 1986, ApJ, **300**, 379

Torres-Dodgen, A. V. 1994, this volume

Torres-Dodgen, A. V., and Weaver, Wm. B. (TW) 1993, PASP, **105**, 693

von Hipple, T., Storrie-Lombardi, M., Storrie-Lombardi, L., Irwin, M. J. 1994, this volume

Weaver, Wm. B. 1990, *BAAS*, **22**, 848

Discussion

Garrison: Do I understand correctly that you used equivalent widths as your data? Why not use all the pixels?

Weaver: The WN and A-star problems used only equivalent widths since those were available. The temperature classifications used raw Reticon spectra. I expect the A-star classification to improve when we use the whole spectrum, and this meeting now inspires me to write a cross-correlator to do that.

Garrison: Also, how does the net handle reddening, telluric lines, etc.?

Weaver: Reddening *would* be handled in the usual way: the network would have to be trained to recognize the effect – just as you would train a human classifier.

von Hippel: You have pointed out that, even if continuum information remains in the spectra, reddening need not be a problem for the neural networks if they are trained with stars with known reddening values and these are given as "classes" to be learned. I believe the problem can also be inverted to give a line and continuum type versus a line type, and that would be a measure of reddening. Your thoughts on this, please?

Weaver: In theory, the more information that you give the neural network, the better. Adding reddening as something for the network to determine is probably much easier for the network than distinguishing among the B stars. The only cost is a few more hidden neurons and another output neuron. Again, the process emulates human learning and the first thoughts about how it should work are guided by thinking how a human would be trained.

Olsen: How does the network react to interstellar lines?

Weaver: We have not (deliberately) shown it interstellar lines. You would train it by showing the network stars with and without I.S. lines. It would learn to ignore them or you could let it identify them as interstellar.

The MK Process at 50 Years
ASP Conference Series, Vol. 60, 1994
C.J. Corbally, R.O. Gray, and R.F. Garrison (eds.)

Automatic Spectral Classification Using Metric-Distance Techniques

Jerry LaSala

University of Southern Maine, Portland, ME 04103

Abstract.

Automated classification methods are widely recognized as a necessary next step in the evolution of stellar spectral classification. Such methods offer the promise of greater homogeneity of classifications as well as the increased speed required by programs to classify fainter stars and by the rapidly increasing volume of data produced by modern digital detectors.

Pattern-recognition techniques constitute one of two general approaches to the problem of automatic classification, the other being criterion-evaluation methods. Pattern-recognition approaches are more closely correspondent to traditional visual classification by human classifiers. Several approaches to pattern recognition are possible, including principal components methods, metric distance techniques, and the use of artificial neural nets.

For the past several years I have been engaged in developing and refining automatic classification programs based on metric-distance methods. In this talk I will present a brief review of the historical background of these methods and the current status of metric-distance classification efforts. I will report on my most recent work, the application of these methods to approximately 250 MK-dispersion slit spectra of solar-type stars obtained and visually classified by Garrison.

1. Introduction

I should like to begin by thanking Chris Corbally for inviting me to speak at this workshop, and thanking all of you participants who have made this such an enjoyable and inspiring meeting.

As Ted von Hippel (*et al.* 1994) pointed out, the last major review of work in the field of automatic classification was that of Kurtz (1984) at the MK Process workshop. In that talk, Kurtz proposed the weighted metric-distance algorithm I shall discuss today. Almost immediately after that meeting Kurtz devoted himself full-time to the study of galaxy distributions; I had worked with him on some of his earlier spectral classification work and decided to continue the collaboration, taking over much of the project myself. The project has become very much a part-time endeavor, due to lack of funding, and progress tends to come in aperiodic delta functions, with occasional progress reports surfacing at various meetings (LaSala 1987, 1988a, 1988b, 1989; Kurtz and LaSala 1991). Chris's invitation to report here occasioned the latest singularity in productivity. I am going to try to fill you in on progress in metric-distance spectral classification since 1983.

First, some background. West (1976) points out that we may divide the approaches to automated spectral classification into two groups, which may be called criterion evaluation and pattern recognition.

The criterion-evaluation approach consists in measurement of various specific spectral features and calibration of these measures into the desired parameters, such as effective temperature and surface gravity, or spectral type and luminosity class. Examples of criterion-evaluation presented at this meeting are the papers of Meyers-Rice and Young (1994) and Malyuto and Shvelidze (1994).

Pattern-recognition methods include cross-correlation, various multivariate statistical methods, artificial neural nets, and metric-distance methods. All seek to emulate the human classifier in making use of patterns within the whole spectrum and comparing the spectrum to be classified with some sort of standard or learning set. At this workshop, von Hippel (*et al.* 1994) and Weaver (1994) presented classification systems based on artificial neural nets, and I shall discuss metric-distance algorithms. The work of Penprase (1994) straddles the division, containing elements of both the criterion-evaluation and metric-distance approaches.

2. The Metric-Distance Algorithm

Each digitized spectrum is regarded as an n-element vector, where n is the number of resolution elements or pixels. The standard spectra **S** represent fixed points in this space. The metric distance d_{xs} between a program spectrum **X** and a standard **S** is given by

$$d_{xs} = \frac{1}{n}\sqrt{\Sigma_{i=1}^{n} \alpha^2 \left[X_i - S_i\right]^2},$$

where α^2 is a weighting factor to be defined. If α^2 is set to 1, we have the ordinary Euclidean metric; Penprase (1994) uses a step function: $\alpha^2 = 1$ at selected features, $\alpha^2 = 0$ elsewhere.

Kurtz (1984), at the 1983 MK Process meeting, proposed defining

$$\alpha^2(i) = \frac{\Sigma^2(i) - \sigma^2(i)}{\sigma^2(i)},$$

where Σ^2(i) is the variance of pixel (i) within the error box associated with some initial approximation, e.g., "G," and σ^2(i) is the variance of the same pixel within the final classification box, e.g., "G2." This definition gives the most weight to those features which have the smallest variance within the final classification box and larger variance among boxes. Thus it weights most strongly those features which discriminate the final classification from the general mix of spectra. Kurtz and I have adopted and continue to use this metric in our work.

Relative advantages of the various approaches are discussed elsewhere (Kurtz 1984; LaSala 1988). Suffice it to say here that criterion-evaluation methods tend to be faster, while pattern-recognition methods are more analogous to the process of visual classification and more in keeping with the spirit of the MK Process. I do wish to emphasize what I see as the important advantages of the particular metric-distance algorithm we have adopted.

To begin with, this approach follows the spirit and philosophy of the MK Process. I was not always sure that this is so, so some of the earlier papers on this work have titles like "Automatic MK-style Classification" and "Automatic Two-dimensional Classification." After considerable reflection and study, I have become convinced that our approach is indeed compatible with the spirit of the MK Process. In addition, our method allows direct comparison of each program star with the classification standards. The weights used to determine the classification can be displayed and examined; comparison of the features given the greatest weight with the features most strongly relied on by human classifiers serves as a check on the validity of the system.

The system should be easily transferable to new instrumental setups by re-observing the standards, which an observer using a new setup must do anyway. There should be no need to re-observe the large numbers of spectra needed to calculate the weighting factors; the weighting factors are a property of the classification system and not of the particular instrument.

Moreover, the metric-distance technique can readily identify peculiar spectra, and can isolate as peculiar features which deviate from the standard by several standard deviations. It can also quantify the degree of peculiarity in terms of the known variance of the feature within the classification box. Finally, our technique facilitates the search for new classification dimensions by identifying features with large variance in the final classification boxes. These are features inadequately discriminated in the current classification system.

3. Rectification

I should like to take a moment to emphasize the importance of rectification or continuum removal for all pattern recognition techniques. There is far more power, in the Fourier spectrum sense, in the spectral continuum and the instrumental response function than in the spectral lines. Even small residual variations, after rectification, can be comparable in strength to the lines, and pattern-recognition programs will tend to focus on these variations as much as on the lines. Thus it is important that such variations be kept to an absolute minimum.

Kurtz and I have tested a number of continuum removal techniques and currently prefer a multi-step process we call "reflattening." We begin by applying the Fourier division technique (LaSala and Kurtz 1985) to produce a rectified spectrum. Then a Fourier-smoothed residual continuum is calculated and subtracted from the rectified spectrum to produce the reflattened spectrum. In addition, anytime we calculate a difference between spectra, we reflatten the difference spectrum to remove any third-order continuum effects. As Figure 1, from Kurtz and LaSala (1991), shows, the results are quite good.

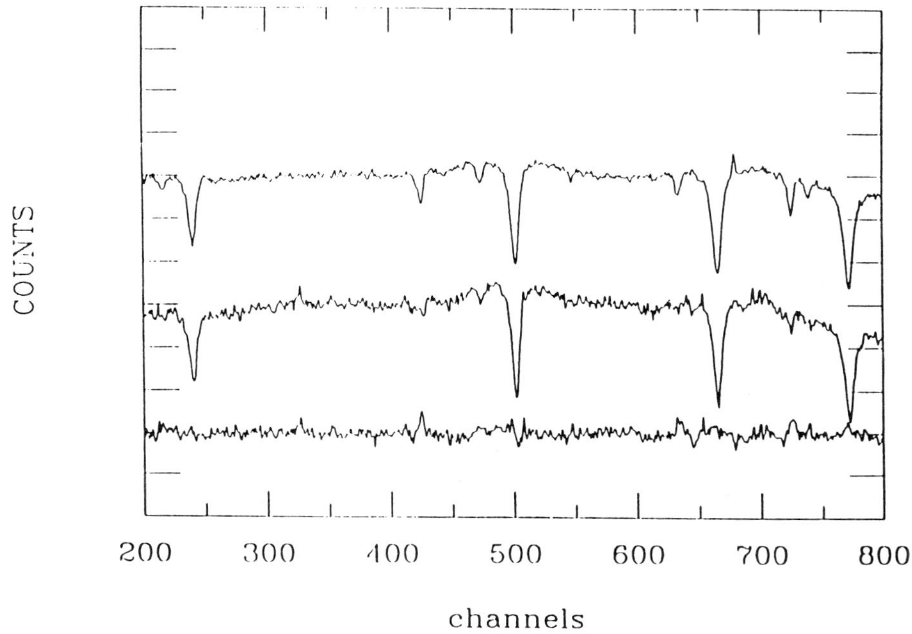

Figure 1. Top: typical B3 V spectrum. Middle: typical B7 II spectrum. Bottom: difference between top and middle spectra.

4. Recent progress

Now let me report on what has been done in the field of weighted-metric-distance classification in the past decade. In that time I have undertaken one minor and two major projects. The first was a trial run in 1985. I obtained several 4 degree objective-prism Burrell Schmidt plates originally taken by McCuskey in the 1940s, and I digitized these on the PDS microdensitometer at the David Dunlap Observatory. Two undergraduate research students, K. Andrews and R. Rae, tested cross-correlation based classification programs on these spectra. The results were comparable to Kurtz's (1982) results with spectra of similar dispersion. This encouraged me to undertake more ambitious projects.

In 1986, Nancy Houk graciously lent me 8 plates obtained as part of the Michigan Spectral Survey (Houk and Cowley 1975, Houk 1978, 1982). I digitized these at the APM facility at the Institute of Astronomy, Cambridge. I obtained a total of about 3500 digitized spectra. To date, Kurtz and I have analyzed and classified the subset of this data consisting of the approximately 350 B-type stars. We presented results at the 1991 Sanduleak Memorial workshop (Kurtz and LaSala 1991). Figures 1-7 are reproduced from that paper. In Figure 1 we see mean B3 V and B7 II spectra, both reflattened, and their difference, also reflattened. Figure 2 shows the square of the difference spectrum; integrating this would give the Euclidean distance between the two spectra. But many of the biggest components of this distance can be seen to be such things as the

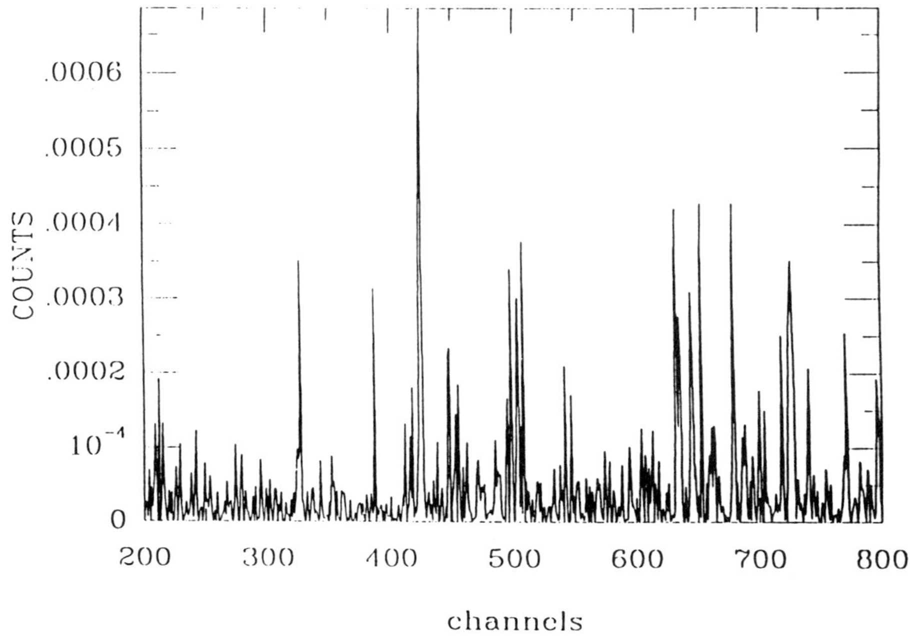

Figure 2. Square of the difference in Figure 1.

cores of the hydrogen lines and instrumental glitches, which we know carry little real information. We improve our discrimination by calculating the variances and weights discussed above. Figure 3 shows σ^2 the variance among all spectra classified by Houk as B7 II. Figure 4 shows the weighting factors α^2 calculated from σ^2 and the overall variance among B stars. We see that the weighting factors emphasize the He lines and the wings of the Balmer lines. Multiplying Figure 4 by Figure 2 and summing gives the squared metric distance between the spectra. We applied these methods to the full sample of B-star spectra, using one step to assign a temperature class and a second to assign a luminosity class. We get a mean error in spectral class of 1.14 subtypes, and our luminosity errors show the same diagonal structure familiar to visual classifiers (Figure 5), showing that the program is having the same difficulties, and making the same mistakes, as human classifiers. We believe that we can improve our mean errors considerably by a second pass through the classification program, with weights recalculated based on the new error boxes, but this work remains to be done. Figures 6 and 7 illustrate the power of the metric-distance technique for identifying peculiarities. Figure 6 is a peculiar B7 II spectrum and Figure 7 the difference between this and the mean B7 II spectrum. We readily see the features which make the spectrum in Figure 6 "weird." By comparing Figure 7 with the σ^2 intrinsic variance among B7 II stars, we can quantify "how weird it is."

The third project, and most recent results, involve slit spectra of higher dispersion. In 1988 Bob Garrison offered me access to his collection of slit spectra for further work. I spent six weeks that summer at DDO digitizing

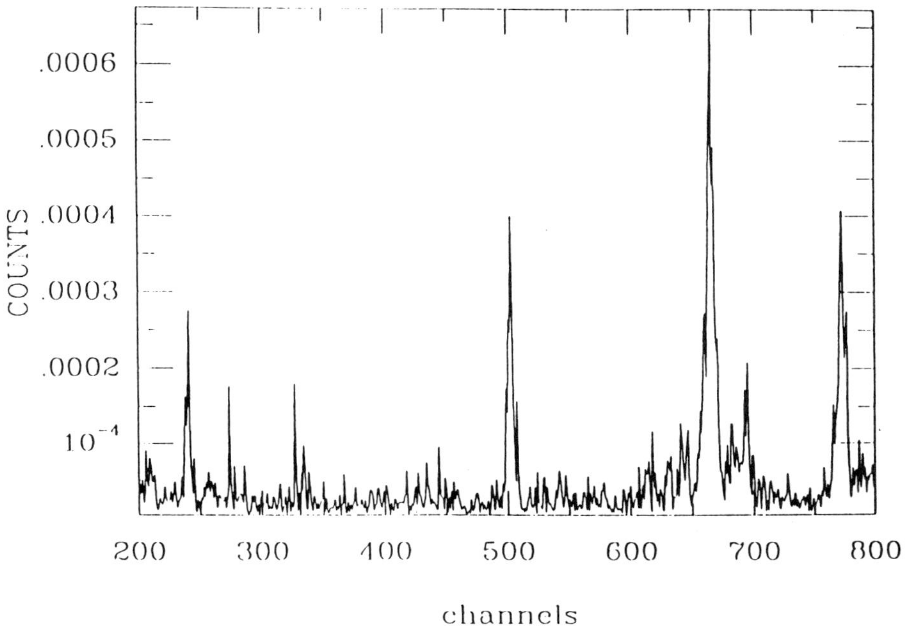

Figure 3. Variance spectrum among all B7 II spectra.

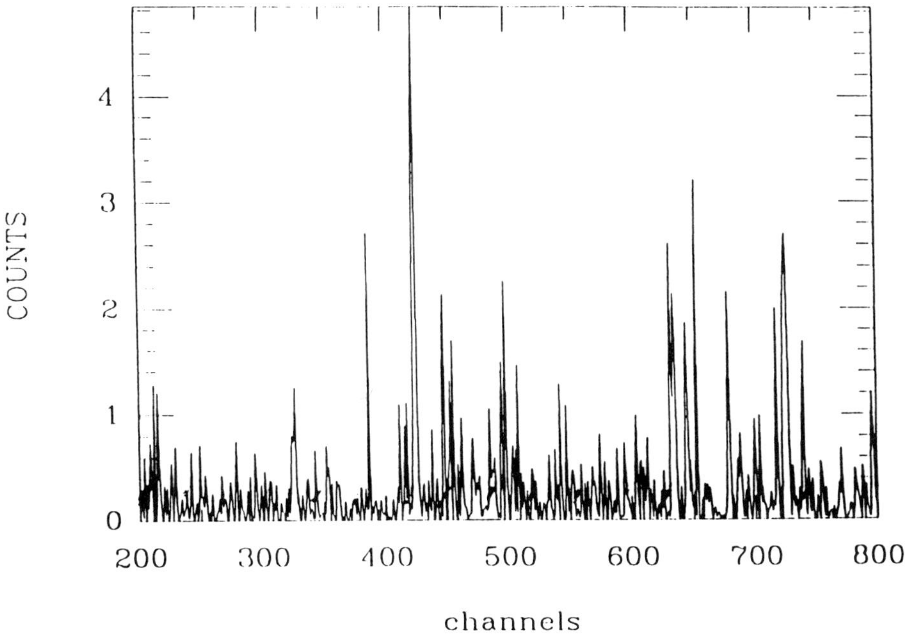

Figure 4. Weighting factors for B7 II classification.

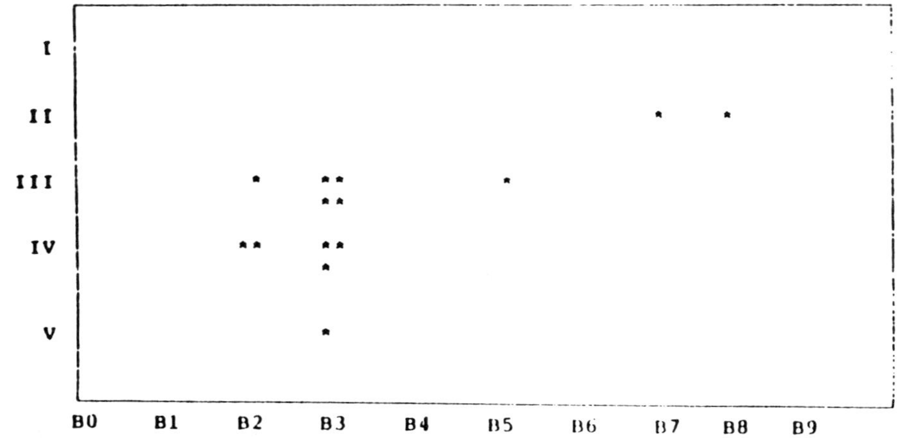

Figure 5. Distribution of program classifications of spectra classified B3 III by Houk.

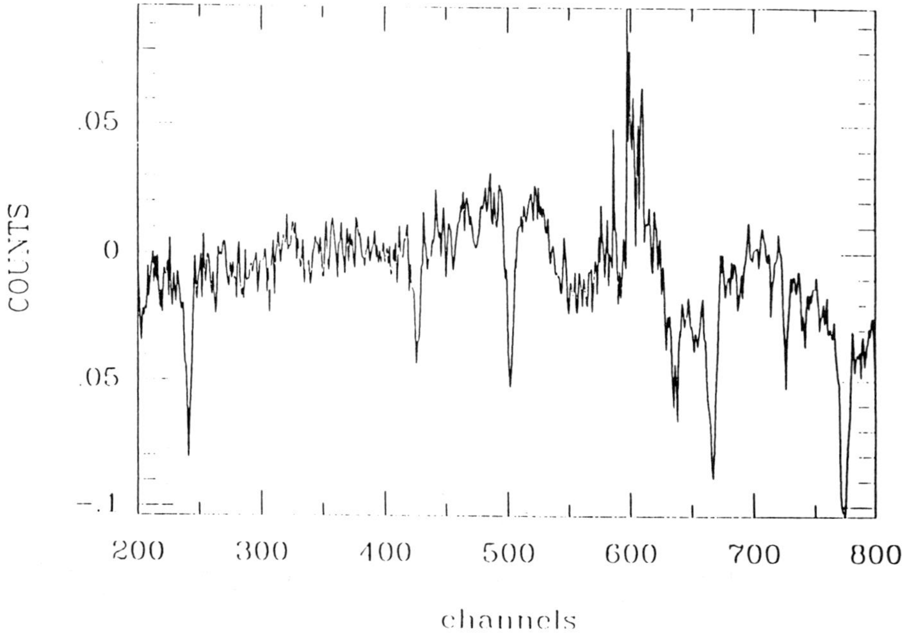

Figure 6. A peculiar B7 II spectrum.

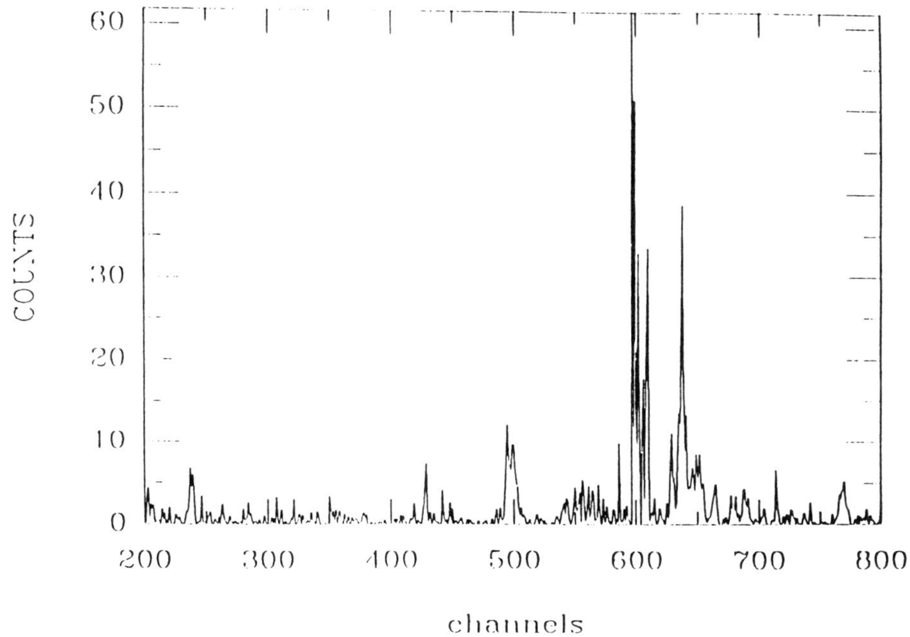

Figure 7. Difference between Figure 6 and mean B7 II spectrum.

selected samples from his extensive data base. I obtained a sample of about 250 solar-type dwarfs, half at a dispersion of 67 Å/mm and half at 120 Å/mm, and a sample of about 100 yellow supergiants at 67 Å/mm. All spectra were digitized and reduced to intensities at DDO; further analysis was and continues to be done at the Harvard-Smithsonian Center for Astrophysics.

This summer I undertook the analysis and classification of the 67 Å/mm solar dwarf spectra, spurred on by the need to give this talk! I wavelength calibrated the spectra and reduced them to a linearized wavelength scale. About 40 spectra had to be temporarily removed from the sample at this point due to unsatisfactory wavelength calibration; these will be recalibrated and used in future work. The remaining 80 spectra were rectified by the Fourier division technique but have not been reflattened in the preliminary work presented here. Means, variances, and classification weighting factors were calculated for each subclass based on the types assigned by Garrison. Figure 8 shows the mean spectrum for type G0; Figure 9 shows the variance σ^2 within subclass G5; Figure 10 shows weights α^2 for subclass G5.

Using the mean spectra as standards, the stars were classified; Figure 11 compares the results of the program classification with the types assigned by Garrison. We notice immediately that the stars classified by Garrison as G5 show unusually large errors in program classification. Inspection of the individual spectra revealed that several with bad wavelength calibrations were inadvertently left in the sample; time did not permit recalculating these results before the meeting. Even with these known errors left in, the mean error in the

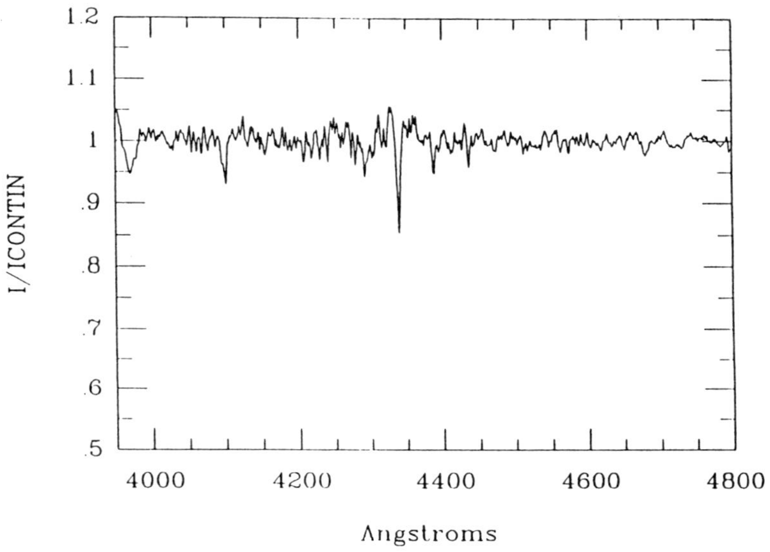

Figure 8. Mean G0 spectrum.

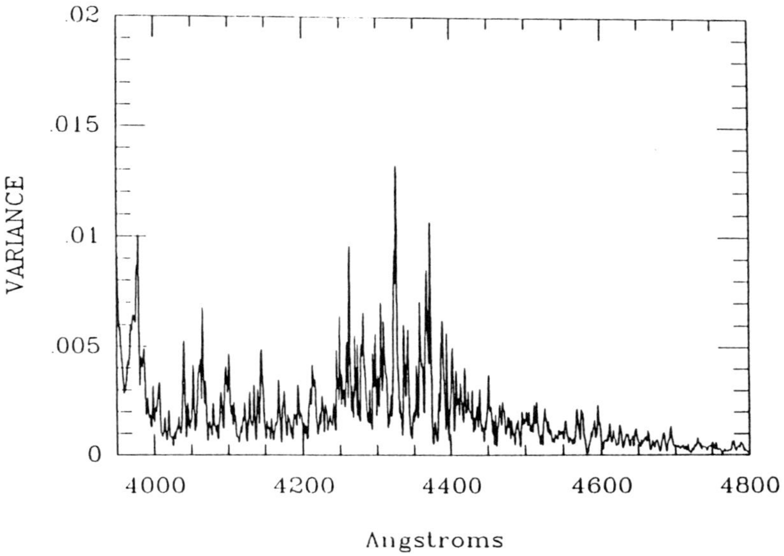

Figure 9. Variance among G5 spectra.

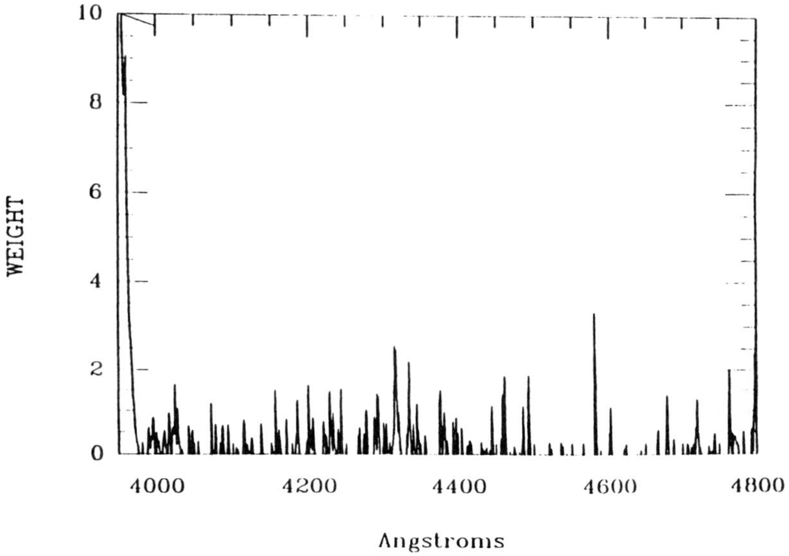

Figure 10. Weighting factors for classification G5.

Program Type \ Garrison Type	F8	G0	G1	G2	G3	G4	G5	G6	G8
G8									2
G6								3	
G5							5		
G4		1		3	1	2			
G3					5				
G2	1	5	1	10			7		
G1			9	1		1	2		
G0		3							
F8	1								

Figure 11. Comparison of program types with Garrison types for G-star spectra.

program classifications, as compared with Garrison's values, is just under 0.4 subtypes. In addition, 4 asteroid spectra not included in the computations of means or weights were classified. Two of these were classified as G2, and two as G1. These mean errors are comparable with Garrison's estimates of his own mean errors of visual classification.

5. Future plans

I emphasize that the results just presented are preliminary. I plan now to go back and recalibrate the spectra rejected from the current sample, reflatten all the spectra, and recalculate all the means, variances, and weights, and reclassify the spectra. I also plan to test the weights thus obtained on the classification of the 120 Å/mm spectra, as a test of my assertion that the weighting factors are transferable from one instrumental setup to another. I also plan to return to the data provided by Houk and apply the classification programs to the full range of spectra in that data set, and to reduce and classify the yellow supergiant spectra obtained from Garrison. In addition, Ted von Hippel and I are planning a collaboration, comparing our respective techniques on the large database of spectra he is obtaining from Houk.

Acknowledgments. I wish to thank R. F. Garrison and Nancy Houk for providing data for this work; and Garrison and M. J. Kurtz for collaboration and many valuable discussions which contributed to the evolution of this project. Partial support for this work was provided through the AAS Small Research Grants Program by a grant from the Margaret Cullinan Wray Charitable Lead Annuity Trust. Partial support for attendance at this workshop was provided by the Vatican Observatory Research Group.

References

Houk, N., and Cowley, A. P., 1975, *University of Michigan Catalog of Two-Dimensional Spectral Types for HD Stars*, Vol. 1 (Ann Arbor: University of Michigan).

Houk, N., 1978, 1982, *University of Michigan Catalog of Two-Dimensional Spectral Types for HD Stars*, Vol. 2, 3 (Ann Arbor: University of Michigan).

Kurtz, M. J. 1982, *Thesis*, Dartmouth College.

Kurtz, M. J. 1984, in *The MK Process and Stellar Classification*, ed. R. F. Garrison (Toronto: David Dunlap Observatory), 131.

Kurtz, M. J., and LaSala, J. 1991, in *Objective-Prism and Other Surveys*, ed. A. G. D. Phillip and A. Upgren (Schenectady, NY: L. Davis Press), 133.

LaSala, J., and Kurtz, M. J. 1985, PASP, **97**, 605.

LaSala, J. 1987, BAAS, **19**, 749.

LaSala, J. 1988a, in *Astronomy from Large Databases*, ed. A. Heck and F. Murtagh (Garching-bei-Munchen: European Southern Observatory), 127.

LaSala, J. 1988b, "Application of Pattern-Recognition Techniques to a Homogeneous Database of Stellar Spectra", paper presented at special session of IAU Commission 45.

LaSala, J. 1989, BAAS, **21**, 759.

Meyers-Rice, B. A., and Young, E. T. 1994, this volume.

Penprase, B. E. 1994, this volume.

Malyuto, V., and Shvelidze, T. 1994, this volume.

von Hippel, T., Storrie-Lombardi, M., Storrie-Lombardi, L., and Irwin, M. 1994, this volume.

Weaver, W. B. 1994, this volume.

West, R. M. 1976, in *Proceedings of the Third European Astronomical Meeting*, ed. E. K. Karadzi (Tbilissi: Abastumani Astrophysical Observatory), 23.

Discussion

Houk: You said you will ultimately use standards. Will you also be using other stars in the classification?

LaSala: Spectra to be classified will be compared directly with the standards. The large body of other classified spectra, for instance the spectra digitized from your plates, are still necessary to "train" the system by establishing the variances and weighting factors. The training, however, should not need to be repeated for each new instrumental setup; the weights once established should be transferrable.

Abt: I wonder whether your definition of weirdness is the same as our definition. For instance, an Ap(Sr) star may have ten times the normal Sr abundance and the normal stars already have strong (i.e., on the flat part of the curve of growth). But an Ap(Hg) star has 1000-10,000 times the normal abundance of Hg but the result is still a weak Hg line. To us, the Ap(Hg) star is more weird than the Ap(Sr) star. By your definition, is the weirdness dependent on equivalent widths?

LaSala: Our weirdness measure depends on the deviation of a feature from the standard relative to the intrinsic variation of that feature within the classification box. So, if a weak line is less weak than the standard by a few standard deviations, that will show up as very weird. This is probably not identical with your definition of weirdness, but it will call attention to the spectra which you will want to examine further.

Von Hippel: How much computer time does it take to classify one stellar spectrum?

LaSala: Our current programs are by no means optimized for speed, but they classify a sample star in about fifteen seconds on a VAX 11-750.

Gray: You made the point early in your talk that your method is more consistent with the philosophy of the MK System because it was measuring, in a metric space, distances from actual standard stars. This is in contrast to the neural network method which learns what an A2 III spectrum is, say, by looking at large numbers of spectra classified as A2 III. Thus it seems to me that the neural network method is not truly based on direct comparison with standard stars. I wonder if you, and von Hippel, and Weaver, could all comment on this.

La Sala: I question whether neural nets can be standard-based in the same way, but I don't know.

von Hippel: Philosophically, our neural networks are taking Houk's classifications as "standards" – maybe tertiary standards on your hierarchy of standards. We will have to demonstrate that the clouds of Houk's classifications are statistically equivalent to the classifications of the primary standards.

Garrison, to von Hippel: Is there any way that you can inject into the neural net technique a high weight for standards, as opposed to using a cloud of stars classified (e.g., A0 V) by Houk? I think that it is important to use the standards as part of the learning net with more weight than all the other stars.

von Hippel: If there were enough standards, we could use just them to train on.

Weaver, to Gray: Neural networks can learn on multiple observations of standards. The variance between different measures of the same standard are usually larger than the star-to-star (cosmic scatter) for that classification box. In fact, we *have* trained on standards only; these results were about the same as those given in my presentation.

Weaver, to LaSala: Your technique seems very sensitive to wavelength scale and zero point. Will this be a problem in classifying unknown stars?

LaSala: It is less sensitive to wavelength scale than it appears. Also, the initial step is a cross-correlation to determine approximate spectral class and wavelength scale. This information can then be used for an automated wavelength calibration by fitting to selected lines if necessary. There should be little problem except for objects so extreme that they cannot be matched with any class. These should be visually inspected anyway.

The MK Process at 50 Years
ASP Conference Series, Vol. 60, 1994
C.J. Corbally, R.O. Gray, and R.F. Garrison (eds.)

Digital Spectral Classification of Stars: An Automated Method

B. E. Penprase

Pomona College, Dept. of Physics and Astronomy, Claremont CA 91711

Abstract. A new method of digital spectral classification is presented, which builds upon earlier metric classification techniques developed by Kurtz and La Sala. The new technique has been applied to a sample of over 110 stars which appear toward high-galactic-latitude molecular clouds, in connection with a study of the interstellar medium. For each of the program stars, a comparision is made with an atlas of standard stars. The comparison is a statistical test which computes a mathematical measure of similarity between program and standard star. A variety of stellar rotational velocities and instrumental background levels is simulated in the comparison to determine more accurately the temperature classification for the star. An initial temperature classification is made based on the Ca II and Hγ line strengths. Then a revised temperature classification is determined based on four indepenent metrics. The luminosity classification is also based on four metrics which are sensitive to both the stellar line strength and the shape of the line profile. The technique offers the possibility of a simple, repeatable technique for classifying digital spectra, which can be applied to faint stars in crowded fields.

1. Introduction

Since the pioneering work of Morgan, Keenan, and Kellman (1943) and Johnson and Morgan (1953), the power of the MK classification system has been amply demonstrated in its ability to reveal the details of stellar evolution and galactic structure. The additional increase in the state of astronomical instrumentation has brought about revolutionary improvements in the quality of stellar spectra, which has allowed spectral classification in ultraviolet wavelengths (Walborn & Fitzpatrick 1990) as well as in the near-infrared wavelengths (Torres-Dodgen & Weaver 1993). The rise of multi-spectral studies of stellar properties, and the predominance of digital detectors on spectrographs has brought about additional opportunities for spectral classification which require techniques much different than the established objective-prism, spectral-classification technique for optical spectra.

The use of digital spectra should theoretically allow for the development of a spectral classification technique which is completely repeatable, objective, and sensitive to both the strongest and weakest lines in the stellar atmosphere. Recent efforts in digital spectral classification generally divide into two basic approaches, neural nets and artificial intelligence techniques (von Hippel *et al.* 1993), and metric based techniques (Kurtz 1988, 1982; La Sala 1988, 1994).

Most previous automated classification techniques have also been 'trained' and tested with the same set of spectra, and the accuracy of the technique is generally evaluated based on a sample of stars with known spectral types. In the current work we present a technique which was developed for the purpose of providing spectroscopic parallaxes to stars which are toward high-galactic-latitude molecular clouds. The classification is based on a comparison with the digital spectral atlas of Jacoby *et al.* (1984), and we test the accuracy of the technique using an independent set of spectra, obtained for the high-galactic-latitude molecular cloud project. The complete details of the observational program and the results may be found in Penprase (1992, 1993). The digital spectral classification method was designed to use the most basic and reliable criteria of the stars in our program, which were mostly in the range from B5-F5.

We have used in our program a combination of the metric technique of Kurtz and La Sala, and an analog of a photometric classification technique described by Davis (1974). Our technique is also unique in that it compares the shapes of the stellar absorption lines, and matches each program star to the closest standard star after allowing for variations in stellar rotational velocity and instrumental background level. The accuracy of our technique was determined by comparing the derived spectral types of our program spectra with the spectral types from the Michigan spectral catalogue (Houk & Cowley 1975). After verifying the accuracy of the automated spectral classification, we then applied the technique to 115 stars which appear behind high-galactic-latitude molecular clouds in an effort to derive distances and extinctions toward the clouds.

2. Observations and Data Analysis

The stellar spectra for our program were acquired during two observing runs at CTIO during March and December 1990, and during one run by Dr. J. C. Blades at the SAAO during January 1990. In all cases we obtained digital spectra of stars in the wavelength range from 3500-5100 Å, at a resolution of 1-2 Å per pixel. For each star, stellar and sky spectra were obtained simultaneously, and a He-Ne-Ar comparison spectrum was used for the wavelength calibration. Each spectrum was reduced with the IRAF system, and the continuum was fitted for the classification algorithm using a low-order spline.

The algorithm for spectral classification was developed in the IDL language and run on SPARC stations at the Space Telescope Science Institute. We obtained a machine-readable version of the digital spectral atlas of Jacoby *et al.* (1982) from the NSSDC, which provided the reference spectra for the comparision with program stars. To simulate the effects of rotational velocity, we have convolved the Jacoby atlas with Gaussians which simulate stars with rotational velocities $v \sin i = 0$, 180, 240, and 320 km s^{-1}. The program spectra were then compared with the expanded Jacoby atlas, and for each star a metric was computed which measured the similarity between program and standard star over the selected wavelength intervals. All comparisons were done after the continua were fitted and the spectra resampled to the same 0.5 Å per pixel resolution. Variations in radial velocities were removed by resampling the wavelength scale in a manner which mapped the hydrogen lines to the same pixel numbers in all images. An example of the type of data used in the program before continuum fitting is shown in Figure 1, which presents the program star HD 25154 (classified as an A5 star by our algorithm), and two of the Jacoby standard stars, HD 235262, and HD 222336. The figure shows clearly the variations in hydrogen and calcium line strengths and profile shapes, to which the algorithm is most sensitive.

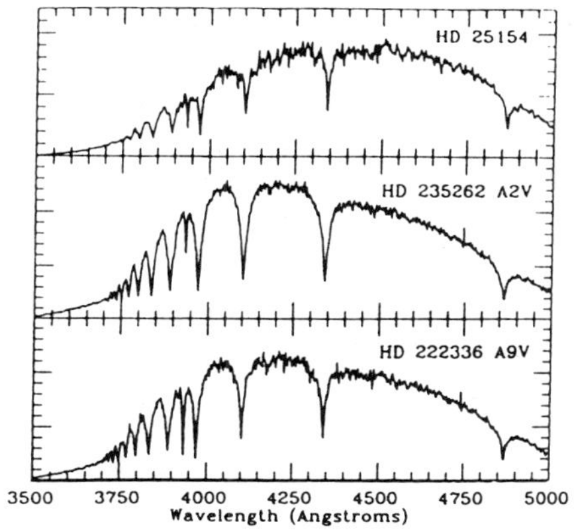

Figure 1: The automatic classification uses the entire spectrum, and tries to closely parallel the human procedure for classification. Here we present one of our program stars, HD 25154, along with two of the comparison stars before continuum fitting.

3. Details of the Technique

The basic technique we used was a metric comparison between program and standard stars. Before comparing the program and standard stars, however, we performed an initial classification using the Hγ and Ca II stellar line strengths to select a restricted range of possible spectral classes for detailed comparison. The equivalent-width technique was found greatly to speed up the algorithm, and it was capable of deriving a temperature classification to within 5 MK subclasses. Figure 2 shows the results of plotting the Hγ and Ca II equivalent widths for the stars of the Jacoby *et al.* (1984) atlas. The crosses show the average values and dispersions for groups of stars separated by 5 MK temperature subclasses, ranging from O5-B0 (on lower left) to K6-M0 (right), demonstrating the basis for our preliminary temperature classification using the Hγ and Ca II lines.

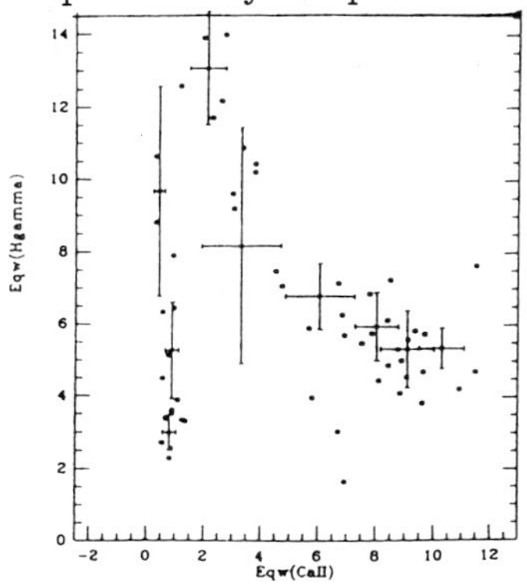

Figure 2: By combining the Hγ and Ca II line strengths, a spectral classification becomes possible. This simple technique was used for a preliminary spectral classification. Here we show the average locations for stars with spectral types (from left) of O5-B0, B0-B5, B5-A0, etc. in the Hγ, Ca II plane.

After a preliminary classification of the star using the equivalent widths of the Hγ and Ca II lines, we then evaluated metrics over several wavelength intervals. The metric we chose was an unweighted Euclidean metric over the selected wavelength interval or

$$M_{kj} = \frac{1}{N}\sqrt{\sum_{i(\lambda_1),i(\lambda_2)} (I_{ki} - I_{ji})^2} \tag{1}$$

where j is the number of the reference spectrum chosen, and λ_1 and λ_2 are the beginning and end of the wavelength interval for the metric. The reference-spectrum library consisted of 644 spectra from the 161 stars included in the Jacoby atlas, which included 4 rotational velocities for each of the reference stars. We evaluated values of four temperature metrics and four luminosity metrics for each program star, and the wavelength intervals chosen are summarized in Table 1.

Table 1 WAVELENGTH INTERVALS USED IN CLASSIFICATION METRICS

Wavelengths (Å)	Lines Included in Metric	Estimator Type
3815.5–4915.5	Entire spectrum	Temperature, Luminosity
3923.6–3943.6	Ca II K	Temperature, Luminosity
4310.5–4370.5	Hγ	Temperature, Luminosity
3995.0–4076.7, 4126.7–4315.5, 4365.5–4836.3, 4886.3–4915.5	"metal lines"— All stellar lines excluding Ca II, and hydrogen lines	Temperature, Luminosity
4165.0–4185.0	Fe II, Ti II blend	Luminosity for B8–F2
4193.0–4213.0	Fe I blend	Luminosity for B8–F2
4121.0–4141.0	Si II	Luminosity for B2–F5
4471.0–4491.0	Mg II	Luminosity for B2–F5

The temperature metrics chosen were the Ca II lines, the hydrogen Balmer lines, the 'metal lines', and then a final metric which considered the entire spectrum. These metrics are those which were found to be most effective with stars from types B0 to F5, and also were general enough to apply to stars of MK classes beyond this range, as well as spectra with low signal-to-noise ratio.

A schematic of the automatic digital spectral classification system is presented in Figure 3, which shows a graphical outline of the entire procedure. The metrics for each of four temperature discriminants were evaluated, over a restricted range of the standard library, and the minimum value of the metric was evaluated to determine the most similar spectrum in the standard library. The restricted range of standards was chosen to correspond to those stars which were within 5 MK temperature subclasses of the preliminary classification provided by the equivalent widths.

There were typically 100 comparision spectra used for each program spectrum. A graph of the values of a given metric then will have a minimum which selects the best matched stars, and whose width also conveys information about the relative certainty of the matching.

The temperature metrics were used to derive a revised temperature classification as shown in Figure 3. From the four temperature metrics we also determined a preliminary luminosity classification. If from the temperature metrics the luminosity classification indicated a class of III or I, we then verified the possibility of the higher luminosity class by evaluating luminosity sensitive

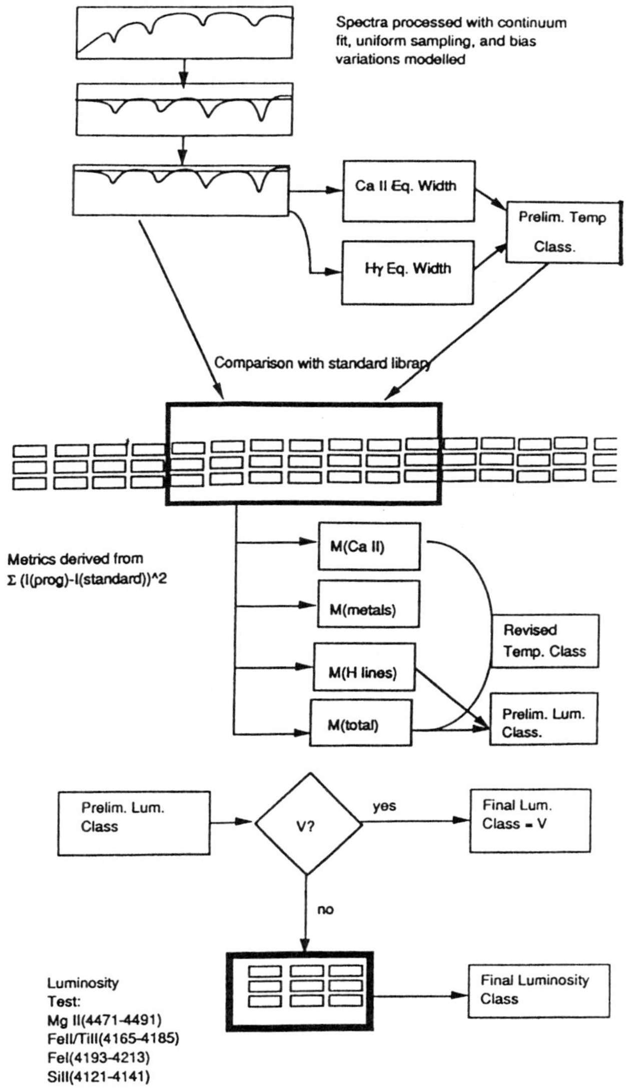

Figure 3: Schematic of the entire classification scheme. The spectra are processed with a continuum fit, and 3 bias levels are tested. A preliminary classification is done with the line strengths, and then the stars is compared with the standard star library with several simulated rotational velocities. The minimum value of the metric is selected to measure similarity between program and standard stars, and large values of minimum metrics are also useful for selecting peculiar stars. After the MK temperature subclass is determined, a set of four luminosity metrics are applied to the stars to verify the preliminary luminosity estimate.

metrics over those comparison stars within one MK temperature subclass of the program star.

The luminosity metrics were chosen from papers by Gray and Garrison (1987, 1989) as well as by Zekl (1982). The luminosity metrics included the Mg II line (4471-4491), the Fe II/Ti II blend (4165-4185), the Fe I line (4193-4213), and the Si II line (4121-4141). If more than two of these luminosity metrics indicated a higher luminosity class, then the higher luminosity class was adopted.

After both the final temperature and luminosity estimates were returned from the algorithm, we then also did a visual check of the resulting spectrum, to provide some human intervention in the process. As most users of computers now know (and also users of astronomical reduction software such as IRAF), it is important to monitor continually the processing of data, and the visual inspection was used to catch gross discrepancies, and problem spectra.

4. Verification of Accuracy of Technique

For the 115 stars in our program, 32 of them had spectral types from the Michigan Spectral Catalog, and this formed the main source for verifying the accuracy of the technique. A comparision between the results from the automated digital classification with the Michigan spectral catalogue is presented in Table 2.

Table 2 COMPARISON OF MSC SPECTRAL TYPES WITH DIGITAL SPECTRAL CLASSIFICATION

HD	SAO	Spectral Type (MSC)	$(B-V)_0$	$(U-B)_0$	M_V	Spectral Type (This Work)	$(B-V)_0$	$(U-B)_0$	M_V
027365	216735	F0.0 V	0.32	0.03	2.70	A8.5 V	0.29	0.09	2.48
028497	149674	B2.0 V	−0.24	−0.81	−2.40	B0.5 III	−0.28	−1.05	−4.75
028856	149712	A1.0 V	0.02	0.01	1.00	A2.5 V	0.05	0.07	1.40
029394	149766	F5.0 V	0.45	−0.02	3.60	F3.0 V	0.41	−0.01	3.60
029482	149776	B9.0 V	−0.07	−0.18	0.20	A0.5 V	0.01	−0.01	0.80
029573	149789	A0.0 V	−0.01	−0.02	0.60	A1.0 V	0.02	0.01	1.00
029624	149799	A5.0 V	0.15	0.09	1.90	B9.0 V	−0.07	−0.18	0.20
029650	149802	A2.0 V	0.05	0.05	1.30	A1.5 V	0.04	0.03	1.15
029709	149812	F0.0 V	0.32	0.03	2.70	F2.0 V	0.35	0.00	3.60
029851	149831	A2.0 V	0.05	0.05	1.30	A0.5 V	0.01	−0.01	0.80
030729	149914	A0.0 V	−0.01	−0.02	0.60	A0.0 V	−0.01	−0.02	0.60
031585	150008	A9.0 IV	0.29	0.07	1.92	A7.5 V	0.24	0.10	2.30
031726	150029	B2.0 V	−0.24	−0.81	−2.40	B1.0 III	−0.26	−0.96	−4.40
031743	150028	A0.0 V	−0.01	−0.02	0.60	B9.5 V	−0.04	−0.10	0.40
031788	150036	A1.0 V	0.02	0.01	1.00	A1.0 V	0.02	0.01	1.00
036224	217346	F5.0 V	0.45	−0.02	3.60	F3.5 V	0.42	−0.01	3.60
036300	217349	F2.0 V	0.35	0.00	3.60	F1.5 V	0.35	0.00	3.38
056073	218545	A1.0 IV	0.00	0.00	0.54	A1.5 V	0.04	0.03	1.15
056183	152619	B3.0 V	−0.20	−0.68	−1.60	B4.0 III	−0.18	−0.59	−2.60
059421	218740	B8.0 IV	−0.10	−0.40	−0.65	B8.0 V	−0.11	−0.36	−0.20
068556	219549	F6.0 II	0.41	0.13	−2.51	F6.5 V	0.49	−0.01	3.70
074146	236157	B5.0 IV	−0.16	−0.55	−1.80	B4.0 III	−0.18	−0.59	−2.60
079061	220914	A1.0 V	0.02	0.01	1.00	A4.0 V	0.12	0.09	1.70
086687	221657	B4.0 II	−0.15	−0.71	−4.46	B2.5 III	−0.22	−0.85	−3.45
093474	179224	A8.0 V	0.27	0.09	2.40	A9.5 V	0.31	0.06	2.63
093721	222342	B9.0 III	−0.08	−0.17	−0.60	B9.5 III	−0.05	−0.12	−0.30
094473	179318	B5.0 III	−0.16	−0.50	−2.20	B3.0 IV	−0.20	−0.69	−2.49
094724	222423	A0.0 V	−0.01	−0.02	0.60	B7.5 III	−0.11	−0.43	−1.35
095065	222458	A7.0 V	0.20	0.10	2.20	A4.0 III	0.12	0.12	0.60
125679	158509	A2.0 V	0.05	0.05	1.30	A2.0 V	0.05	0.05	1.30
135485	159132	B3.0 V	−0.20	−0.68	−1.60	B4.0 III	−0.18	−0.59	−2.60
148379	226813	B2.0 I	−0.17	−1.00	−6.30	O5.5 V	−0.32	−1.19	−5.60

We can see from the table that a good agreement exists among the spectral classifications. To present quantitatively the comparison, we have plotted in Figure 4 the results of the temperature classification from the four metrics against the 'actual' temperature class from the Michigan catalog. At the top of Figure 4 is the final derived temperature subclass from Table 2 plotted against the actual class from the Michigan spectral catalog. We see from the plots that any of the temperature metrics individually could be used to derive some degree of accuracy, and that the final agreement between the two classifications is to within 1.5 MK temperature subclasses from B0 to F5, which covers the majority of the stars within our program.

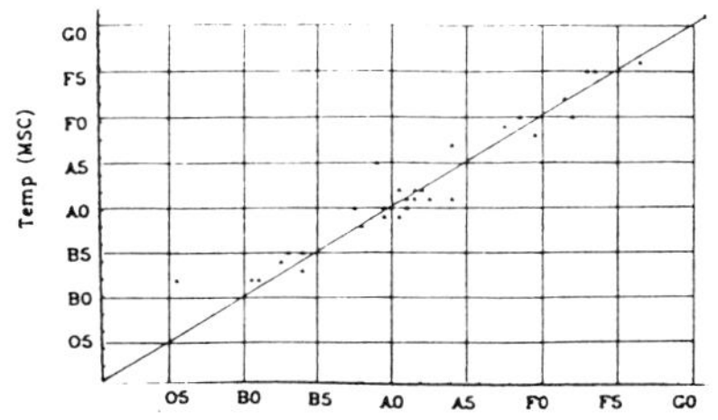

Figure 4: (top) Plots of final estimated MK temperature subclasses (x axis) for our program stars against the Michigan Spectral Catalog types (y axis). We also show the plots for the Ca II metric (middle), and the metal line metric (lower), which indicate the different estimates provided by the independent metrics.

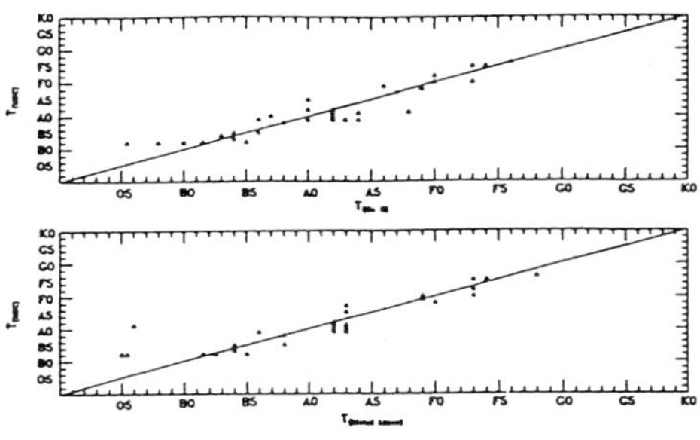

Serious discrepancies exist for the hottest stars, and the method needs to be refined to account for the different criteria necessary for the hottest and coolest stars. Nevertheless, we believe that a conclusive demonstration of the feasibility of the method has been accomplished, and with an independent source of classification we have obtained accuracies of 1.5 MK classes.

Acknowledgments. This work was performed as part of the doctoral thesis at the University of Chicago, and under the predoctoral fellowship program at the Space Telescope Science Institute, Baltimore, MD. I would like to thank my advisor at Space Telescope, Dr. Chris Blades, for all his help, as well as my University of Chicago advisor, Dr. L.M. Hobbs. Nolan Walborn and Robert Garrison also provided useful comments on the development of the technique.

References

Davis, R.J. 1974, ApJ, **213**, 105

Gray, R.O., and Garrison, R.F. 1987, ApJS, **65**, 581

Gray, R.O., and Garrison, R.F. 1989, ApJS, **69**, 301

Houk, N., and Cowley, A.P. 1975, *University of Michigan Catalog of Two-Dimensional Spectral Types for the HD Stars*, Vol. 1., Ann Arbor: University of Michgan Press

Jacoby, G.H., Hunter, D.A., and Christian, C.A. 1984, ApJS, **56**, 257

Johnson, H.L., and Morgan, W.W. 1953, ApJ, **117**, 313

Kurtz, M.J. 1982, *Ph.D. thesis*, Dartmouth College

Kurtz, M.J. 1988, *Astronomy from Large Databases*, Munich: European Southern Observatory

La Sala, J. 1988, *Astronomy from Large Databases*, Munich: European Southern Observatory

La Sala, J. 1994, this volume

Morgan, W.W., Keenan, P.C., and Kellman, E. 1943, *An Atlas of Stellar Spectra*, Chicago: University of Chicago Press

Penprase, B.E. 1992, ApJS, **83**, 273

Penprase, B.E. 1993, ApJS, **88**, 433

Torres-Dodgen, A.V. and Weaver, B. 1993, PASP, **105**, 693

von Hippel, T., Storrie-Lombardi, L., Storrie-Lombardi, M.C., and Irwin, M. 1993, MNRAS, preprint

Walborn, N.R., and Fitzpatrick, E.L. 1990, PASP, **102**, 379

Zekl, H. 1982, A&A, **108**, 380

Discussion

Abt: What happens when you measure a metallic-line (Am) star in which the three criteria give inconsistent results? Do bells ring and lights flash?

Penprase: The technique has the capability of identifying magnetic and other peculiar stars, by using the actual value of the minimum metric. In cases where the spectrum matches none of the comparison stars, the minimum metric has a relatively high value, and so it is possible to identify peculiar stars in this way.

Walborn: I think that this is a very nice practical development and application of automatic classification. However, the Ca II will be completely dominated by the interstellar contribution earlier than mid-B, so that parameter is not applicable there. The extensive Walborn and Fitzpatrick digital atlas of OB spectra is available electronically, and could help.

Penprase: Yes, this is a definite source of variability for the hottest stars, and the deviations seen in Figure 4 may be caused by this. The Ca II lines do indeed need to be deemphasized for these hottest stars.

Roman: How well do the parameters track each other?

Penprase: The different metrics are generally correlated, although it is important to note that they do arrive at slightly different MK temperature classifications, and we use weighted average to combine them.

Parthasarathy: Cool stars have emission in the cores of Ca II K-lines. Using the Ca II K-lines in the automated method of classification may give rise to uncertainties in the spectral types of cool stars.

Penprase: This is an interesting point, although I am not sure whether the emission in the cores would substantially alter the derived temperature classification, since the shape of the entire line profile is considered in computing the metric.

La Sala: How did you decide on your weighted averages? Did your calculations of distances (residuals) weight different features differently or was the metric purely Euclidean?

Penprase: Within the metric the computation was Euclidean, and for the weighted average of the temperature metric, we gave more weight to the Ca II and total spectrum metrics (2×) than to the other metrics.

Weaver: Could you comment on your approach to luminosity?

Penprase: We used the Mg II (λ4481), the Fe II/Ti II blend (λ4175) the Si II line (λ4131) and the Fe I line (λ4203) as luminosity indicators. The stars were assumed to be class V unless the preliminary luminosity estimate indicated a higher class. If two or more of the above luminosity indicators signalled a higher luminosity class, that class was adopted. This was found to be a satisfactory technique, and a further refinement would be to tailor the luminosity indicator more closely to the temperature class of the program star.

Gray: You couldn't have got these indicators from my papers with Garrison (1987, 1989). I found that, while the blend at λ4175 is certainly luminosity sensitive and a case could be made for Mg II and Si II showing some luminosity sensitivity in supergiants, the Fe I λ4203 blend is so luminosity *insensitive* that it can be used for comparison with λ4175.

MacConnell: Now that you have finished your thesis, will you continue with the automated classification work?

Penprase: Well, my new position at Pomona College should be an ideal one for continuing this work, since I can take observations at the College observatory, and also I can get students to help me with the work.

The MK Process at 50 Years
ASP Conference Series, Vol. 60, 1994
C.J. Corbally, R.O. Gray, and R.F. Garrison (eds.)

MK Classification in the Direction of the Galaxy Anticenter

K. B. Chargeishvili

Abastumani Astrophysical Observatory, Al.Kazbegi ave.2a, Tbilisi, 380060, Rep. of Georgia

Abstract. MK classification is performed for 6037 stars to $12^m_.5$ in a 140 sq. deg. area in the direction of the Galaxy anticenter. It is shown that, for the purpose of revealing peculiar stars, the capabilities of the objective-prism spectra taken on Kodak IIIa-J plates are equal to those obtained by the slit spectrograph on Kodak IIa-O. This fact is significant from the standpoint of the MK Process.

558 stars of different peculiarities are discovered. 86 emission stars and 391 O-B2 stars are revealed. In the direction of the Galaxy anticenter no fragment of any other Galactic arm, except for Orion, is observed. An abundance of O-B2 stars is noticed in the direction of positive Galactic latitudes. It can be said that a similar effect is marginally present for stars of all spectral types. Ap and especially Am stars do not show such concentration near the Galaxy plane as do the ordinary stars of the same spectral type. The fine structure of interstellar absorption is investigated.

1. Classification at Abastumani Observatory

The interest in spectral classification at Abastumani astrophysical observatory began early in the sixties when the 70-cm meniscus telescope was put into operation. The telescope is equipped with 8° objective prism (dispersion 166 Å/mm near Hγ). Such an instrumental combination together with excellent sky transparency at Abastumani have made it possible to obtain high-quality data. For more than 30 years since then the criteria of the MK System have been developed and refined under the direct guidance of Prof. Kharadze and Dr. Bartaya at our observatory (Kharadze, Bartaya 1960; Bartaya 1979a). Figure 1 clearly points to the reliability of our classification. Here our data are compared with those of Morgan and other well-known classifiers (Jaschek *et al.* 1964).

Classification has been done at Abastumani Observatory in the selected areas of the sky which have served to solve different problems of stellar astronomy in cooperation with other observatories. The classification work in 115 Kapteyn Selected Areas of the Northern Sky up to $11^m_.5$ is especially noteworthy. Two catalogues have already been published: for KA 2-43 (Bartaya 1979b) and for KA 44-67 (Bartaya, Kharadze 1992). The catalogues for other areas are being prepared as well. All these works were based on photographic observations taken with Kodak IIa-O plates. The Abastumani criteria were developed just for this emulsion. In addition, Kodak IIIa-J plates were used for the work that I am going to tell you about in detail.

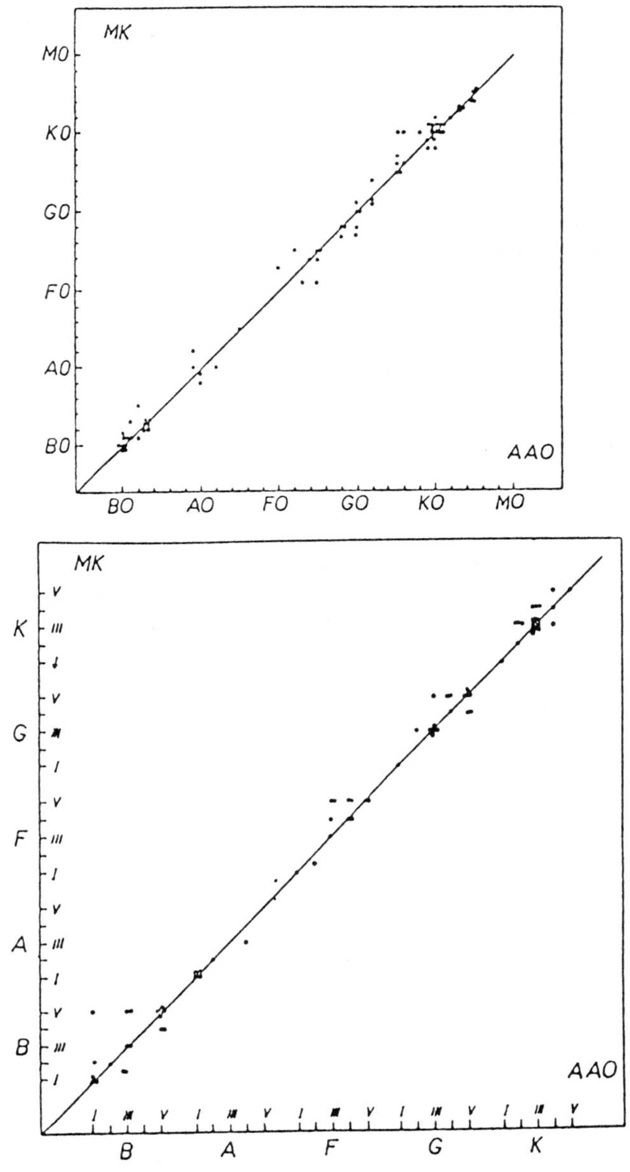

Figure 1. Comparison of Abastumani classification data with those of Morgan and other well-known classifiers.

2. New techniques for the current work

This work deals with a spectral analysis of a 140 sq. deg. area in the direction of the Galaxy anticenter. Kodak IIIa-J was used in this work for the first time in the practice of our observatory. Under the conditions of hypersensitization and a 60 min. exposure time, penetration was increased by a magnitude with this emulsion. This is important for research in stellar astronomy. It must be noted that while sensitivity is important, it is not the only advantage of this emulsion with respect to Kodak IIa-O. Since it remained fine-grained regardless of a 60 min. exposure time, the spectral quality did not grow worse but was improved. Besides, Kodak IIIa-J emulsion extends further to the long-wave region. This fact provides more information from our data for red stars.

Against the background of the new capabilities for this emulsion, the criteria which had been worked out for Kodak IIa-O had to be revised. It was found that these criteria could be mainly applied to Kodak IIIa-J as well, but in the latter case we deal with higher quality information data allowing improvement in the classification criteria. This especially concerns O-B5 stars for which the criteria are weakly visible. The wings of hydrogen lines in B- and A-type stars are well-defined and, accordingly, the stars of luminosity III and V can be distinguished reliably.

In particular, the increased capabilities of this emulsion for revealing peculiar stars should be noticed. In this task our possibilities are *equal* to those of the slit spectra taken on Kodak IIa-O with 125 Å/mm dispersion. This was confirmed by the fact that Kodak IIIa-J made it possible to find almost all peculiar stars listed in the Atlas by Morgan, Abt, Tapscott (1978). These peculiarities were very seldom or not at all distinguished on Kodak IIa-O plates.

As an illustration of the increased capabilities of the Kodak IIIa-J emulsion we can take a region, KA 49, which was classified by Bartaya using Kodak IIa-O emulsion. As a result of our new classifications, 40 peculiar Ap and Am stars were revealed, while Bartaya had found only 5 such in the same area. The possibility of revealing peculiar, non-MK System stars has enabled us to make our modest contribution to the further development of spectral classification from the standpoint of the MK Process.

3. The new catalogue and its analysis

After we had implemented the above-mentioned techniques, a catalogue of spectral types and luminosity classes for 6037 stars to $12^{\mathrm{m}}\!.5$ was compiled in our 140 sq. deg. area (Figure 2) in the direction of the Galactic anticenter (Chargeishvili 1988). In the classification process 558 peculiar stars were found; out of these 549 are original discoveries (Kharadze, Chargeishvili 1990), 86 are emission objects (Chargeishvili 1989a), and 391 O-B2 stars are found as well.

The Galactic anticenter direction is of interest from the standpoint of a lot of stellar astronomy problems. It is essential for investigating the problem of spiral structure. At present, none of the spiral structure indicators give a real picture of the Perseus Arm in this direction.

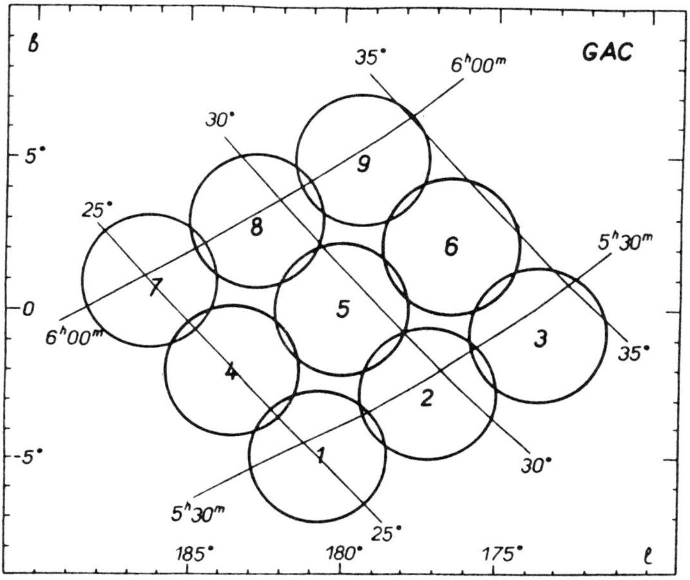

Figure 2. Circles 1,2,...9 correspond to the areas under investigation in the direction of the Galactic anticenter.

3.1. The OB stars and Galactic structure

Out of 391 O-B2 stars found by us, only 78 turned out to have photoelectric *UBV* values. Therefore, we cannot speak about their densities in the anticenter direction, and we shall restrict ourselves to considering the projection of their positions onto the Galactic plane. In this way we can compare our results with similar data by Kharadze and his coauthors in the Perseus direction (Kharadze *et al.* 1985).

Figures 3 and 4 show the projections of O-B2 stars onto the Galactic plane in the direction of the anticenter and Perseus Arm respectively. Figure 4a is borrowed from the paper by Kharadze *et al.*(1985). In these figures all the B stars are discussed. The data are taken from the catalogue by Bartaya (Bartaya 1979b). To make the comparison easier, we plot Figure 4b which shows only the O-B2 stars according to the data of the same catalogue. From Figure 4 the density decrease between the two locations of the inner edge of the spiral arms, Orion and Perseus, is clearly seen. This is of interest since we present the rare case when observational data of single stars make it possible to distinguish the above-mentioned Galaxy arms so distinctly. The dashed line shows the location of the inner edge of the Perseus Arm, which is based on investigation of the velocity fields of young stars (Gerasimenko 1983). As it is seen, the results obtained from two different data sets are in good agreement.

An extension of the Perseus Arm in the anticenter direction, if it really exists, might be in the interval of 2-3 kpc according to references (Schmidt-Kaler 1966) and Figure 4. But, as is seen in Figure 3, there is no discontinuity

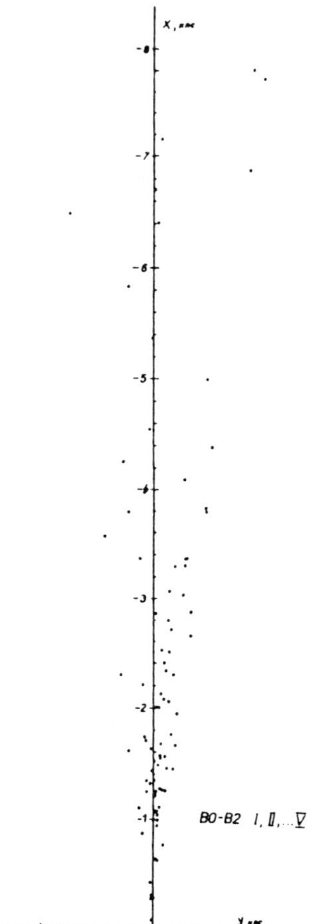

Figure 3. Projection of locations of O-B2 stars on the Galactic plane in the direction of the anticenter.

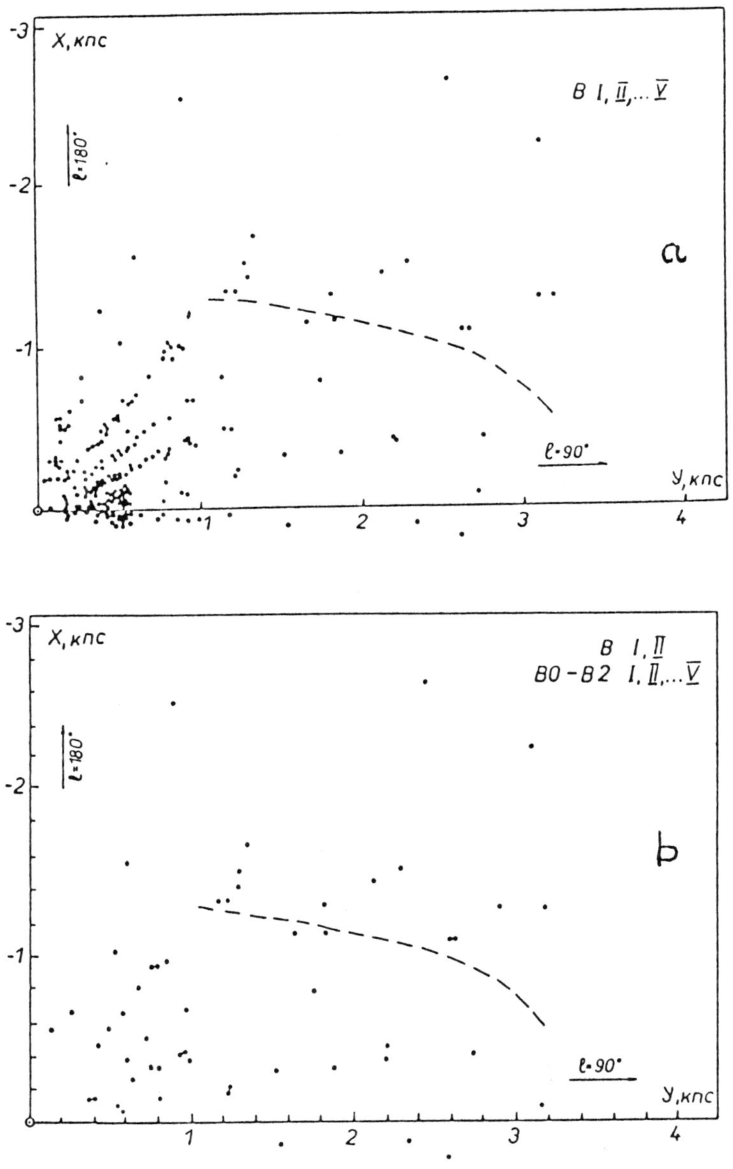

Figure 4. Projection of locations of hot B stars in the direction of the Perseus Arm.

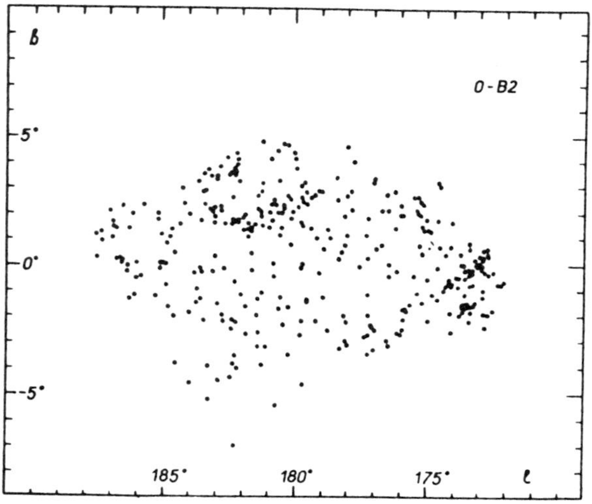

Figure 5. Surface distribution of O-B2 stars in the Galactic anticenter region under investigation.

in this interval (just as observed in Figure 4). This fact makes the existence of the Perseus Arm fragment in the anticenter direction less certain (Chargeishvili 1992).

At the same time, taking into account that the volume of the investigated area increases with the distance, from Figure 3 a conclusion can be drawn that, when moving away from the Sun, the density of O-B2 stars falls. This points to existence of the local spiral arm, the Orion Arm. These results are consistent with those of McCuskey (1967).

If, in the near future, the measurement of *UBV* photoelectric magnitudes is possible for all the 391 O-B2 stars found (Chargeishvili 1988), discussion of such data will throw more light on every problem that has arisen.

Figure 5 shows the surface distribution of O-B2 stars. As can be seen, the majority of them are located above the Galaxy equatorial plane. Both for hot O-B2 stars and the Galaxy gas layer, a similar effect had been observed earlier.

3.2. Distributions with Galactic latitude

We considered it to be of interest to see the distibution of different spectral-type stars with Galactic latitude (Figure 6). The region studied by us $|b|=\pm5°$ does not represent a disk of such thickness that the density decrease should be appreciable even for B and A stars; but we find that such a decrease is observed. The fact that Galactic anti-center (GAC) 3 and GAC 8 differ in stellar abundance can be explained by a higher transparency in these directions (Chargeishvili 1992). At the same time, GAC 3 coincides with an O-association. While bearing in mind the above remarks, there is a definite, though slight, increase in the abundance of stars of all spectral types for the positive Galactic latitudes.

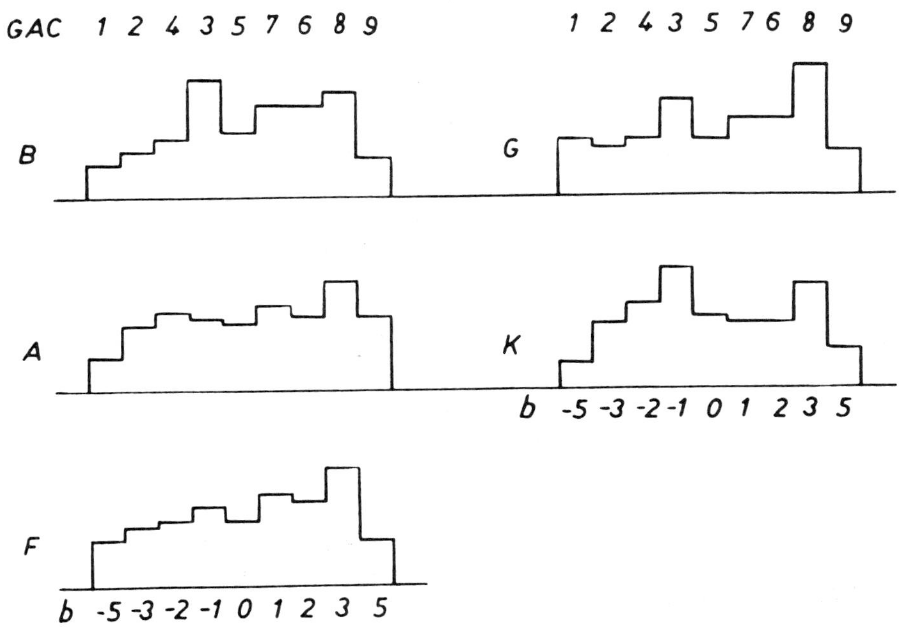

Figure 6. Distribution of stars according to the Galactic latitude within particular spectral type.

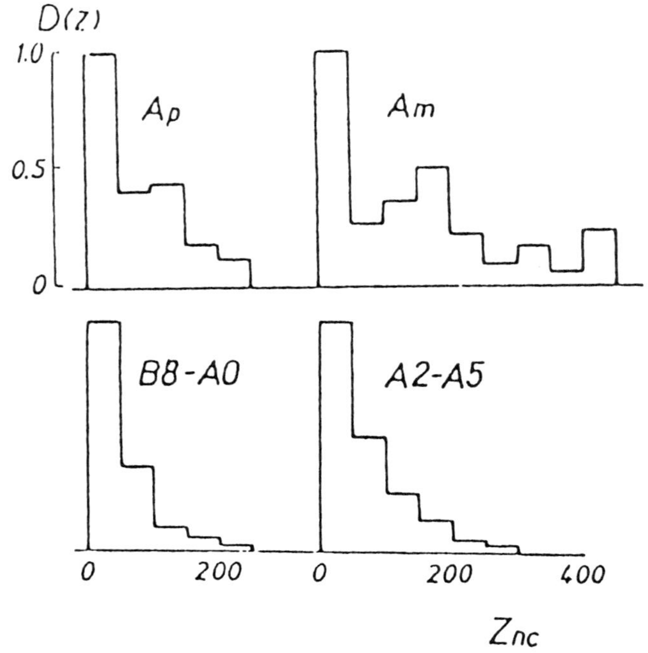

Figure 7. Spatial distribution of Ap and Am and their suitable normal stars in B8-A0 and A2-A5 spectral intervals perpendicular to the Galactic equatorial plane.

3.3. Ap and Am stars

The Ap and Am stars represent the six hundred or so peculiar stars revealed by our study. Through our data these stars are given for the first time in full coverage to 11^m. The Ap and Am stars are basically concentrated in the B5-A7 spectral interval, with Ap stars from the B8-A0 and Am stars from the A2-A5 spectral ranges. Their percentages in these intervals are 9% and 26% respectively. The Ap and Am stars are characterized by less of a concentration near the Galactic plane compared with the corresponding normal stars (Figure 7). This especially applies to Am stars (Kharadze, Bartaya 1977; Chargeishvili 1989b).

References

Bartaya R.A. 1979a, Bull.Abastum.Ap.Obs., **50**,3-237

Bartaya R.A. 1979b, Bull.Abastum.Ap.Obs., **51**,3-315

Bartaya R.A., Kharadze E.K. 1992, Bull.Abastum.Ap.Obs., **72**, 3-270

Chargeishvili K.B. 1988, Bull.Abastum.Ap.Obs., **65**, 3-239

Chargeishvili K.B. 1989a, Bull.Abastum.Ap.Obs., **66**, 19

Chargeishvili K.B. 1989b, Bull.Abastum.Ap.Obs., **66**, 25

Chargeishvili K.B. 1992, Bull.Abastum.Ap.Obs., **70**, 41

Gerasimenko T.P. 1983, Pisma v Astron.Zh., **9**, 721

Jaschek C., Conde H., de Sierra A.C. 1964, *Catalogue of Stellar Spectra Classified in the Morgan-Keenan system.*, La Plata

Kharadze E.K., Bartaya R.A. 1960, Bull.Abastum.Ap.Obs., **25**, 139

Kharadze E.K., Bartaya R.A. 1977, Astron.Nachr., **298**, 111

Kharadze E.K., Bartaya R.A., Pavlovskaya E.D., *et al.* 1985, Bull.Abastum.Ap. Obs., **59**, 91

Kharadze E.K., Chargeishvili K.B. 1990, AJ, **99**, 379

McCuskey S.W. 1967, AJ, **72**, 1199

Morgan, W.W., Abt, H.A., Tapscott, J.W. 1978, *Revised MK Spectral Atlas for Stars Earlier Than The Sun*, (Yerkes Observatory, University of Chicago, and Kitt Peak National Observatory)

Schmidt-Kaler Th. 1966, IAU Trans., XIIB, 415

Discussion

Walborn: These 400 or so OB stars in the Galaxy will provide a very important candidate list for comparing the metallicity of Galactic stars versus those in the Magellanic Clouds. I hope that you will publish a catalogue and finding charts for these stars in a readily accessible place.

Chargeishvili: We are very pleased by your interest in the OB stars that we found. Data and charts for these stars are given in our catalogue *The Catalogue of Spectral and Luminosity Classes of 6037 Stars in the Direction of the Galaxy Anticenter* (Chargeishvili 1988), and this will be sent to you.

The MK Process at 50 Years
ASP Conference Series, Vol. 60, 1994
C.J. Corbally, R.O. Gray, and R.F. Garrison (eds.)

Present State of the Work on Automated Spectral Classification by Means of Objective-Prism Spectra

V. Malyuto

Tartu Astrophysical Observatory, Estonia

T. Shvelidze

Abastumani Astrophysical Observatory, Georgia

Abstract. The present state of the work on automated spectral classification by means of objective-prism spectra is briefly described. Faint program stars having reliable basic physical parameters are used in line with standards to improve calibration of the criteria.

1. Introduction

Some years ago, a complex program of studying the main meridional section of the Galaxy was initiated by some astronomers in Kiev, Tartu, Abastumani, and Vilnius with the aim of improving our knowledge of spatial and kinematic characteristics of stellar populations. The use of absolute proper motions of stars (with respect to galaxies), photoelectric *UBVR* and Vilnius photometry together with automated quantitative spectral classification for large stellar-statistical samples form the most essential features of the program. Areas within 30° from the main meridional section of the Galaxy are involved.

To classify stars, objective prism stellar spectra of intermediate dispersion (D = 166 Å/mm at $H\gamma$), obtained with the 70-cm meniscus telescope at the Abastumani Astrophysical Observatory, are used. The field diameter is 4°50′ and the limiting photographic stellar magnitude is about 12 mag.

Our automated quantitative spectral classification of F-K stars applies criteria evaluation and is based mainly on two software backages: the SDR package for spectrometric data reduction and the CTATEC-2 package for the linear regression model used in classification. Underlying the SDR package is the transformation of spectral densities into the values of quantitative classification criteria (Malyuto and Pelt 1981). Underlying the CTATEC-2 package (Malyuto and Shvelidze 1989) is the definition of the coefficients of a multiple linear-regression model "criteria values *versus* main physical parameters" based on the standard stars. Spectral classes, luminosity classes and [Fe/H] values are estimated for programme stars by inverting the model and inserting criteria values.

2. Selecting the spectra and digitizing the observational data

The spectra, widened to 0.15-0.25 mm, are recorded on Kodak IIaO photographic emulsion. For their digitization we apply the PDS machine installed at the Tartu Astrophysical Observatory. The slit area corresponds to 5×100 μm on the plate, one displacement step equals 5 μm, and the recorded linear distance is 10 mm (2001 pixels). The specific package for automatic stellar spectra detection and identification from objective prism plates has not been elaborated yet. Instead we select the spectra visually and use manual positioning on the PDS machine. To fix the approximate starting point for spectral recording, we use the line H Ca II + Hϵ (λ 3969 Å), easily identifiable visually in all the spectra.

3. Spectrophotometric data reduction

The SDR software package includes the main program together with a set of subroutines written mainly by J. Pelt. The package has been written in the FORTRAN 77 language and runs on a UNIX-based computer.

The package contains the following steps:

a) filtering of the spectra from the emulsion grain noise,

b) transforming densities into intensities,

c) fitting the continuum by means of the cubic spline technique,

d) standardizing pixel numbers,

e) measuring 53 classification criteria (the ratios of equivalent widths confined by the continuum line, the residual spectral intensity distribution and the wavelength borders).

4. Spectral classification of stars

The multiple linear-regression model "quantitative criteria values *versus* main physical parameters" applied to about 100 standard F-K stars specifies the CTATEC-2 package. By inverting the model we classify the stars. From repeated measurements of classification criteria (from 2 to 6 spectra per star) the r.m.s. differences between the standard and calculated values of basic physical parameters of standard stars were found. They equaled ±0.09 for spectral classes, ±0.43 for luminosity classes, ±0.22 for [Fe/H].

At the Tartu and Abastumani Observatories, our automated spectral classification is in routine use for classifying automatically large samples of stars. The B-magnitude is now obtainable from the spectra through a specific subroutine by the application of the standard mathematical B-filter; the obtained accuracy is about $0\overset{m}{.}1$.

The results of our spectral classification in two fields near the North Galactic Pole have been applied to studies of the stellar populations together with proper motions with respect to galaxies and Vilnius photometric classification in neighbouring fields (Andruk, *et al.* 1993). Preliminary results satisfactorily agree with the published data.

5. Recent developments

Our regression model used in classification was based on spectral data for about 100 standard stars of normal metal content with reliable MK classification, together with a few metal-deficient standard stars ($B < 7^{m}_{.}0$). The discrete values of spectral and luminosity classes compiled from the literature were transformed into the spectral and luminosity codes which were presumed to be continuously changing values. We suspected that such an approach might introduce additional uncertainties in stellar distances resulting from our classification. Therefore we decided to include into regression more direct physical parameters: $\log T_{\rm eff}$, M_v and [Fe/H]. However, for those stars which are MK standards the mentioned physical parameters have been derived from only MK data through calibration "spectral and luminosity classes *versus* $T_{\rm eff}$ and M_v" (Straizys and Kurilene, 1981). Figure 1 shows the H-R diagram for standard stars (122 in number).

The sample presented in Figure 1 suffers from the shortage of stars in the F0-G2 region as well as from the total shortage of metal-deficient stars in the sample. The lack of photographic plates prevents us from observing additional standard stars. However, many program stars in the areas might be used to calibrate the criteria (in line with the standards) if reliable $T_{\rm eff}$, M_v and [Fe/H] values for these program stars were known from other sources. Fortunately, two fields photographed by us are located within 5° from the NGP and a lot of papers devoted to the NGP region are available. We have used the following sources:

a) DDO photometric classification of G-K stars listed by Hartkopf and Yoss (1982) together with corrected and additional DDO classfication kindly sent to us by Dr. K. Yoss;

b) Photometric classification of F-stars in the Strömgren photometric system compiled from different sources (mainly from Knude 1989, Hill *et al.* 1982, and from the SIMBAD data);

c) Vilnius photometric classification of F-K stars observed and processed by Dr. S. Bartashiute.

The total number of program stars involved in calibration of our spectral classification criteria is about 140. The H-R diagram for the standards and program calibration stars is presented in Figure 2. The latter stars are much fainter ($8^{m}_{.}0 < B < 12^{m}_{.}0$) than the standards ($B < 7^{m}_{.}0$) and they were photographed with widely differing exposures (less than 1 min and about 20 min for standard and program stars respectively). Suspicion arises that the criteria values may be randomly and/or systematically distorted by the effect of exposure. To check it, we selected the stars of about the same luminosity and of normal metallicity and plotted the criteria against $\log T_{\rm eff}$. Different symbols are used for standard and program stars. An example is given in Figure 3. It has been concluded that the mentioned effect is rather small or absent.

In Figures 4-5 similar examples are given for effects of metallicity and/or luminosity on the criteria. We suppose that the inclusion of standard and program stars in the sample of calibration stars will improve the classification results and extend the range of $T_{\rm eff}$.

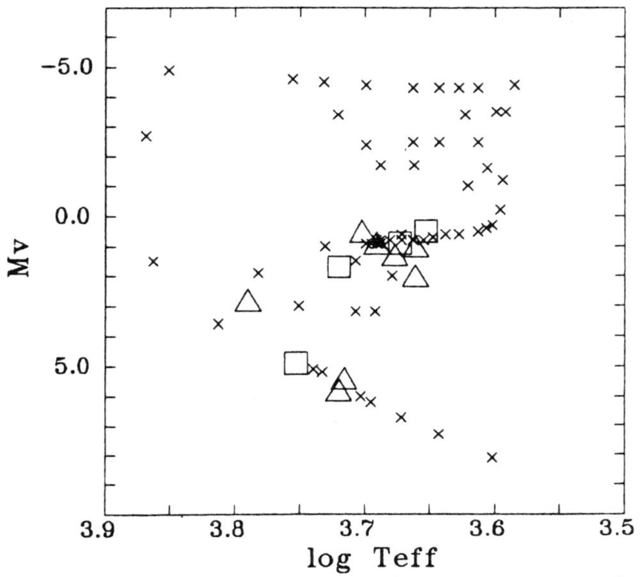

Figure 1. A sample of the standards. The designations used: crosses are normal metallicity stars ([Fe/H]> −0.4), triangles are stars with the intermediate metal deficiency (−0.4 >[Fe/H]> −1.0) and squares are stars with low metal content (−1.0 >[Fe/H]> −3.0)

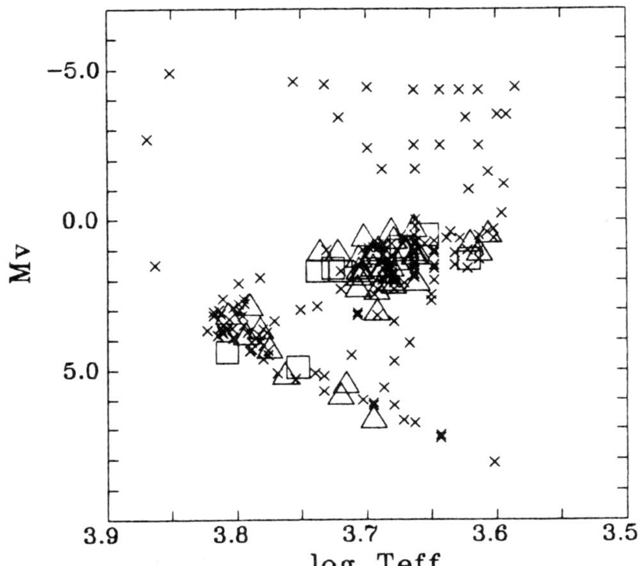

Figure 2. The same as in Figure 1 but for the standards and program calibration stars

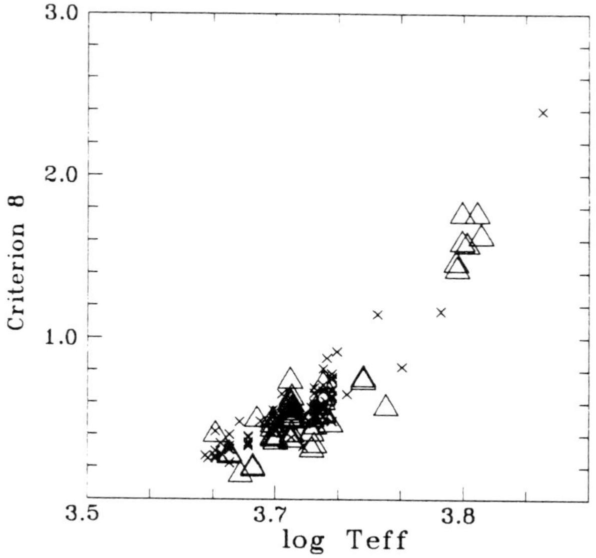

Figure 3. Criterion 8 versus $\log T_{\rm eff}$ for the standards (crosses) and program calibration stars (triangles) with $3^{\rm m}\!.0 > M_v > -2^{\rm m}\!.4$ and with normal metallicity ([Fe/H]> −0.4).

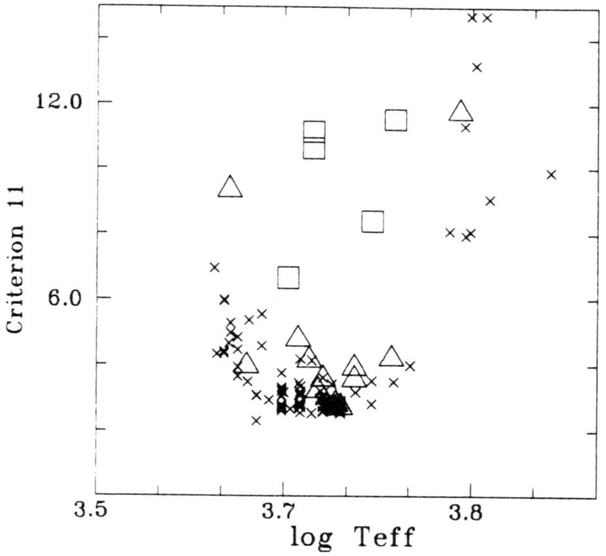

Figure 4. Criterion 11 versus $\log T_{\rm eff}$ for the standards and program calibration stars with $3^{\rm m}\!.0 > M_v > -2^{\rm m}\!.4$ and with different metallicities. The designations used: crosses are stars with [Fe/H]> −0.4, triangles are stars with −0.4 >[Fe/H]> −1.0 and squares are stars with −1.0 >[Fe/H]> −3.0.

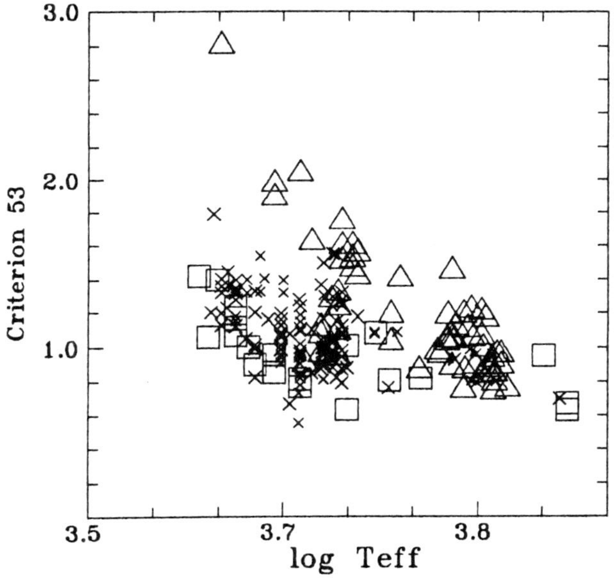

Figure 5. Criterion 53 versus $\log T_{\rm eff}$ for the standards and program calibration stars with normal metallicity ([Fe/H]> −0.4) and with different luminosities. The designations used: crosses are stars with $-3^{\rm m}_{.}0 > M_v > -2^{\rm m}_{.}4$, triangles are stars with $M_v > 3^{\rm m}_{.}0$, squares are stars with $M_v < -2^{\rm m}_{.}4$).

At present the spectra for about 1000 stars in four areas (three plates per area) have been digitized. The spectra will be re-classified with a new regression model and applied to studies of the stellar populations of the Galaxy with the use of proper motion data measured by Dr N. Kharchenko of Kiev Observatory. Another six regions of the program have been observed through our objective prism on Kodak IIaO and IIIaJ emulsions and will be treated later.

References

Andruk V.N., Bartashiute S., Kharchenko N.V., Malyuto V.D., Shvelidze T.D., 1993, "A Study of Stellar Populations in the Main Meridional Section of the Galaxy", Proc. 3rd DAEC Meeting, Oct. 12-16, 1992, Meudon, France (in press)

Hartkopf W. and Yoss K.M. 1982, AJ, **87**, 1679

Hill G., Barnes J.V. and Hilditch 1982, Dominion Astroph. Obs. Publ., **16**, 111

Knude J. 1989, ApJS, **81**, 215

Malyuto , V., Pelt, J. 1981, Prepr. the Acad. Sci. Estonian SSR, **A-1**, 1

Malyuto, V. and Shvelidze, T. 1989, ApSS, **155**, 71

Straizys, V. and Kuriliene, G. 1981, ApSS, **80**, 353

Discussion

Houk: What is the dispersion or resolution?

Shvelidze: It is the same as in Chargeishvili's paper, 166 Å/mm.

Weaver: Are your quoted errors in temperature and luminosity based on the statistical fit to the standards or on testing those fits to "unknown stars" that are independently classified?

Shvelidze: Our quoted errors are based on the statistical fit to the standards.

Abt: There is a potential error in any studies that use bright standards for faint unknown stars. The beautiful HST pictures shown by the Nolan Walborn of the R136 region in the clouds show that many of the "stars" are small groups. The bright standards are ones in which all composite spectra have been eliminated. But if one looks at stars 10 times farther away, there will be 100 times more composite spectra. Therefore one has to explore the increased effects of multiplicity.

Shvelidze: I agree that effects of multiplicity on faint unknown stars have to be investigated specially. This problem is common to both photometric and spectral surveys. However the problem is less serious in spectral studies like ours because a preliminary visual inspection of the spectra may enable us to detect certain composite spectra.

The MK Process at 50 Years
ASP Conference Series, Vol. 60, 1994
C.J. Corbally, R.O. Gray, and R.F. Garrison (eds.)

Faint, Cool Supergiants in the Southern Milky Way

D. Jack MacConnell

Astronomy Programs, Computer Sciences Corp.
Space Telescope Science Inst.

Robert F. Wing

Ohio State University

Edgardo Costa

Universidad de Chile

Abstract. We report on the classification features in cool, luminous stars taken on CCD frames in the near infrared.

1. Introduction

Since 1985 we have been collecting intermediate-band photometry in eight colors (the Wing system) and fairly low-resolution CCD spectra ($\mathcal{R} \sim 1000$ at 8000 Å) in the red of hundreds of candidates for cool supergiants marked on near-infrared objective-prism plates of very low dispersion. The candidates are typically of mag $I \sim 10$ or fainter. Most of the photometry has been taken with the 1.0-m Yale telescope and the spectra with the 1.5-m telescope at Cerro Tololo. The objective of the program is to improve the completeness in the census of cool supergiants along the southern Milky Way, to obtain their distances, and hence to use them in applications to problems of galactic structure. We gave a thorough review of the rationale of the program and the methods of the search and classification in MacConnell, Wing, and Costa (1992). Here we restrict our attention to normalized spectra in two wavelength regions of: 1) MK standards with types from Keenan and McNeil (1989) and 2) a sample of program stars which have been measured in the Wing system, comparing them with standards of the same temperature type.

2. Observations

The eight-color photometry is sufficient to classify the program stars K3.5 and cooler into a temperature class, based on the strength of the γ (0,0)-system of TiO with band-head at 7054 Å, and a luminosity class, based on the strength of the CN (3,1) and (0,0) bands near 8120 Å and 1.1 μm, respectively. Each of these molecules is a good discriminant of the relevant parameter – TiO increases with decreasing temperature, and CN grows with increasing luminosity – except that

CN strength varies with temperature and must be corrected for the presence of TiO, particularly at 8120 Å. In cool supergiants, the (3,0) and (4,1) CN bands in the 6900 – 7100 Å region are strong, so the temperature type must be corrected for their presence. The reduction routine written by Wing compensates for this.

We have been obtaining the spectra of as many of the photometrically-observed candidates as possible in order to: 1) classify those earlier than K3.5 which the 8-color system cannot classify, 2) eliminate S stars and other peculiar stars which the system was not designed to identify, and 3) serve as a check on the photometric classification. In order to classify a program star from its spectrum, we use its photometric temperature type to choose a set of MK standards of the same type with which to compare it. We then look for the closest match in the region of the luminosity-sensitive features, CN and Ca II.

Figures 1 – 4 show the effects of luminosity in these two regions for cool MK standards over a range of temperature. The most-luminous star in each of these figures is plotted with a solid line. In these and the succeeding plots, we have chosen the pseudo-continuum points at about 7810 and 8640 Å. The effect of luminosity is most striking in both regions for the early K stars as is seen in Figures 1a and 1b. For the early M stars (Figures 2a and 2b), the discrimination is not as marked, especially at Ca II, although the equivalent widths probably increase with luminosity. At M3 (Figures 3a and 3b) the difference in CN at 7900 Å is still quite clear as it is at Ca II 8662 Å, but the situation is less clear at the other two Ca II lines. At M4, the discrimination is poor and even seems to be the reverse of what is expected for CN beyond 8020 Å and in all three Ca II lines. Whether this is due to stellar variability or to some artifact of the continuum placement is not easy to determine. Notice the slope of the spectrum in Figure 4a; this is due to a strong TiO band with head at 7589 Å. It seems that neither region should be trusted for classifying stars this late.

Figures 5 – 7 present the plots for three program stars compared with standards of similar temperature and luminosity class over the range K3.5 to M2. Comparing just the CN plots, it is clear that AM 26 is less luminous than the other two program stars which is consistent with the photometry. The distinction is less clear in the Ca II lines, however.

3. Conclusions

Spectroscopy of cool supergiant candidates serves as a useful adjunct to the 8-color photometry. Although the strength of the infrared Ca II triplet has long been considered an important luminosity discriminant, it appears that the strength of CN in the 7900 Å region is more reliable and easier to use. This is particularly true for the M stars. As always in dealing with spectra of cool stars, the placement of the continuum in the normalization step and in the comparison with standards is crucial.

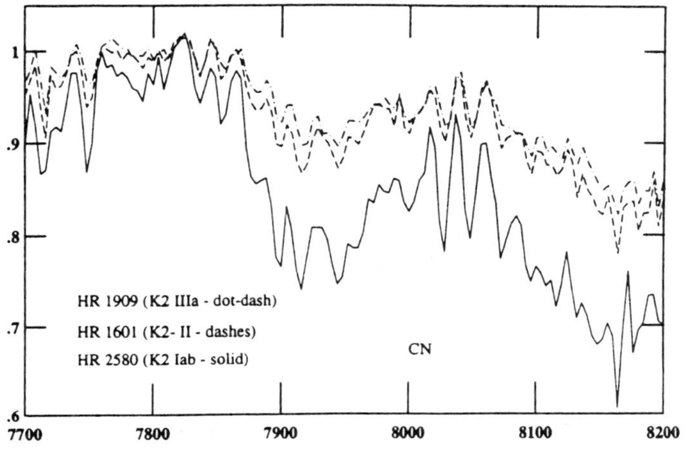

Fig. 1a

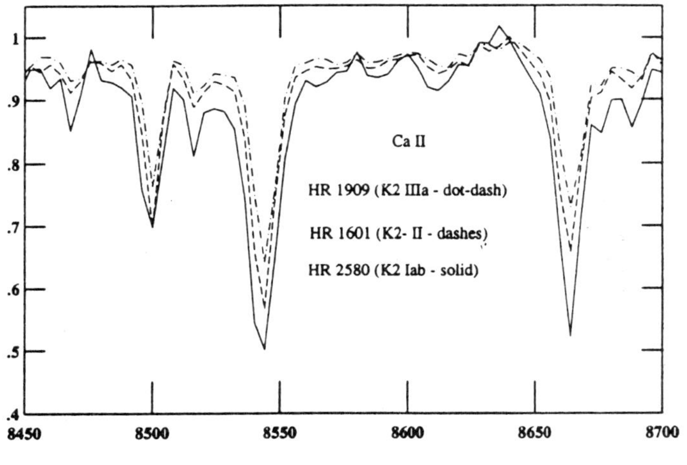

Fig. 1b

Figure 1.–4. Luminosity effects in the region of the luminosity-sensitive features, CN and Ca II, for different M stars.

The Figures 5–7 show 3 program stars versus standards over the range K3.5 to M2.

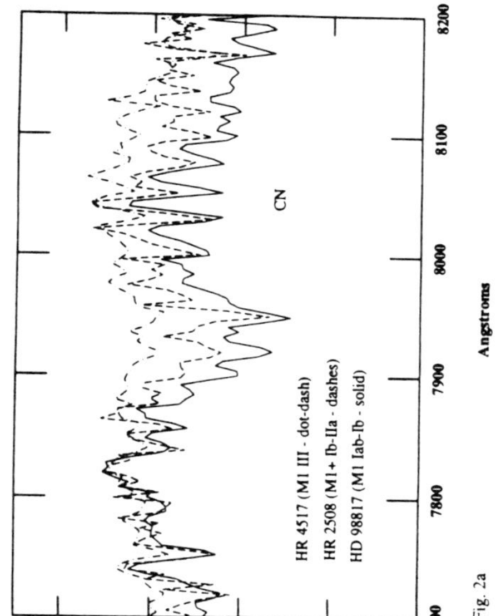

Fig. 2a

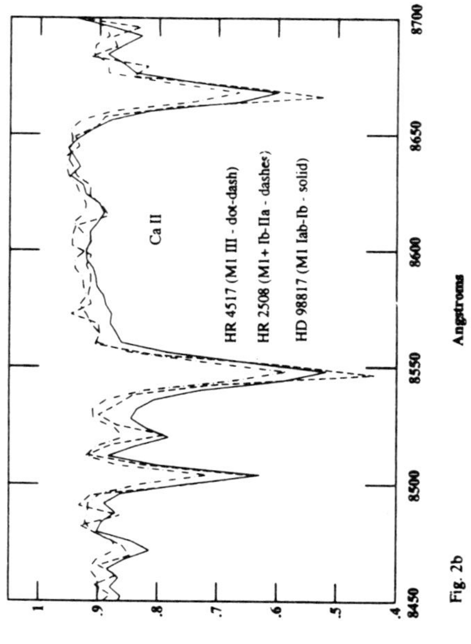

Fig. 2b

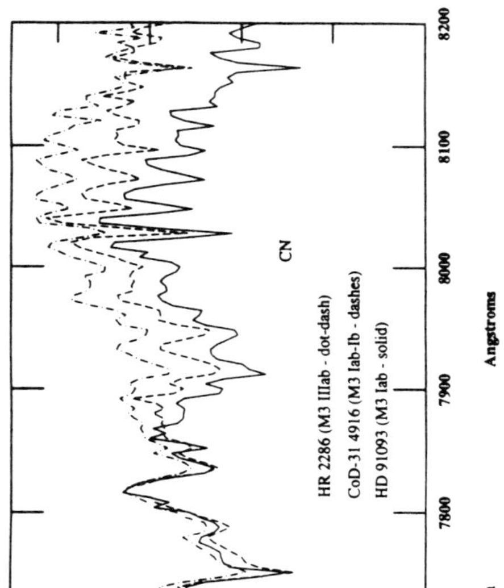

Fig. 3a

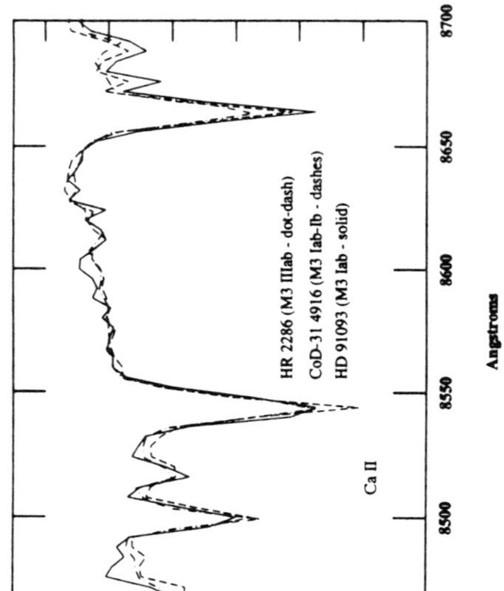

Fig. 3b

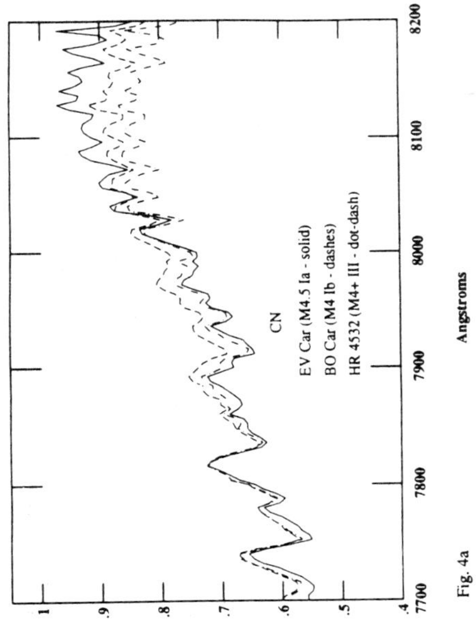

Fig. 4a

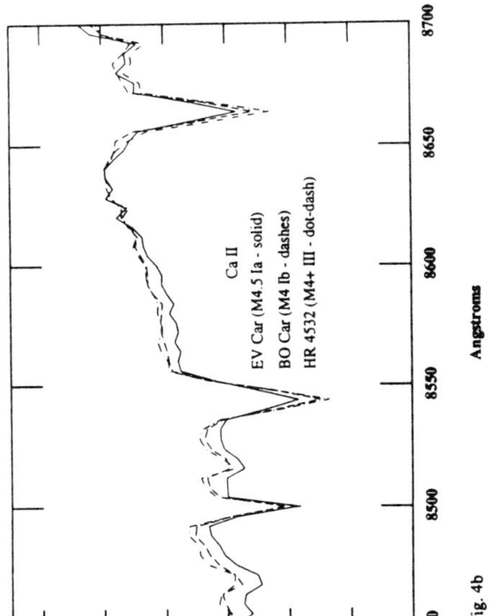

Fig. 4b

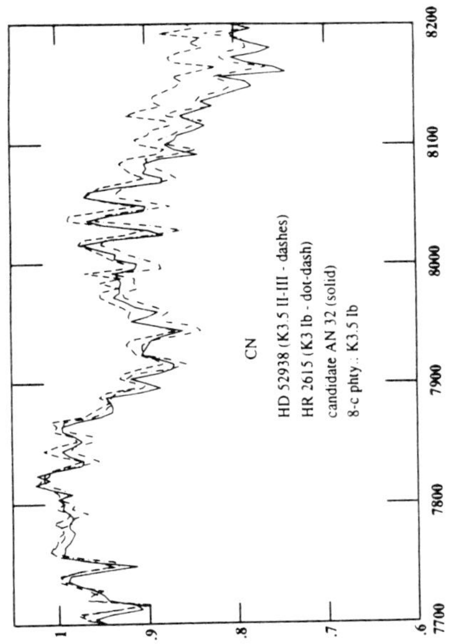

Fig. 5a

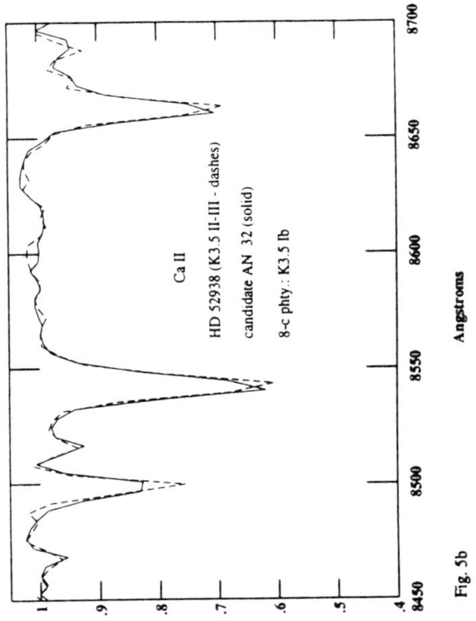

Fig. 5b

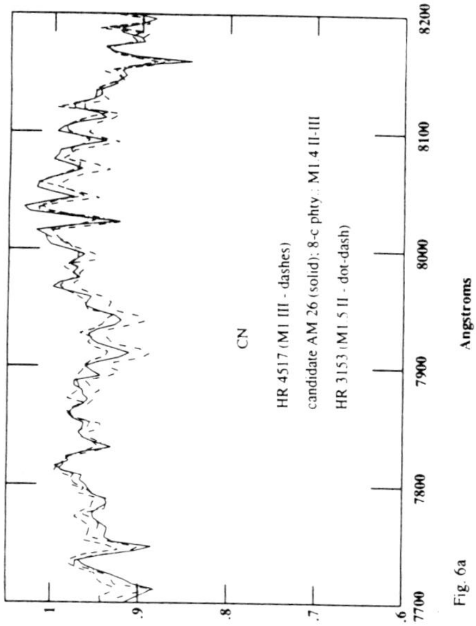

Fig. 6a

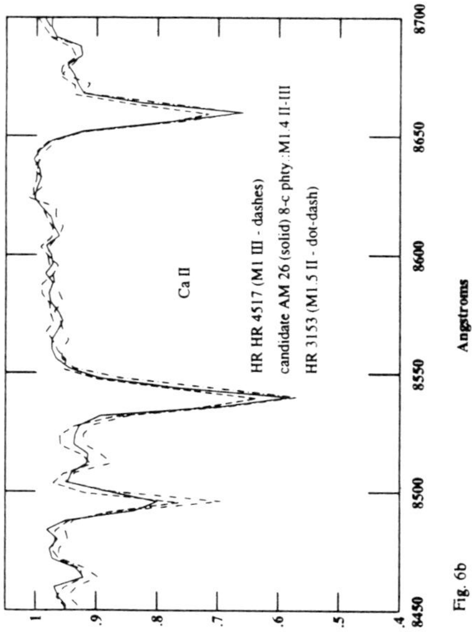

Fig. 6b

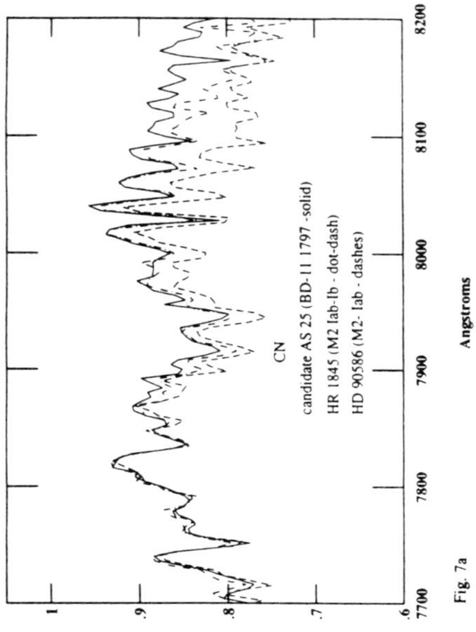

Fig. 7a

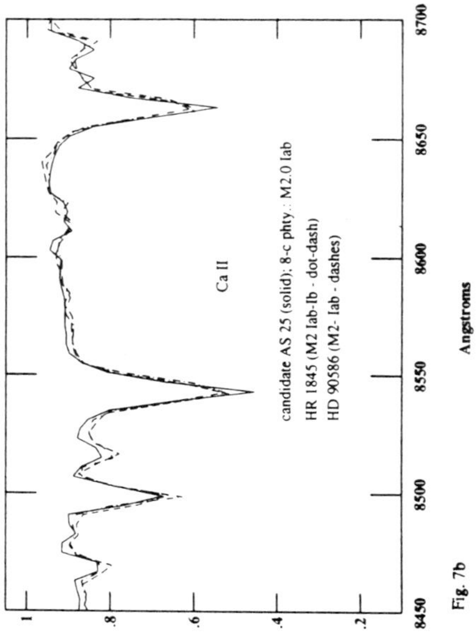

Fig. 7b

References

Keenan, P.C. and McNeil, R.C. 1989, ApJS, **71**, 245
MacConnell, D.J., Wing, R.F., and Costa, E. 1992, AJ, **104**, 821

Discussion

Garrison: The Ca II triplet is usually more sensitive to temperature than to luminosity. Do you have a sample of the temperature sequence for M supergiants?

MacConnell: I have a quite complete grid of about 85 Keenan-McNeil standards K0 and later and of luminosity class III and higher. I classify in luminosity from the spectra only after I know what the wing temperature class is, so I know with which Keenan standards to compare.

Lloyd Evans: Is there any possibility of confusion between CN and LaO on your spectra? This question is prompted by the finding of Catchpole and myself that one third of a sample of S stars found on IR objective prism plates by Westerlund and Olander are not S stars but (mostly) M supergiants.

MacConnell: Our spectra also cover the region around the ZrO band at 6474 Å, so we can exclude the S stars. In fact, we have found two new S stars – they can appear very similar to reddened, cool stars on the objective-prism plates, and the Wing photometry alone may not distinguish them.

The MK Process at 50 Years
ASP Conference Series, Vol. 60, 1994
C.J. Corbally, R.O. Gray, and R.F. Garrison (eds.)

Spectral Classification from Low-Resolution CCD Long-Slit Spectra [1]

H. Hensberge

Royal Observatory of Belgium, 3 Ringlaan, B-1180 Brussels, Belgium

W. Verschueren

University of Antwerp (RUCA), Astrophysics Research Group, Groenenborgerlaan 171, B-2020 Antwerp, Belgium

J. Rogl

Technical University, Gusshausstr. 25-29, A-1040 Vienna, Austria

Abstract. Quantitative measurements on high S/N, low-resolution CCD spectra contain sub-pixel accuracy information that is not fully exploited by classification through visual comparison. The presence of discretization effects at a level that exceeds greatly the S/N is identified as the limiting factor in applying visual classification to our CCD data. The development of a quantitative classification procedure relative to MK standards is discussed, with emphasis on early-type stars.

1. Project

This work is part of a long term programme on young stellar groups. Spectral classification and Walraven photometry are used, together with astrometric data, for identification of cluster membership and for an evaluation of basic astrophysical quantities. Our principal object is the young cluster NGC 2244, in the Rosette H II nebula.

The spectral classification is based on long-slit spectra obtained with CARELEC (Lemaitre *et al.* 1990) between early 1987 and late 1990. The long-slit observation mode ($\sim 5'$) permits accurate background correction in the emission nebula. In addition to around 250 cluster target stars, a grid of some 300 MK standards and reference stars (including a few peculiar stars) has been observed to allow a homogeneous differential classification process.

These CCD spectra have a significantly lower 2-pixel resolution (15.5 Å) than prescribed for MK classification, but have a high S/N and cover a broader wavelength region, from below the Balmer jump to beyond Hα. The data discussed here have typically a S/N of 200 at $\lambda > 4000$ Å, which is set rather by systematic noise than by photon noise.

---

[1] Based on observations collected at the Observatoire du Haute-Provence

We present our progress in the development of a digital classification procedure that uses quantitative parameters, including preferentially those suggested as guidelines in the MK System. The requirements for the procedure are on the one hand the full exploitation of the high S/N, rather than automation, and on the other the compatibility with the MK System. For the latter reason, the procedure is developed directly from the MK standards and reference stars. Tests on artificial noise-free data ascertain the insensitivity to discretization.

Our up-to-date experience indicates that the accuracy of the measurements depends primarily on the avoidance of low-resolution discretization effects. Through critical examples, the achievement of sub-pixel accuracy is emphasized and the correlation of basic quantitative parameters with the MK sequence is shown for early-type stars.

2. Data Reduction

The major advantage of applying optimal data extraction at high S/N is the gained insight in the error budget. We used Horne's (1986) algorithm with a more complete error model (Hensberge and Verschueren, 1992). The availability of spectra obtained from optimal and straightforward extraction, and the estimation of theoretical and empirical errors on every spectral point enable us to evaluate the reliability of any feature in an individual spectrum. In addition, the two-dimensional cross-order profile, residuals, extraction weights, and the background model are saved in a 'control image' for check purposes. Theoretical error estimates include photon noise and read-out noise in raw spectral data and flat-fields, and include uncertainties in background subtraction, in flat fielding, and due to CCD defects; empirical errors were estimated from the residuals with respect to the derived smooth cross-order profile. Flat-field and wavelength calibration frames were taken with all scientific frames, in the same telescope position.

The wavelength calibration is made in a robust way, considering together the calibration line positions on all frames, in order to disentangle discretization influences on measured line positions from spectrum shifts and dispersion changes. A quadratic fit is adequate, with frame-to-frame and run-to-run variations limited to shifts in pixel space and small changes in the linear dispersion coefficient. The calibration accuracy is better than 0.1 pix. Rebinning to wavelength is not performed, in order to keep the data points uncorrelated, and most of all because of the induced systematic rebinning errors at low resolution.

Normalisation to a 'local weak-line level' is made using an impersonal procedure in order to avoid systematic errors between neighbouring spectral types, rather than "optimising" the normalisation for individual spectra.

3. Test Cases

3.1. Hydrogen

Tests made on a set of artificial lines, ranging from point spread function (PSF) profiles to lines with damping wings as strong as in A dwarfs, and discretized in different ways, reveal that *at our resolution*, in the absence of blends,

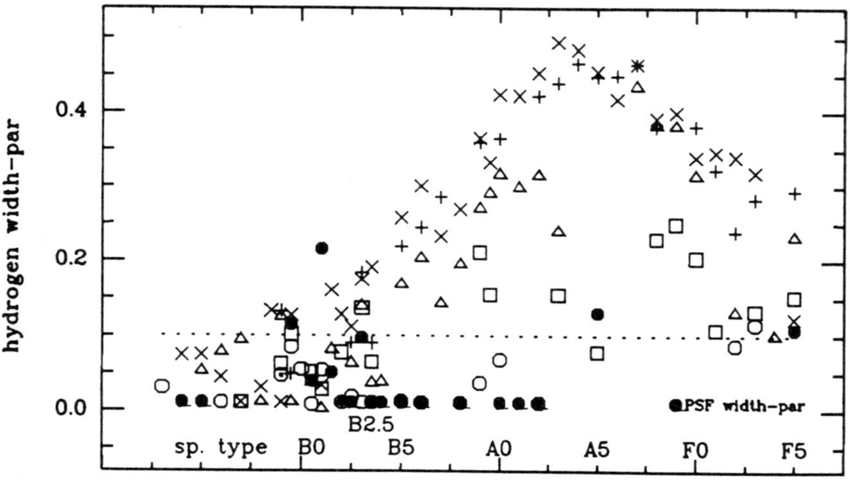

Figure 1. Global hydrogen $width_{par}$ (formula (1) applied to $H_{\alpha,\beta,\gamma,\delta}$) as measured on MK standard stars of type F5 and earlier. Only above the dotted line, the intrinsic line broadening exceeds significantly the PSF width. Symbols refer to luminosity class : x(V), +(IV), △(III), □(II), o(Ib), ⊙(Iab), •(Ia).

width and core intensity indicators quasi-free of discretization errors can be constructed as a combination of geometrical and arithmetical averages of pixel intensities relative to the central pixel intensity, and taken symmetrical with respect to the pixel closest to the line center:

$$width_{par} = 1.8\sqrt{\frac{I(-n)}{I(0)} \cdot \frac{I(+n)}{I(0)}} - 0.8\frac{I(-n) + I(+n)}{2I(0)} \tag{1}$$

where I(n) refers to measured intensities at n pixels distance of the central pixel, i.e., the one with minimum flux. Width parameters are evidently more robust against blending when measured at intensity levels that dominate the intensity of weaker blending lines, and at line depths that are one order of magnitude larger than the uncertainty on their values. Therefore, n = 2 is used for broad lines, while for narrow lines, a more reliable measurement is obtained from the 3 central pixels only (n = 1); both n = 1, 2 are used for lines of intermediate width.

The intensity of the line core is measured by correcting the central intensity as follows:

$$I_{core} = I(0) + \frac{(I(+1) - I(-1))^2}{I(+1) + I(-1)} - (0.56 + 3(width_{par})^3) \tag{2}$$

Results for the MK standards and reference stars are displayed in Figures 1-2.

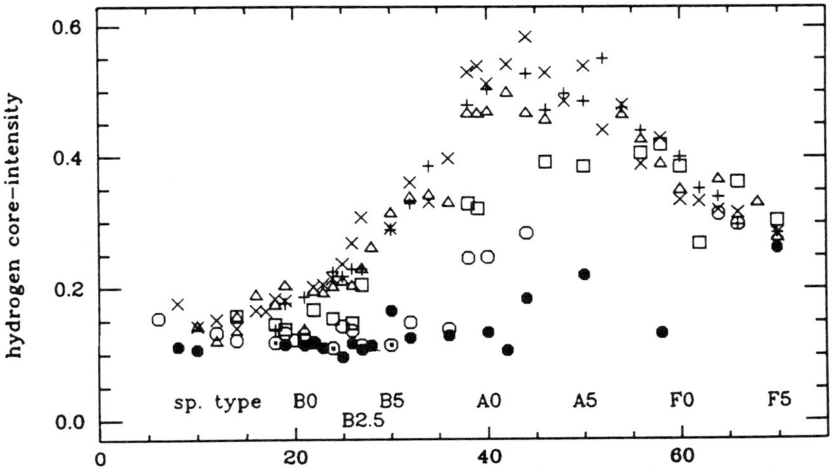

Figure 2. Global hydrogen core intensity (formula (2) applied to $H_{\alpha,\beta,\gamma,\delta}$). See Figure 1 for the interpretation of the symbols.

3.2. Unresolved features with both components used in the MK System (intensity-ratio criterion).

Many MK criteria involving line-intensity ratios can not be used directly at our resolution since one or both lines are part of a blend, or both lines form a common blend. Therefore, we investigated to what extent information can be retrieved in case the constituents are unresolved at roughly one pixel separation.

Looking for the most robust parameter that correlates uniquely with the line-intensity ratio of the 2 blending lines, we selected the gravity position of the blend, with respect to the calibration-predicted position of one of the constituent lines. However, for two reasons, the final estimate should be computed from the measurement of several center-of-gravity estimates, each determined using a different, small number of pixels. Indeed, first of all, noise (systematic, e.g., blending and random) at a larger distance from the center destabilizes the gravity estimator. Secondly, the fact that a measurement interval consists of an integral number of pixels and thus is generally not properly centered at sub-pixel level, induces discretization effects. The induced error changes sign with intervals containing an even versus uneven number of pixels, and diminishes for larger intervals (Figure 3). Hence, the bias is largely eliminated using:

Feature-position parameter =

$$\frac{(x_{n+1} - x_{n-1})^2 x_5 + (x_5 - x_3)^2 x_{n+1}}{(x_{n+1} - x_{n-1})^2 + (x_5 - x_3)^2} \qquad (3)$$

with n=5 or n=3, the latter if the wings of the feature are obviously blended.

Figure 4 shows this feature-position parameter for the He4471/Mg4481 blend for the MK standards of type A5 and earlier. The correlation with temperature is obvious, largely independent of luminosity. The scatter amounts only to a small fraction of a pixel. The correlation of the intensity of the feature with

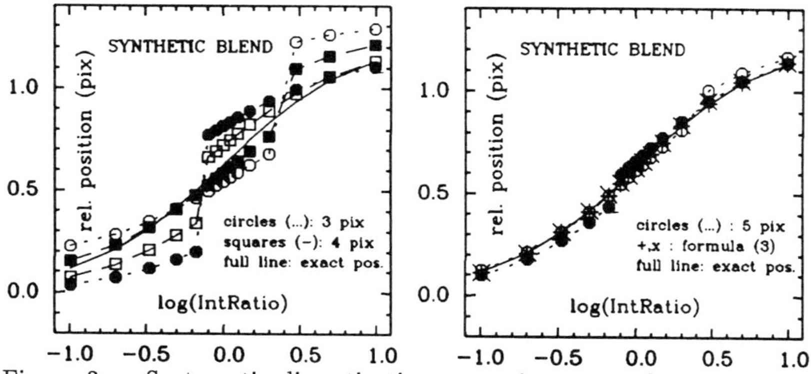

Figure 3. Systematic discretization errors in center-of-gravity measurements on a test case (2 noise-free Gaussian lines with fwhw=1.9 pix and separation 1.25 pix). The error depends on the intensity ratio of the components (abscissa) and on the pixel fraction at which the true gravity center is situated (full line). The measured position (ordinate) is expressed relative to the position 0.0 of one of both constituents. Open versus closed symbols refer to those discretizations that produce the largest systematic errors. Left panel for measurement over 3 versus 4 pixels, right panel for a 5-pixel interval and for the prediction of formula (3).

the MK spectral type (not shown here) still contains additional classification information.

3.3. Individual features with PSF profile

From a set of discretized PSF profiles, each with a different shift in position relative to the pixel center (step 0.05 pixel or less), we select the template corresponding to the discretization predicted through the wavelength calibration. The intensity parameter

$$I_{feature} = \frac{\sum[I(obs) * I(PSF)]}{\sum[I(PSF) * I(PSF)]} \tag{4}$$

is found to be fairly robust against discretization and blending outside the feature core, if one sums, at our resolution, over 2 to 4 pixels. Ratios of intensities of nearby features are least susceptible to uncertainty in the local weak-line level. When used on weak lines, preferentially several of these measures are combined into one global luminosity or temperature parameter. Such a global strategy has been succesfully applied to the estimation of the strength of the diffuse interstellar band spectrum (DIB), in this case however taking into account the intrinsic width of each feature. The DIB's were then eliminated, so that e.g. the measurement of Hβ in reddened stars is not influenced by the DIB in/near its red wing.

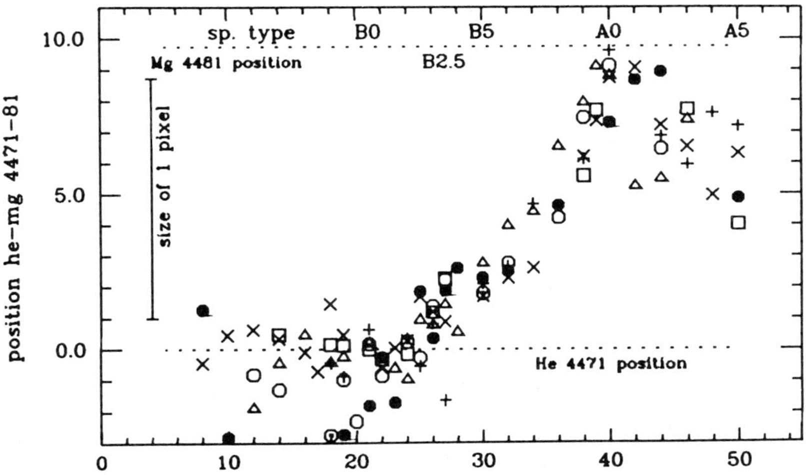

Figure 4. Feature-position parameter (in Å) relative to He λ4471, as measured on the He λ4471 / Mg λ4481 blend in MK standards of type A5 and earlier. See Figure 1 for the interpretation of the symbols. The line intensity ratio of both components is in the MK System a temperature indicator for B1-A0 stars.

3.4. Criteria not used in MK classification

An obvious case of a parameter not used in the MK classification scheme, but useful at our resolution and wavelength coverage, concerns the detection of CP2 stars in the 520 nm region. The broad 520 nm depression, used commonly as a photometric detection criterion (Maitzen, 1976), is easily recognized. Moreover, the loss in photometric detection sensitivity at A5 and later can be avoided spectroscopically (Hensberge, 1994).

4. Future work

We showed that important classification indicators can be described in terms of quantitative parameters which are at high S/N fairly robust against discretization effects and blending. The search for a proper combination of these parameters (most of them having a weight different from zero only over a limited spectral range) that quantifies the over-all (dis)similarity of the spectra, is in progress. Specific indicators should draw attention to peculiarities in a spectrum. Our extensive set of MK standards and reference stars permits us to optimise the compatibility with the MK System.

References

Hensberge H. 1994, in *Proc. 25th meeting and workshop of European Working group on CP Stars*, ed. I. Jankovics and I.J. Vincze (in press)

Hensberge H., Verschueren W. 1992, in *Proc. 4th ESO/ST-ECF Data Analysis Workshop*, p.13 (ed. Grosbol and de Ruijsscher)

Horne K. 1986, PASP, **98**, 609

Lemaitre G., Kohler D., Lacroix D., Meunier J.P., Vin A. 1990, A&A, **228**, 546

Maitzen H.M. 1976, A&A, **51**, 223

Discussion

Abt: Why does the composite strength decrease in the later A's? Is it due to a contribution of 4471 Fe I?

Hensberge: Exactly, it is due to blending by Ti and Fe lines.

Weaver: How carefully must you do the wavelength calibration?

Hensberge: When using positions of blends, like for the He 4471/Mg 4481 feature, as indicator for the relative strength ratio of the components, the accuracy of the wavelength calibration is directly reflected in the measured parameter. Hence, the calibration accuracy should be a small fraction of the pixel size, say better than 0.1 pixel at least. In order to obtain this accuracy, the most crucial issue is to make sure that the measured positions of the calibration lamp spectral lines themselves are unbiased by discretization effects.

The MK Process at 50 Years
ASP Conference Series, Vol. 60, 1994
C.J. Corbally, R.O. Gray, and R.F. Garrison (eds.)

Aitoff Projection Plots for Stars in Volumes 1–4 of the Michigan Spectral Catalogue

Shannon K. Austin

Georgia Inst. Technology, Atlanta, GA 30332

Nancy Houk

Univ. of Michigan, Ann Arbor, MI 48109

James R. Sowell

Georgia Inst. Technology, Atlanta, GA 30332

Aitoff equal-area projection plots of stellar distribution in galactic coordinates l, b are displayed, for HD stars between $\delta = -90°$ and $\delta = -12°$. Surface distributions are shown because, for most of these stars, no accurate magnitudes and colors are available, making distance calculation so uncertain as to not be worth doing. The poster paper includes O stars, supergiants of each temperature class, normal giant and dwarf stars of grouped temperature classes and some early-type and late-type peculiar stars. O stars and most of the peculiar-star plots are not shown in Figures 1–4, but copies may be obtained from the authors upon request, and most may be published later in the Astronomical Journal.

Houk thanks the NSF for support through Grant No. AST-8913683.

Discussion

von Hippel: Why is the distribution of the Am stars so different from the A0 V stars?

Houk: The Am stars may have a Ca H & K line type of A0, but the real temperature type of the stars is more nearly equal to the hydrogen line type, or even the metallic-line type; these are usually around A7-F0. The Am distribution is very similar to the normal F0 V distribution shown in Figure 3.

von Hippel: So, the Am stars are fainter in absolute magnitude than the early-A type stars and with them we do not punch out so far into the halo of the Galaxy.

Abt: The magnitude effect, which von Hippel is talking about, should be strongest in the plots you showed for the A0 V stars. The stars brighter than V = 6 are nearby and should show a more isotropic distribution than those at V = 9, which should show a greater concentration to the galactic plane.

Drilling: How many stars does one point represent in your plots?

Houk: Each point represents a single star.

Drilling: What is the significance of the plots of luminosity in your poster?

Houk: If you are referring to the graphs of CN stars (not shown in published proceedings), they are plots of the actual luminosity, determined by Sr line/Fe line ratios versus a 'pseudo-luminosity' obtained by comparing the strengths of the CN break with those of normal stars of different temperatures and luminosity. Thus the strong CN stars have a CN break equivalent to the normal CN break in class II or Ib stars, for example, of the same temperature and luminosity type. The plots showed that for type G8 the majority of CN stars classified were strong CN rather than weak CN stars and that this imbalance becomes even more pronounced as one goes to the later temperature types of K0, K1, and K2. Outside of these temperature types it is difficult to detect strong or weak CN stars at MK classification dispersion.

MacConnell: I think one must take care not to over-interpret the plots since the HD plate material may have been less homogeneous than we would like. Even with care and modern techniques, a photographic survey can have rather different limiting magnitudes from one field to another.

Houk: Yes, that is true, but in this case I do not think the region is low density because of the limiting magnitude, since for earlier and later spectral types there is no low density region.

McCarthy: Do you have any comments about the non-HD stars on your plates.

Houk: I am careful only to look at and classify the HD stars.

Garrison: Classifying the non-HD stars is the task for LaSala and von Hippel, not Houk. (Laughter)

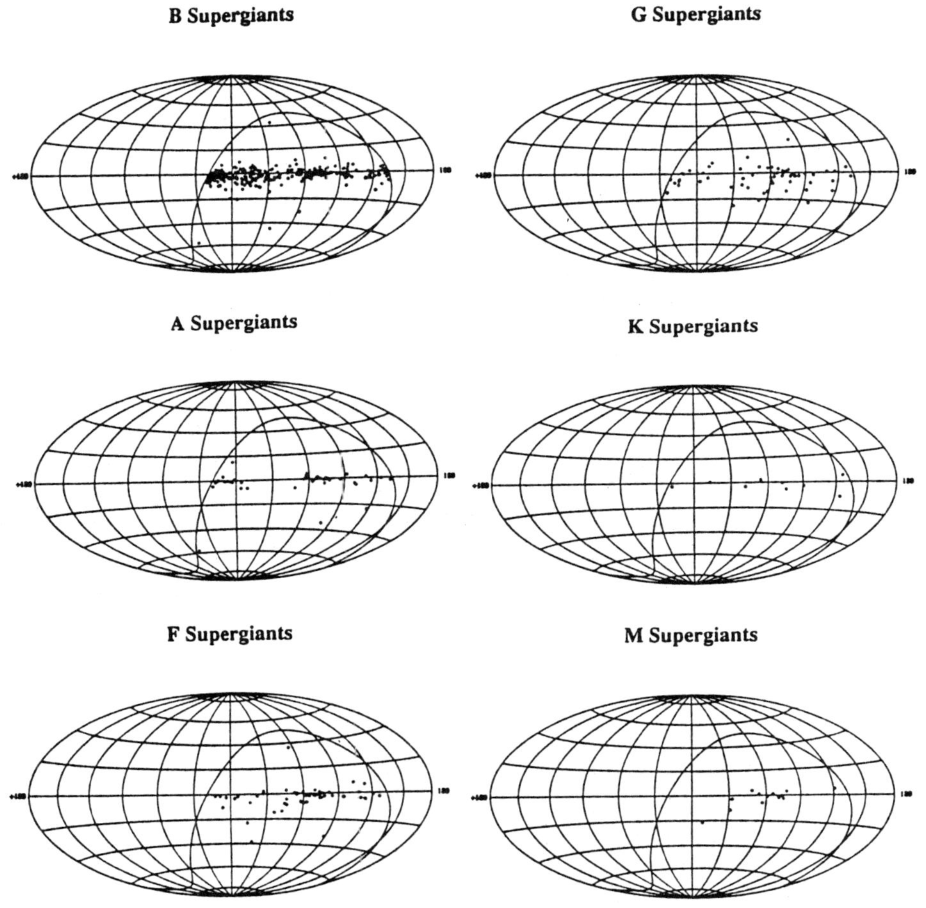

Figure 1. These Aitoff equal-area projection plots show the distribution of supergiants, in galactic coordinates, from Volumes 1-4 covering the area from $\delta = -90^\circ$ to -12°.
The galactic center ($l = 0^\circ$, $b = 0^\circ$) is at the center with l increasing toward the left. Some Magellanic-Cloud supergiants are included.

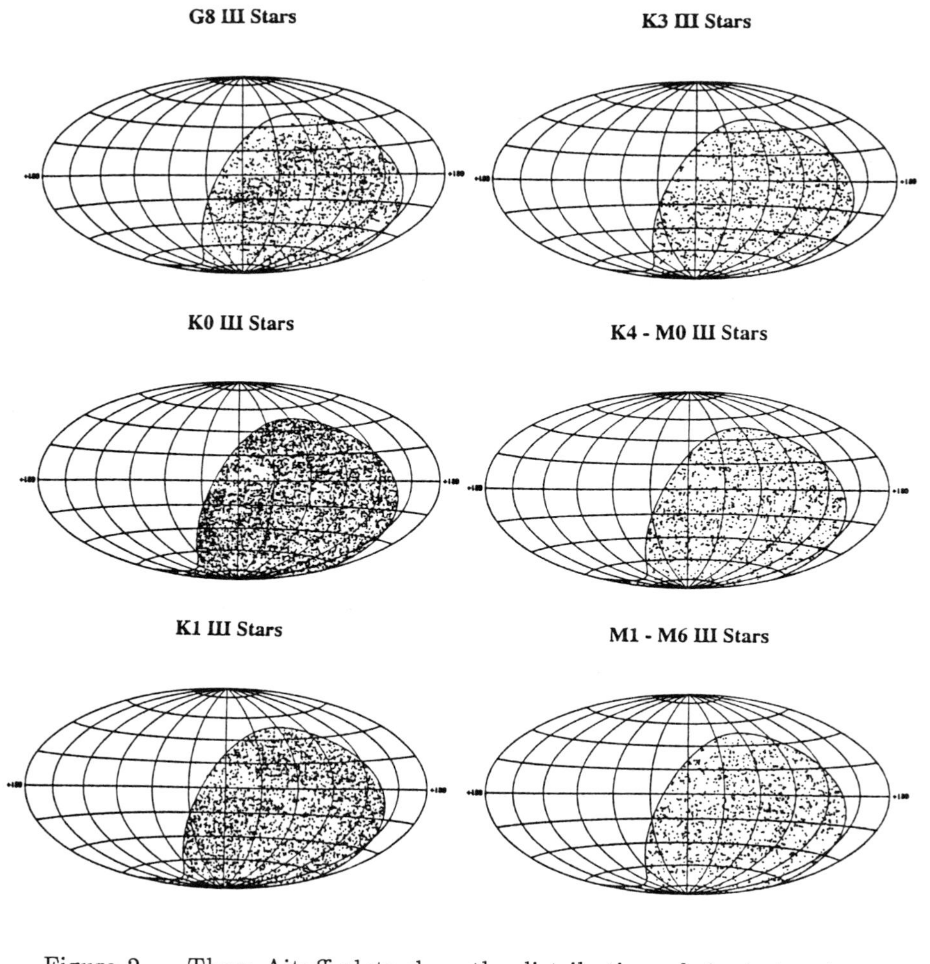

Figure 2. These Aitoff plots show the distribution of giant stars in galactic coordinates, covering the area from $\delta = -90°$ to $-12°$. The galactic center ($l = 0°$, $b = 0°$) is at the center with l increasing toward the left.

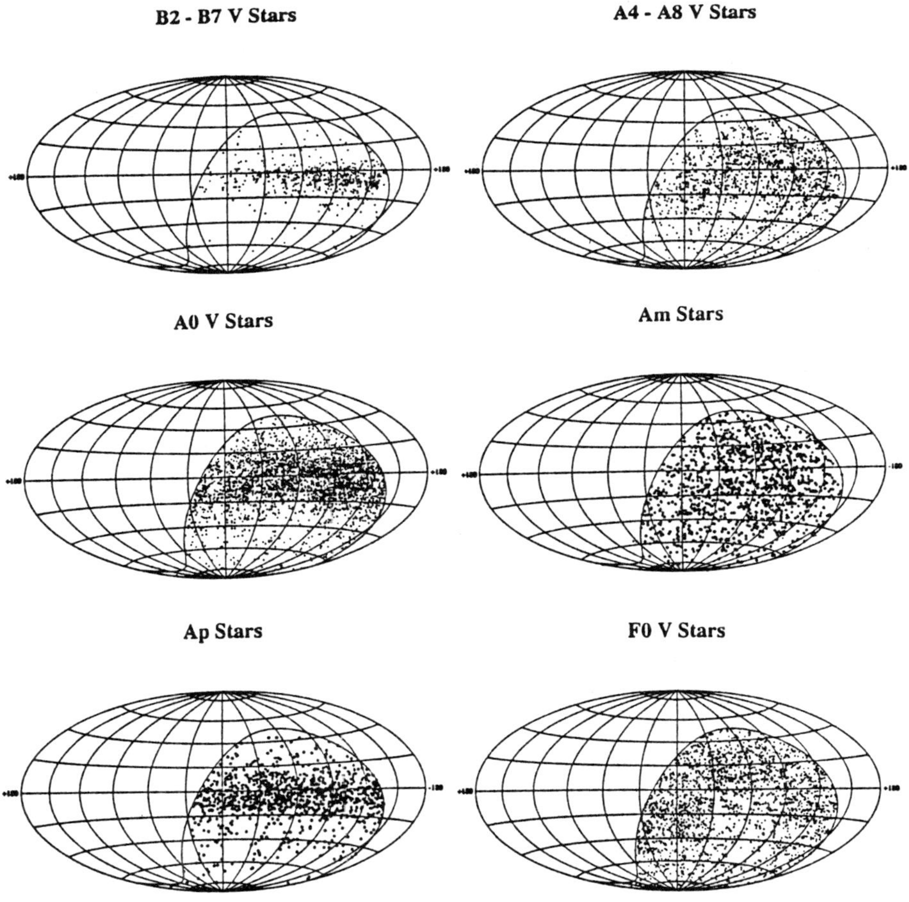

Figure 3. These Aitoff plots show the distribution of early-type to mid-type dwarf stars covering the area from $\delta = -90^\circ$ to -12°. The distribution of Am stars, similar to that of F0 dwarfs, is also shown.

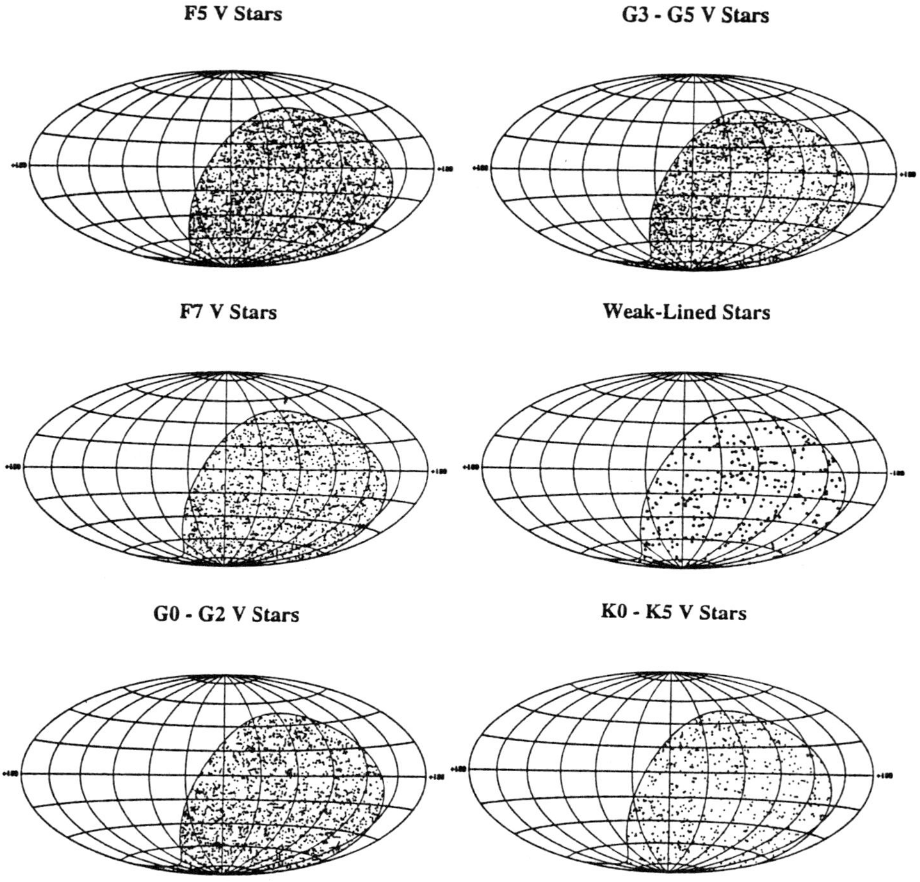

Figure 4. These Aitoff plots show the distribution of mid-type to late-type dwarf stars covering the area from $\delta = -90°$ to $-12°$. The distribution of the metal-weak stars is also shown.

PART VIII.

GENERAL DISCUSSION

GENERAL DISCUSSION

1. Introduction

Garrison: One of the reasons why Chris Corbally asked me to chair this final morning's discussion may have been to keep me quieter than usual, or perhaps more diplomatic, which can be difficult, but I'll try. We have had a good deal of even paced discussion during this most pleasant workshop. Let us continue in the same spirit, perhaps not solving any of the significant problems, but at least bringing them up for further collective consideration. We can start off on any of topics that have been suggested, for instance, ...

2. Discrimination levels

Garrison: Through Morgan and Keenan and company, some of us together have achieved a certain level of discrimination in classification; and that level of discrimination is really what makes the MK System valuable. When we go to lower resolution or when we go to other wavelength regions, we may be able to achieve a slightly higher discrimination level. But one of the things over which we have to be a little bit careful is that the higher discrimination is sometimes a false one. For example, MK classification assumes that you look at as big a piece of the spectrum as you can and that you integrate all of the features of the spectrum. If you take one feature or one line ratio and get a quantitative measurement of it, you are going to get a higher level of discrimination, but it may not fit with the rest of the spectrum.

Let me illustrate that with $\lambda 4121$ to $\lambda 4144$, which is the helium triplets to singlets ratio. At B2, you can discriminate ten luminosity classes quite easily if you use only that ratio. But if you look at the whole spectrum, there's a little fuzziness that comes in: one criterion might tell you that it is a B2 III and another criterion might tell you that it is a B2 III-IV, and so forth. So we need to be careful that, in interpreting higher levels of discrimination, we don't just measure one line ratio and get a hundred discrimination levels, and then call that a better classification; it may well not be. Just remember that the MK Process involves looking at the *whole* spectrum in the end, and that the line ratios are guides. They are useful – and some line ratios are more useful than others – but they should be used intelligently.

I don't want to talk the whole time, so let us move on to another topic. I'm sure that everybody has a lot of opinions about ...

3. Classification notation

3.1. Notation for extensions

Garrison: This suggestion for a topic was probably meant to draw our attention to the need to do some coordination and worrying about what notation to use when implementing new dimensions and extensions to the MK System. What if somebody sets up a scheme in the ultraviolet, or the infrared, or the X-ray, or the radio region, or whatever? How do we distinguish each of the classifications so that, when you have a bibliography of MK types, it is clear that a particular classification has been done in the infrared and that is has been made at 20 Å resolution? I mean, if someone uses the far infrared or a very low resolution to infer an MK type, how can we know that fact when looking, say, in the *Bright Star Catalogue* and be able to distinguish it from a *bona fide* MK type?

For a start you could use small letters of the alphabet, and instead of an "F5 Ib," you might call it an "f5 ib." The problem is to think up enough different notations for each new system. So when you are thinking in terms of extending the MK System, you should worry about how to distinguish your new type from an MK type or from another extension's type.

I find that this is very important when I'm working with my complete bibliography of MK classifications. I have to go through the original publication myself and put some mark in the computer database, which tells me that this is not a visual spectral type but a photometric type, etc. Some people, for example, give a photometric type that is from the *uvby* system and they'll say, "Okay, this star has the colors of a B3 V star." If they put that notation down in the table, others will feel that, just given the table, this star is really supposed to be a B3 V MK type. You know, that is really misleading, and so we have to be careful how we use notation in tables. If anybody has any suggestions for notation that could distinguish this, I would be really appreciative for my database.

Rountree: Well, two things: first of all, in the publications that George Sonneborn and I've prepared, one of which was published in the *Ap.J.* already and a second paper which will be coming out next year, I've used italics to indicate MK types derived from ultraviolet spectra, although I truly believe that these types are very close to the MK System since we used MK standards. In the atlas I found it difficult to maintain that italic notation. It is hard to have the typesetter keep printing things in italics, and often we did not do that except when the distinction was truly necessary.

Secondly, my experience is that a number of these notations have been used temporarily in the past and have gradually been absorbed into the standard MK notation. To give an example from my graduate school days, Rudy Shields was working for Morgan at the same time that Bob and I were, and he used slightly different standards for some of the B stars. Morgan felt that Rudy's types were not quite on the MK System and so he insisted that Rudy use small letter notation for them. This persisted for a couple of years but fortunately never made it into print. By the time Rudy got ready to publish, Morgan had relented and Rudy was allowed to use the standard MK notation – which is a good thing, because otherwise it would have caused a lot of confusion. (*Garrison*: I'll bet Rudy was well calibrated as well. I used to work with him.) So I agree with you that some distinction is necessary, certainly for photometric types or types that truly do not use MK standards, but one has to be careful not to create too many, that's all.

Roman: I have a few points. Firstly, I was going to suggest that one simple notation would be parenthesis for types that are based on systems other than the standard MK region.

Secondly, I'm not sure that, in all of the work extending to new regions, we have been able to do as careful a job as we might to correlate the types that we derive in these other regions to the MK types. It may not be possible to do a one-to-one correspondence for all types, because I suspect that different regions will have different levels of discrimination in particular groups of spectra. For example, it may very well be that the infrared will do a better job of discriminating among the cool stars than the MK System can. We need a finer grid of types, but they still can be anchored in the MK types.

Finally, I have in mind, not so much types from other regions, as the extensions of the MK System to indicate metallicity and other "peculiarities" in the spectra. We are beginning to get very complicated notation, and I'm wondering if anything can be done to simplify the system.

Corbally: I like the idea of parenthesis rather than italics. In print you can obviously see italics, but when you transmit digital data, you can't transmit italics. It's got to be something that is in ASCII format.

I would add a thought about the complexity: some of the complexity should go into a note rather than into the type that one gives for a particular spectrum. However, in databanks these notes tend to be lost, so there could be an asterisk in the databank's entry to say that this star has a note, and this would help one's obligation to go back to the original paper and to look at the classification.

Notation becomes a really important question to me when I'm remembering that all these classifications from other extensions to the MK System will all end up in Strasbourg. What will we get? Perhaps the place to start is the databanks, to try and get divisions in what they consider should go under "spectral type," so that they will have different categories for an infrared type or an ultraviolet type or whatever.

Houk: My problem is that I've been using parenthesis for uncertainty. So I'm wondering if maybe different classification systems could be distinguished by brackets, or square brackets, or french brackets, or something other than parenthesis – which is fairly commonly used for uncertainty, is not it?

Garrison: "Uncertainty", as usually used in Morgan's work, is denoted by a colon. So, a colon is put after the particular parameter that is most uncertain. For example, if it was the temperature type that was uncertain because of some peculiarity, he would call it a B2: III, or if it was the luminosity that was uncertain, B2 III:. That notation has been found to be very useful.

Morgan also seemed to use a question mark when the spectrum was not good. The colon usually indicates uncertainty because the star is uncertain about what it is: but the question mark is uncertainty simply because the spectrum we have was not good enough. So I've always followed that notation myself.

Houk: In that case I would have question marks all over the place. Anyway, I was just saying that the parenthesis are already preempted.

Roman: I make a plea not to use square brackets. They are not standard and not all machines print them. I use an IBM mainframe and it will not accept the square brackets that I get from other people.

Gray: This is a suggestion that I would like to propose. You can think about it. In this meeting we are beginning to get an inkling of the different parts of the electromagnetic spectrum – the UV, visual, red, infrared – where these new classification systems parallel to the MK system are going to be set up. We have seen that the IUE ultraviolet is one place where extensions or parallel systems have already been set up. I think that we are beginning to see that the red, the near-infrared, and also the H and the K infrared bands are fruitful places where sometime in the future we are going to have parallel systems set up. The stew has to perk and boil a little bit more before we are really certain what these regions are going to be. But when they finally settle down, I think that we might solve this question of "from what region do these MK or similar MK types come" by just simply appending a lower-case letter to the beginning of the spectral type. So, for instance, if we have a classification in the IUE ultraviolet, we simply put uB8 III, and then it is understood by everyone that the classification comes from the 1200 to 3200 Å region.

Garrison: Could it be two letters like "uv"?

Gray: You could have "uv", but why have two when one would do? For instance, when we finally settle on the red part of the spectrum that we think is a fruitful place for spectral classification, we could have just a little "r". For the near-infrared we could use "n" or "ni", and for the H and K bands it is pretty clear that we could use "h" and "k", although that might be confused with hydrogen lines and K lines. That would avoid parenthesis, and italics, and all sorts of non-ASCII characters.

Torres-Dodgen: I like your suggestion of a small first letter, because parenthesis and asterisks are already being used for O-type stars and those would create a lot more confusion.

Garrison: What about a possibility of a parenthesis with a subscript IR or subscript UV. (*Cries from the floor*: Subscripts are hard to print.) Okay, okay. But I have a problem with h-region and k-region types because of H and K calcium and because of hydrogen. We already use those for Am stars for which we give a small "h" for the hydrogen type and a small "k" for the K-line type. But we can think of something, I guess, and draw up a table of suggested nomenclature.

3.2. Notation for third dimensions and peculiarities

Roman: I would like to come back to the other notation problem I mentioned: that of other dimensions besides luminosity and temperature. Is there any way to simplify or standardize that notation?

Garrison: Yes, basically the problem with the third dimension is that it's different for every region of the HR diagram. So I personally have been leaning towards Keenan's notation, especially his simplified notation. In his original notation he gave practically every element an indication, but now he uses iron as a summary. As long as other species like cyanogen and so forth are weak to the same degree as the iron is weak, he will just use the iron as a weak-line characteristic. I like that. I've done a lot of G- and K-type spectra lately and a lot of weak line spectra, and it works well for me.

I've not tried doing this for the Ap stars or the Am stars, but I wonder if something like that might be a refinement that we can introduce. At the present time, what we do for the Ap stars, as you know, is almost that. For example, we will give the best estimate we can of the spectral type – and we know that it is not terribly good – and then in some cases we will just put silicon in parenthesis with no strength indicator or in other cases we will give three types. We will give the hydrogen-line type, the K-line type, and the metallic-line type.

We have also done that for the metallic line stars. It is a little bit confused still but, now that we have had a lot more experience with Ap stars, we might consider going one step further for the Am stars and having a similar nomenclature to Keenan's in the late-type stars. That is, we might use iron, for example, as a kind of a typical indicator for the metallic line stars and give it a strength indication: but then we would have to have set up standards. Keenan *has* set up standards in his atlas. You can actually see what "Fe-4" is, and you have got a standard star that represents that class. So he has done pretty well that way.

Levato: Concerning the Ap stars, what we are doing now is loosing some information that we have in the spectrum. In our work with Helmut in clusters, we have the impression that certain kinds of peculiarities increase with age. We have not quantified this with an index, but it is far from trivial. For instance, we will have an A0 or a B9.5 or an A1 spectrum, which will have a particular intensity for the silicon lines, and the intensity scale is different for each of these temperature classes. Also, among the Am stars, you have some spectra in which the scandium lines are weak and others not. So, someone must establish the grid of indices and say, for example, "This is an Ap Si+2", and show the standard; but no one has done this yet.

Garrison: This is a task which needs to be done by somebody who has a great deal of experience with Ap stars and who has a good sample of specimens on which to build this experience. It is a big task, but I think a necessary one.

3.3. Notation for variable stars

Hensberge: For stars like these peculiar stars there may be another, more relevant problem: these stars may be known to be periodic spectrum variables. So the biggest loss in information may not be related to knowing how strong some peculiarity is, but knowing the time of observation for the spectrum. Because, if you find in your literature three different entries for the peculiarity of a star, without any time of observation, you loose the information about whether the star is probably a spectrum variable or not. That may be very interesting information for people doing high dispersion work on these stars. So, my suggestion is to include the time of observation, at least in a note and even for non-periodic variables.

Garrison: Yes, I agree that time variability is a serious problem as well. I ran across that with the Be stars, and I'm sure Janet also has a lot of feelings about that. We give quite a complicated type for the Be stars that is descriptive about what a spectrum looks like right now, but the problem is that five years or ten years from now it may not even *be* a Be star. Janet has suggested a Be star nomenclature which is quite useful, but unfortunately the stars are variable. So unless the classification has got a date attached to it, somebody will take a spectrum, expect it to look like the earlier classification, and if it does not, then they will say, "Oh well, she was wrong" – which is not true. That is a real problem, and one for the Ap stars, too, which can vary on long time scales. If we only have one spectrum or two spectra, we may not even see that variation.

Roman: I think the situation with spectrum variables, whether Be's or peculiars or anything else, is very analogous to the radial velocity situation where the radial velocity is usually quoted as an average (or, if you know it, the K value) and has some indication that it is in fact variable. Then you usually have to go to another catalogue to get the individual measures as a function of time. In a similar situation, you may not want to give *detailed* definition of silicon strength or of strontium strength, but simply indicate that this is a silicon star and where to go to find out more details of what happens over a period of time.

Garrison: Yes, probably about the only solution we can really make is simply to indicate what we see; but there are some stars in those categories that are not variable and there are a few that are variable. It is the ones that are variable, if we are doing a survey and have only one spectrum, that get to be a problem. We certainly don't want to triple our effort by taking three spectra in every survey, two close together in time and one further apart, but that may be necessary for exact survey work.

Drilling: So, what I've done with photometric, not spectroscopic observations is to put the averages in the published paper but to refer to, e.g., a NASA database, and then put the times of observations in that database so that they are not lost.

Lloyd-Evans: One actually encounters some very severe problems in the classification of variable stars. Some stars are relatively well behaved. For instance, on moderate to low dispersion spectra of classical Cepheids you don't encounter too many problems; but once you start to look at stars with shock waves, you have a very different picture. The situation there seems to be that when you have shock waves, which for the first approximation you can take as being present in stars that have, at some part of their light cycles, strong Hα emission, then you get what sometimes we call "veiling effects" and you get spectra which really don't fit at all well into any classification scheme. So you'll find that the red spectra of RV Tauri stars have titanium oxide bands while their blue spectra indicate a much higher temperature.

In the carbon Miras, for instance, some very peculiar things happen. During an event usually following light maximum, in the blue spectral region not the red, the C_2 bands and the λ4737 bandhead in particular can disappear completely and be replaced by comparatively sharp emission features, whereas the λ5165 bandhead and the bandheads to the red of that appear relatively normal. At the same time in the red, the normal relationship between the C_2 bandheads and the sodium D feature – which you might consider set up from the more normal stars of small amplitude or from constant stars – that normal relationship breaks down completely and you get stars which have weak C_2 and yet weak sodium D, which contrasts with the more normal stars in which you'll find weak C_2 going with a very strong sodium D.

What I've done in particular for some of the hotter stars, the RV Tauri and Type II Cepheids, was to follow the example of George Preston and his collaborators. I must say, I was partly forced into this by having spectra of very low resolution where you could not see the normal MK features. Anyway, I gave what amounts to an absolute strength for each feature by comparing my spectra with a set of standard star spectra and saying, "Well, this spectrum has CN like a G2 supergiant, but it has a calcium K-line like an A5 dwarf." So you have got something that is not at all an MK classification but a collection of absolute feature strengths. Ultimately one hopes that somebody will come along and do as Dimitri Mihalas did for the examples that were quoted the other day and give us an astrophysical explanation.

To finish off, it is worth noting that there are rather exotic spectra around, such as for some stars selected from IRAS in which it is quite clear that circumstellar material is important. These are definitely not standard MK spectra and such exotica will start turning up in other surveys going down to faint magnitudes.

Garrison: That is a good point, and I think the veiling for M stars is a serious problem. The way that Keenan always has gotten around that, and in which I have too, is by using band ratios, and you find that you get fairly reliable types that way.

Parthasarathy: I should like to add as another problem that there are Hα emission surveys which just indicate that the star is an Hα emission object but give no spectral type information. It would be very useful to have the spectral types and luminosity classes for all the Hα emission-line objects, because recently we came across one star which has changed its spectrum in the last twenty years. Fortunately, the spectrum was taken by David Clicken in 1971, so we do know what it was like twenty years ago. But that kind of information is seldom available.

A further problem is that in the case of binaries there is also no systematic spectral type information. For example, the several hundreds of Algol-type binaries in the binary star catalogues have no systematic classification at the MK dispersion.

Garrison: Well, you point to the underlying problem that we are a finite group and we have finite time, so we can't do all these tasks; but I certainly agree with you that it would be very desirable to do so. I had a student some years ago, Austin Gulliver, to whom I just suggested that he look at the twenty classical shell stars, because nobody had looked at them since Merrill in 1956. Nobody! He found a lot of new and interesting information simply from those twenty stars, not a very extensive group.

Corbally: Dr. Parthasarathy may be interested that I've got a good number of spectra of the wider Algol systems which I took in collaboration with Richard Wade and Ron Polidan. I am slowly working through them, but I am wondering whether they may really need a new dimension for their proper classification. For instance, I've looked at some of the B-type spectra against MK standards and, generally, they are showing weaker helium than the standards do. One needs new eyes, as it were, and the leisure to be able to think about the phenomena. So, this may be a fruitful area for an extension to the MK System.

Houk: Is it really totally unfeasible to get the remarks into machine readable form and into databases? I do that for my catalogue.

Garrison: That certainly would help. We could make a recommendation to the database keepers that this is something they should do.

Weaver: We wandered a little bit into the topic of variability, and one thing that is often not done, when people are looking for certain types of peculiarity, is to list the stars for which they did *not* find those peculiarities. Nelson Irvine did a survey a few years ago of three or four hundred potential Be stars, and he only listed those in which he found Be characteristics. There was nothing I could do to persuade him to publish in any way the ones that he found with no Be characteristics. Of course, later those are going to be Be stars. You would like to know of the times when they were not. So even if the editor beats you into not publishing all the data, there are lots of other databank areas where you could put it and where, if people really want to study something, they can retrieve it.

3.4. Notation for photometric types, and references

Olsen: I have a few comments concerning photometric types. I've been guilty of putting MK types into tables since many years ago, but I don't do it any more, not very often at least. I will now ask photometrists always to be sure that, when they translate their photometry into a spectral type, they call the result a prediction. Because photometry is just mentioning a few things in the spectrum and it has nothing to do with a spectral type. Even if you are pretty sure that you know what the spectral type is, you should still call it a prediction. It is not even an estimate, because you have no spectrum. You are actually predicting what you think it will be when someone goes out and looks at the spectrum.

Another thing for what it is worth: I've used the old Mount Wilson notation in my tables because I was sure nobody would take a type out of the table and mistake it for a Mount Wilson classification. But I think they are very good. You put 'sg' for subgiant, and you call what you think is a solar type star a dG2 and not a G2 V. The notation also uses cF4 for what you think is an F4 supergiant, and so on. If you had those kinds of predicted types in a table, surely people who read it would not put into some database but would say, "There might be something interesting there," and go on and look at the spectrum.

Garrison: I think that does help. At the time of making up my huge database – which is about 100,000 index cards, not including Nancy's even, that were put separately into the computer – I found that, when going back to the papers with the original data, it was at least useful if at the top of the heading of the column they had "MK predicted" or something that could immediately signal to me that here was the problem or the source. So I urge people always to go back to the original source to get your data. Likewise, I urge you not to use Buscombe's catalogue which has absolutely no references in it and seems to be forcing you *not* to go back to the sources. That's deadly stuff, and I've had discussions with Buscombe again and again and again, but I can't get him to add the references. "No student help" or "I've not got enough money" is his excuse. "Well, don't do it at all then," I say. Similarly, when you use the *Bright Star Catalogue*, use it as a finding list or something, but don't use it as a quote or as the ultimate source. The suggestion of using "predicted" at least helps if it's in the head of the column.

Ah yes, Wayne, you have done a lot of bibliography work. You probably have some really good suggestions.

Warren: Well, one thing that Dorrit and I have decided, based on what we have heard at the meeting, is that we are going to try to put references into the *Bright Star Catalogue*. (*Applause!*) It's an immense job, and it'll have to be done piecemeal as we go along, but we ought to start it now.

Crawford: I shall make a comment, too, on the photometry where we have it a lot easier because we are looking at less parameters. Basically, all we do is add a colon if we don't understand why the thing is varying and add a "v" if we do know why. (That is, we have precise data, but it can certainly be variable.) We also do what Chris suggested and add an asterisk if we are referring to a note. However it can actually be more interesting than that because, when we went from *UBV* to 4-color photometry, we added a whole other column. I think that spectral types are getting so complex now with other dimensions that, rather than trying to find notation to put everything you are seeing within the MK type, perhaps you really should think of adding another column for another dimension. That could work, and one way to do it is just to use the asterisk to refer to a note, which refers you to the original source, which *has* the other column.

Garrison: In all this there is certainly a compromise that has to be made, because we can describe the spectrum in infinite length and that is not a classification: that is a description. So we have to be careful to keep things simple and yet be precise; and that is not always easy.

4. Extensions to the MK System

4.1. Methodology

Roman: We have heard a fair number of papers at this meeting about MK types derived at wavelengths other than the standard K-line to a little past Hγ that the MK atlas and the MK System is based on. What I've not heard is anyone who has tried to derive or set up the MK Process in a wavelength region other than the standard one. Now this is a far cry from simply taking spectra of the MK standards and producing an atlas, and I don't think you want automatically to decide to set up the MK Process in every possible spectral region. What I would like to see is the setting up a program of looking into various spectral regions and finding out the advantages and disadvantages of that region, and perhaps at more than one resolution, for stars of particular spectral types. Then go back to the very best of these combinations, and for it set up the MK Process from scratch, looking at what changes with temperature, what changes with luminosity, what changes with other parameters. Establish a grid of standards for that particular set of characteristics, and then, after that has been done carefully, try to tie it into the "MK" system so that – and I think this is important – you don't have a set of pseudo-MK types that are sufficiently different from the true MK types to cause confusion. This is a very major effort, and maybe I've missed some of the work that has been done, but I've not seen that started yet.

Walborn: In the context of the ultraviolet atlas, I've not really done that with the O stars, because I did not find it necessary, since I found such good correlations. But, in fact, that is what Janet Rountree was doing in her study of the low-luminosity B stars. She was setting up criteria for actual *independent* classification in the UV.

Liebert: That is what Pat Boeshaar did in her thesis on M dwarfs as well.

Garrison: It is an enormous amount of work, as Roman says, but it is something that is fruitful because in the interface between the new system and the old system you have information.

Weaver: We have thought about that with the near-infrared spectra. In the end, we decided that it was pretty dangerous to create a new system in the way that it was done for the galaxies and then ignored.

Roman: I think we have to set up the system independently and then link it.

Weaver: We were very sensitive to places where it might not track. We don't want to spend a lot of time setting up a system that would not be used.

Roman: That is why I said that before you set up the system, you have to test it to see whether in fact it is going to be a useful system.

Garrison: Yes, and one of the things that I'm always struck with is that the evolution of the MK System in the blue-violet region was accidental, in a way, because the prism spectrographs cut things off at 4,000 Å and the emulsions were not available above Hβ. So that defined the system: you did not have any choice really. But in fact, for the general reference frame, for the whole kit and caboodle, it is really the best wavelength region, and let me summarize this.

For specific problems obviously the near-infrared is great, such as if you want just to get temperatures of M stars. But the near-infrared spectrum of M stars is dominated by molecular bands. So you have to go down to the blue-violet region, if you can, to see whether the star is metal weak or metal strong. It is very difficult to determine that in the infrared even for M stars, which are strong in the infrared.

In the far-UV the problem is that you run into the wind lines and, as Nolan has pointed out, let's possibly use them, if we use them carefully. It gets less and less useful as you go down into the X-ray region.

The problem with O and B stars in the near infrared is that there is very little light. Also, circumstellar shells dominate in the far infrared, so you are not measuring the star in that case. I take the point when Bruce says that for OB stars which suffer from strong interstellar reddening, infrared spectra are useful.

This summary is not to be taken as literally forbidding the use of classification in any other spectral region, but the point is that the visible is basically the best region for the *overall* classification of stars. Then for specific problems, like reddened OB stars or M giants or M dwarfs, there are other regions that are far more useful. So you should define your classification scheme and your set up to solve your particular problem. but for the general reference frame I still find the visible quite useful.

4.2. Coordination of extensions

Garrison: This topic is a tough one and I don't know quite how to introduce it, but somehow we need to coordinate some of these extensions. We are doing it right here at this meeting, and this is a great way to do it. We have also got an e-mail address list here. Maybe we can talk and confer with each other especially when we are introducing new nomenclatures and ask if another knows whether anybody else is trying to use this at the same time for something else. I think coordination is an important consideration.

White: The History of Astronomy Group has usefully coordinated people who are involved in that enterprise in order to create a subgroup, called a conference. It is an e-mail conference, which you can set up yourselves. A question just cast out into the ether goes to everybody in the conference and responses go the same way. It needs one person to be the administrator of the conference or the specialized bulletin board, if you will, but it is a very useful way to pass around a lot of information very, very quickly.

Garrison: We might also consider setting up a monthly newsletter, perhaps electronically to make it easy and cheap, or something that just comes out periodically whenever we get two or three things that need to be disseminated into the community. Maybe just bringing a difficult problem up and saying, "Hey, this is a problem," is enough.

Corbally: Already there is the *Standard Stars Newsletter* and that could be slightly redefined, such as becoming e-mail rather than hard copy in form.

For those unaware of it, I might add that this is the newsletter of the IAU's Standard Stars Working Group. I mainly do the production side and Martin McCarthy the editing side. Its scope is wider than the MK standards and so it's concerned with anything to do with standard measures for stars: radial velocities are included, as are good determinations of abundances, etc. It comes out when there are enough contributions for it to go to press. If you are not on the mailing list for it, then all you have to do is give me your address and I'll add you. We have cleaned up our list recently, but remain very interested in people who do want to receive the *Newsletter*.

Garrison: The *Newsletter* is a possibility for remaining in touch over new nomenclatures. However, one of the problems with making something only e-mail is that then it gets lost for archival purposes. Don Fernie and I are considering changing the David Dunlap Doings to electronic version but, if so, we shall produce hard copies for libraries.

Just to mention it, the Standard Stars Working Group, of which I'm chair, is a joint working group of Commission 45 and a number of other commissions. We have made the case that this is not just a single-problem focused working group, but that this is an ongoing working group. Because the radial-velocities commission, the stellar-atmospheres commission, and all the other commissions involved have standard stars, so this is an attempt to coordinate them and pull them together into the one working group. So if you are not a member of that working group, please come in. It's not exclusive or intended to be in any way.

Warren: There is a bit of a precedence for this general problem of deciding on nomenclature. As you all well know, there's been a terrible difficulty in the past with the nomenclature of identifications for objects. In IAU Commission 5 we started a working group on nomenclature for celestial objects. We gathered about a dozen people who are experts in nomenclature and worked out a system. The system was accepted and published in the major journals.

Similarly for MK types, it needs a working group to decide on the best notation, then the advertisement of this notation and the asking everybody to contribute reactions, and finally the reaching of a common decision. This should take place within the IAU Commission 45 on Spectral Classification, which would promulgate this standard notation after it's been decided upon. That's one way to solve the problem, though I know it means a lot of work.

Garrison: That's a very good suggestion. Maybe we can think about this after we go away from the meeting and come back to it again next summer when we have our commission meeting. We actually need to give a preliminary announcement that we are going to discuss that and ask for suggestions. (*to MacConnell*: Jack take note. Thank you!) We have the President of Commission 45 right here, so we have just co-opted him.

5. Recruiting into the philosophy of the MK Process

Garrison: I'm concerned that if there are going to be people who understand classification problems in order to develop the MK Process, refine it, and extend it, we need to pull more young people into the philosophy, whether or not they become specialists. Now, there is a really nice way of getting people used to classifying, which Richard Gray has developed. He said he is willing to share the software with anybody. It is a simple plotting program, called "MK", which runs even on a 286-PC and which is great for MK classification. If each and every one of us who is a teacher at a university can set it up among the various undergraduate lab projects, then we would at least get some interest stimulated among the young students, and we may find one or two whose personalities and philosophies are just about right to go into work like this, and so they just may get turned on. Even if they don't, they will have seen that this way of classifying is not just old photographic work based on old, old techniques; we have got new, digital techniques, which embody the traditional principles but are still quite exciting. So that is one route we can go, and maybe other people would suggest others.

Houk: May I make a comment about the old fashioned alternatives? Some still do like to fiddle around with photographic spectra. We have sold a lot of our atlases for practically nothing to people who really use the atlas. Many people buy the atlases in pairs: they use one atlas for standards and cut up the other atlas for unknowns. It works out pretty well.

Garrison: That is a very good suggestion for it uses a set that's already produced. You are right, too, that the visual appearance of the spectrum does turn a few students on. I've had a lab set up at Toronto for quite a long time with a whole set of photographic spectra as standards and then a few unknowns that they are to classify: each year I changed the unknowns. It works, but I don't use it any more, because I'm using "MK" now.

McCarthy: Many good educational materials are channeled through the IAU Commission on the Teaching of Astronomy, as you are probably aware from its regular bulletins. So that is a very good resource for us to build up and from which to learn wisdom.

The other thing I wanted to say was that Mrs. Brück at Edinburgh has a very good series, which we have used at the Vatican Observatory Summer School, and it is based on photographic spectra that they have made. We could keep in mind the kind of projects that they have used successfully at Edinburgh.

Warren: As primarily a photometrist who is interested in getting into digital classification, I would say that Richard Gray has done a wonderful thing by writing such software to be accessible to everyone.

I would add that, for a number of years, we have had a large number of wonderful photographic atlases that can be used right along side of the comparator to classify, but photographic work is now going out: we are having to face that. We have seen a number of digital atlases appear, but what we really need in conjunction with using Richard's program is a high quality digital atlas, in the visible region, of all the MK standards.

Garrison: Okay, I've got the data. It is just that putting out an atlas is, as Nolan and Janet know, a big job. It detracts sometimes from other things, but maybe somebody else can produce one.

Walborn: This is a sort of a sociological comment about young people in spectral classification. There are a number of quite bright young people, very active today, who are not here. I think Anna is the only representative of the Colorado school here. There is Ed Fitzpatrick, Joel Parker, Phil Massey, Katy Garmany, etc. A suggestion is that maybe in future gatherings of this kind we ought to think of casting a wider net. I think there is somewhat of a blood dynasty here, and we tend to be the same people talking to each other and maybe preaching to the choir sometimes. Those now absent came in from different schools, so they are not *bona fide* original MK people: but they work in the field, they contribute massive amounts of data, and it is a value for us and for them that we work together, that we hear what they are doing, and that we try to transmit to them some of our philosophy. I've worked quite a bit with Ed Fitzpatrick and Joel Parker, and I think I've transmitted some of that philosophy to them. So it is a suggestion for perhaps the "sociology" of future gatherings of this type to cast a wider net.

Garrison: Yes, I agree with you completely, but the problem may be a little different because we *did* try to get some of the people and they were doing other things. For example, Mike Kurtz is a young person working in the field of automated classification: his job is in another kind of work and he could not leave it right now. I think that Davy Kirkpatrick had the same problem. He wanted to be here, but he had other commitments that he just could not get out of. Young people have less freedom to move than the more established researchers do.

McCarthy: In response to Nolan, we might think of a working bag-lunch at one session of the American Astronomical Society at which we could advertise that we are going to talk about standards in spectroscopy. That would bring some of these younger people who would more likely be there than if they had to make a journey to this workshop.

Garrison: One of the carrots we can hold out to these younger people is that, at least at Toronto, I find that a very high percentage of my former students are working in astronomy still. Not all of them are working on spectral classification. They are working on other parts of astronomy, but they learned a lot about spectral classification and they have applied it to other fields. One of the things that I'm aware of as graduate coordinator at Toronto, where we have 30 students, is that all students coming in want to be cosmologists and they want to do large scale structure: and you know, this preference is because they don't hear about spectral classification, or spectroscopy, or any of the other interesting things until they get into the program. Then these cosmologists get the first postdoc, and maybe the second and the third, but they are not getting permanent jobs: they are going begging. So a job in stellar astronomy is one thing you can offer the new students. For example, I have now three of the new Masters students who have gotten that word – not by me, for the students pass the word around amongst themselves – and are quite interested in going in directions contrary to that of the whole crowd. We can all capitalize on that a little bit.

Liebert: I've been listening all morning, and there have been a number of themes that have been building up in my head. I'm wondering why only two of our graduate students, who have to walk just eight blocks to get here, are attending. That is not very many compared to the number that normally would attend a scientific meeting held in Tucson, and – here I'm going to be very frank with you – you have not been successful in projecting the kind of dialogue between classification and theory that is the theme of this workshop. The students of today, if interested in stars at all, are interested in the parameters of stars. I believe that spectral classification *is* useful in reaching the parameters, but you have not made that case, and that is the problem.

Garrison: I appreciate those comments, Jim. They come from somebody that is interested, but not totally involved in MK classification like we are.

Boeshaar: I am chair of a private college, undergraduate physics department now, and I've noticed a change in the students who go into graduate school from the time I started at Ohio State in 1970, where people came in from theology, mathematics, astronomy, history. When we look at the students going into graduate school today, they are well founded in physics and they view things very differently than when I started. As one of Philip Keenan's students, I must admit that he emphasized the physical applications/interpretations of classification: e.g., asking if we understood the importance of the excitation level from which an atomic transition arose; or, based on current models (Te, L, abundance) of barium stars, what might we predict for certain atomic or molecular line ratios.

But Jim Liebert is right. Somehow, the ideas behind the MK Process and the inherent excitement in our work have *not* been conveyed to the students of today. I realize that we have heard in this workshop many wonderful presentations regarding the theoretical understanding underpinning the classification process, but I also heard some of the more vocal participants downplay the physics, and I was not alone in this. So why should any student with a physics background want to do classification for a career? I've seen several young astronomers develop or refine very important spectral classification schemes as part of their theses (Diane Turnshek, Rick Crowe, and Davy Kirkpatrick) and then discontinue the effort. Often their actual classification observations were done as a side project, when observing conditions were marginal. Is it possible to interest graduate students today in research involving spectral classification, other than as a sideline endeavor? How does one go about getting increasingly scarce telescope time to do classification observations, other than for the brightest stars? I should point out that the *one* student poster at this meeting only highlights these questions. It has been over 50 years since the introduction of the MKK system. Our understanding of the physics of stellar atmospheres has come a long way since then, as has the wavelength range over which we can obtain spectra. We need to emphasize the importance of using all of this information in any classification scheme, just as Keenan and Morgan incorporated the best empirical data available to them.

I disagree with an overheard comment that students all seem to be flocking into extragalactic astronomy, though Philip Keenan is always accusing *me* of doing so. Jim Liebert has some very bright students investigating cataclysmic variables, while one of Tony Tyson's recent postdoctoral fellows did his Ph.D. thesis in Canada on globular cluster stars. I currently work on deep CCD surveys and brown dwarf physics, though I also realize the value of good spectral data and the MK Process. But classification and its process are a sideline to my current research. What graduate students want is an intriguing, unsolved physical problem to attack (don't we all?), not just fine tuning details in an established paradigm. Somehow, working on the MK System "appears to be like classifying butterflies," in the words of one student. Like Jim, I think that the dearth of student interest in this workshop speaks legions.

There used to be lots of students interested in the MK Process and its applications in the 1970's. The MK Process was taught as part of my graduate courses. Most students knew *how* the process worked and how to use it. Now I'm routinely asked to classify stars for colleagues, even if I don't have any special knowledge of that portion of the HR diagram. What has happened in the intervening years? Has our focus become too narrow? Judging by many papers at this conference, I don't think so, but the *students* do! So how can we communicate how relevant the MK Process actually is to astronomy today; the interplay of physical theory and pattern recognition; and the excitement generated in discovering/predicting a pattern in the behavior of spectral features of a class of stars, of actually pointing to the information theoreticians need to model an atmosphere properly?

Garrison: Those are excellent points, although my own impression is that always there have been just a few people going into spectral classification, inspired by a few great teachers. I mean, Morgan did his work alone: there were not a lot of classifiers around in those days. So I don't think that we can ever expect to have floods of young people turned on by spectral classification. It takes a certain personality.

Walborn: It occurs to me that an example of this problem is all of the ferment and intensive activity in stellar and globular cluster astronomy today over the apparent abundance anomalies of sodium, aluminum, and the CN and CH molecules. I look at some of these papers and I'm not at all convinced that they have the temperature and the three dimensions sorted out. Yet again, the real people doing all that work, with the possible exception of Tom, are not here; we are not hearing from them and they are not hearing from us.

Garrison: That is certainly an area where we need to disseminate our skills.

6. Spectral standards

Garrison: We have quite a few topics suggested on standards and I shall try to introduce them. For example, we could discuss to what extent we should use external information, like theory, in developing new dimensions and extensions; the coordination of standards, so that we don't have ten groups of people all saying, "Hey, this is a standard, that is a standard," then somebody else comes up from the infrared and says, "No, that is a standard." We can discuss faint candidates for large telescopes, exchange of atlases, digital atlases, and so on.

6.1. The interaction with external information

Walborn: I would like to put forward for a little more discussion the point about the independence of spectral classifications. This is absolutely fundamental, and I think we all understand it, so we are sometimes preaching to the converted here, and that is zero order truth. But reality is more sophisticated and complex sometimes, and we need to go to the next level of thinking about this issue. We see that, in fact, in the development of the MK System or any classification system, subconscious knowledge of external information is not only permissible, but vital and essential. Just the concept of the two dimensional system would be instructive. Where did that come from? That was not something that was dreamed up out of the blue just from looking at plates; it was something that developed in the general wisdom of astronomy as the process matured. So I'd like to urge anyone who is interested in these problems to look at the paper I published in 1979, not just because I published it, but because there is a lot of interesting information there (see in *Spectral Classification of the Future*, Vatican symposium, eds. M.F. McCarthy, A.G.D. Philip, and G.V. Coyne, p.337).

I presented there some lessons from the past and thoughts about the future in spectral classification. In particular, I recounted what you can read in the *Ap.J.* (1911) of the original, very interesting series of discussions about how a spectral classification system should be set up. You can see the errors that were made by people who did not fully appreciate this point about maintaining independence, including one by Hertzsprung himself who insisted that the system should be tied to the colors. Antonia Maury's active mind went too far, and she talked about the main sequence being an evolutionary sequence and about reordering the letters to show that, although she admitted there were some uncertainties as to whether the stars were evolving up or down the main sequence. Schlesinger and, most importantly, Henry Norris Russell come through just like a clarion stating these principles of maintaining independence. In fact, Morgan was strongly influenced by Russell, as he acknowledges in his 1937 paper which is well worth reading.

I want to read these two brief paragraphs of what Russell said: "In my opinion, the classification should be based exclusively on a study of the spectra, i.e., of the *line and band absorption*, without reference to color, intrinsic brightness, and the like, much less to theoretical considerations." Now listen to this: "External considerations should be admitted only (1) in the search for differences, *perceptible in the spectra themselves*, which might otherwise escape notice; (2) in determining which of numerous small differences are entitled to specific rank." That is really the point I'm trying to make which Russell appreciated, and Morgan did, and which I think is a fact in the development of spectral classification.

Russell continues: " ... the Draper classification seems to be all the better because the letters are not in alphabetical order. This helps to keep the novice from thinking that it is based on some theory of evolution. ... I would add the suggestion that a comparative study should be made of the spectra of stars of very different total luminosity but the same spectral class (Hertzsprung's "giant" and "dwarf" stars). If any definite and constant *spectroscopic* differences exist, they will be of value in classification." And finally, " ... When agreement on the details of classification is reached, a series of type stars should be chosen, preferably several for each subdivision, and these should be taken as its permanent definition." That is what Russell said in 1911 and I find it rings true across the decades.

For example, an astrophysical point was brought to my attention at a meeting in Colorado during 1989. Although I probably knew better, I had just been assuming that an ionization ratio, let's say Si III/Si IV, should be safe as metallicity changes. That it is not affected, right? Wrong. If, for that ratio, one of the lines shifts to a different part of the curve of growth, that will not be true. We may still describe the line ratio and give it a type, but if we are not aware of what we are doing and then preach to astrophysicists about the purity of what we are doing, that is not going to be very convincing to them.

The other example which we already discussed is about the detection of He I at high resolution in the O3 stars. If it turns out that O3 stars which do not show He I at high resolution are the same ones which show the weak N IV and N V features in the visible, then it might be necessary and desirable to introduce an O2 spectral class to reflect this.

Garrison: That is a very considered set of examples of how we can interact helpfully with theory without going overboard. A thing which I always worry about is taking the current stellar atmospheres theory and changing spectral types, for example, because somebody says that one star has a higher temperature than another. I did an analysis once of Cayrel's table of [Fe/H] for several stars that are well known. It turns out that the spread in [Fe/H] and T_{eff} and $\log g$ that is given by the theoreticians, if you check out the complete range that they give, is about three times the range of types from different classifiers using the UV, visible, or IR. So we have to be a little careful about how much we believe the theoreticians in each case, but I agree with Nolan completely that there is wisdom with them, and that we have to share mutually, and it is a two-way street. It was pointed out by Richard Gray that we need to talk with the astrophysicists and they need to talk with us. At the present time, I think we look more to them than they do to us. So one of the reasons we set up this workshop was to encourage them to take notice that classification is a powerful tool for astrophysical insight. But that mutual give and take is important as long as we just don't "take" without question.

6.2. M dwarf standards

Weaver: I've done a lot of work on M dwarfs with both objective prism and digital spectra. So I was really shocked by a remark of Pat Boeshaar. It is appalling to think that the MK System, from the primary standards point of view, does not include most of the stars in the universe. That is, if you take volumes by cubic parsec, the MK System would not classify these stars in a well defined way from primary standards, and Bob's list, of course, made that clear. I guess the latest dwarf standard was K5 V on your Anchor Points. (*Garrison*: But that is partly because they could not reach those.) That is right; but that argument now no longer holds, and so the MK System has to be finished. Someone has to address, without too overwhelming a bureaucracy, the most frequently found stars in the universe, the M dwarfs.

Garrison: Well, you, and Anna, and Pat ought to get together and use some sufficient resolution to be able to do that. Sure!

Liebert: The M dwarfs don't fit very well into the MK System because somehow the third dimension is much more important for the M dwarfs. This may be because the effects on the spectrum are much more extreme since metallicity causes molecular band ratios to change a lot from TiO to hydrides. If you draw a diagonal line in the HR diagram, it so happens that I work on the other side of that main sequence line, probably a lot more than anybody else in this room, and the MK does not apply to it very well at all. In fact, for things like the subdwarf O stars that John Drilling talked about, we have inherited more from the Mount Wilson system, which I consider to be far inferior; we have picked up the capital 'D' and the little 'sd' from that system, and for the whole subdwarf M problem we sometimes use 'sdM'. So some of the interesting problems of stellar astrophysics lie outside of the MK System and don't seem to be addressable in an easy way.

Garrison: I find that there is a tendency for people to identify us and a few others that are not here as "the team." So if they have a classification problem that they want solved, they say, "You, guys, do it for us, will you please?" However, we are also busy, and we would like to do a lot of different things and sometimes can't get around to them. You and Ed are experts with white dwarf spectra, and now that you have heard all of what we have had to say, you can do that job, not me.

For example, there was a symposium in 1985 at Lake Como and a young galaxy person got up and she said something like, "I was observing with this big telescope and I wanted to classify some of the stars, but I could not because I could not get standards of faint stars." She wanted me to do it for her, but I just said to her, "Well fine, do it." So one of the responses that we can make is to invite people to do some of this work because we are not going to be able to do it all.

Boeshaar: I find my experience to be the same as Bob's when I hear someone say, "Hey Pat, where are your standards? I need some M dwarf standards. I have no time. I can't get big telescope time to do this." My work on standards is done whenever I get time as part of another project, or as a fishing expedition. To get time at CFHT for that fishing expedition, my one page proposal has more physics in it than I would ever show right here. I really have to justify it. I could not get it by saying I wanted to extend the MK System. Given the dearth of telescope time, we should at least strive to educate the astronomical community about the need for good standard data.

6.3. Faint secondary standards

Walborn: I want to discuss the history of standards for large telescopes, but it is also related to the strategy and sociological issue discussed earlier. Some of the ideas we have heard during this workshop I put under the heading of "bad ideas." One of the main reasons why they are bad ideas comes from my having managed four cycles of program selection for the Hubble Space Telescope. The average program allocation is ten spacecraft-time hours. If you put in a proposal to spend a significant fraction of that time observing either bright filtered standards or faint secondary standards, I can guarantee you that your grade will be 5.00, your ranking will be 999, and your priority will be "U" for unapproved. So while we may like the idea of observing zillions of standards every time we get a run on a large telescope, it is not going to happen. Unless we want to beat our heads against walls, we have to do things differently. I'm not saying that you can't take any bright standards (filtered), but certainly not a lot of them.

I see problems with the concept of this grid of secondary standards: as Bob already said, it is difficult to implement. I presume you would have to have a complete grid every twelve hours around the equator to be accessible throughout the year. For the OB stars, of course, that is unfeasible because they don't live on the equator but move in a different circle. Furthermore, if you talk about very faint Galactic OB stars, generally they are going to be heavily reddened, whereas the stars you may be interested in studying are in another galaxy, where they are lightly reddened and maybe have a different metallicity. So, if I really saw my name on that project, I would like to publicly resign because I don't think that is a useful project, at least for the work that I'm interested in doing in spectral classification.

Here are some constructive suggestions of how to do things better. I think we have to get away from the idea of observing lots of standards with our large telescope on our short limited run; and that is really what we have done already anyway. You don't observe the whole MK atlas when you go to observe. You observe some key stars and then you refer to the atlas to tie it in and interpolate. If you insist that you have to have all your standards on the same instrument, build a small telescope – maybe you have one already – with the same f-ratio as the large telescope, and put your same instrument on both telescopes. Do a whole grid of standards, and then observe a minimal number of tie-in objects with a large telescope.

Related to what Jim Liebert was saying, the trouble is that the MK System is applicable only to a small subset of stellar astronomy today. That is one of the reasons why we are losing a lot of the most active and enthusiastic stellar astronomers. They are working on cataclysmic variables and faint subluminous objects in which the MK System really is not relevant, although the MK Process may be relevant. So we have to be more specialized and more astute in how we apply the process to be relevant to specific problems in modern astronomy and astrophysics. For instance, if you are working in the Magellanic Clouds (or in M31, or M33), you should define standards and localize regions which are applicable to the problem, and refer your observations to those standards, just as we do in the MK Process. If they are in a localized region on the sky, then you can observe them with fiber optics and get lots of them on one exposure.

My own optical atlas was done with the Schectman two-dimensional system on the one-meter telescope at CTIO. At that time, there was a similar two-dimensional system on the four-meter. The idea was to get the standards with the one-meter and then do the program with the four-meter, observing a small number of tie-in standards. Alternately, in the time we had this August, we observed stars in 30 Doradus. One setup with fiber optics in Carina got us a whole ream of standards in one short exposure. So if you follow those kinds of strategies, you are going to do a lot better with peer review committees than if you say you have to observe 50 standards, one by one, with a four-meter telescope. You are just not going to do it.

Before I leave this topic of setting up an atlas, I wanted to say that is what Janet and I have done for the ultraviolet. The IUE database should be viewed as a reference system for HST observations. You have to observe some stars in common, obviously, to get some tie in, for there may be very different detectors on the two telescopes: but the data are quite similar and one is never going to be able to observe the full grid with HST.

The bottom line is that we have to maintain the principles and precision of the MK Process. We also have to be a little adaptable and flexible to change in technical and sociological conditions so that we don't limit ourselves, and the power of our technique, because of insisting on traditional procedures.

Garrison: I agree with everything except your characterization of this as a radical idea. I find that these are the ideas we have come up with at this meeting, and obviously the suggestion of the faint standards was made some ten years ago and it has not been done, so *de facto* we are recognizing all of those problems with doing the project and I have to agree that it is simply not a workable idea. One of the advantages of electronic and computerized digital data is that we can now go and set up a library of standards which can be used with other telescopes, and the only thing that we have to be very cautious about is the differences between spectrographs, and to make certain that the resolution of the standard star spectra in the library is high enough so that they may be convolved with the instrumental profile of the spectrograph that is being used for the project. You can check this convolution with only a very few standards. So what you have said about setting up things with the IUE, and so forth, is something I for one can agree with completely, as long as – and I agree especially with your last line – we pay attention to the principles and are careful about the process.

Gerbaldi: In agreement with Jim Liebert, I am more on the side of people measuring parameters than upon people doing weird MK spectral classification. But we are speaking of two different things which cannot be opposed. We must not fight one against the other, saying "You are right," "You are wrong," "Do this," "Measure parameters," "Why are you classifying things with the eyes?" Look at the Geneva system, which is a well defined one. We have measured, roughly speaking, four thousand stars and obtained very good physical parameters for many, many years. Since for more or less the same time Nancy Houk has been classifying more than one hundred thousand stars, we are not at all on the same level. With such specification you gain a broad idea of the population of stars in the Galaxy, and this then determines in which area you would like to obtain more refined parameters. We must not fight one against the other. We must convince the scientific panel giving time on the telescope that both jobs are really useful. I think probably it is a better way to work together.

Garrison: That is precisely the principle of complementarity. I guess you have stated it in a very clear way. Thank you very much.

6.4. Atlases in the digital era

Rountree: I should like to comment on a couple of Nolan's points, although I think my remarks are complementary to Bob's, to mention complementarity. First of all, there are two distinctions that need to be made. There is a difference between photographic atlases and digital atlases; and secondly, there is a distinction, at least in our present sociological climate, between data obtained from space and data obtained from ground-based observatories.

The distinction between photographic atlases and digital atlases is this: the MK atlas was never intended to be classified from. I'm amazed that people have implied that it was so intended a couple of times in the course of this discussion. It was intended to illustrate the differences among the various spectral types, and the idea was always that the person who wanted to do classification would go out and get his own standards on photographic plates. He would then compare his unknowns with his standards on plates. He would never compare his unknowns with a print in the atlas. That was not the idea. So, in the era of photographic plate work, it was simply not possible to produce an atlas of standards which could then be shared by various other people. At the very most you could share your plates with somebody who was using the same telescope: and not very many astronomers were willing to do that, unless they were at the same observatory, because of the fragility of the plates and the difficulty of shipping them from place to place. It is quite understandable that there were not mutually available atlases or collections of standard stars. Now in the digital era, something good has happened: you actually can classify from an atlas. If you take your data and reduce it in the same way that the atlas was reduced, you can literally line up the spectra and you can use the atlas for classification. That is the purpose of my ultraviolet atlas, and it is now possible to think about making atlases of standard stars in all wavelength regions.

Now, what is the difference between space data and ground-based data? It is not just the wavelength region: the difference is that space data belongs to everybody. After it has gone through its proprietary period of usually six months to a year, it belongs to the whole world and anybody can dig out of the IUE archives, for instance, all the standards that he or she wants and can reduce them in any way he wants. This is not true with ground-based data, and this is a very sore point, I'm told. I myself have not been involved, and my understanding is that at NSF, where the people live who run the national observatories, there is periodically a movement to declare all data obtained at national observatories to be common property after a certain proprietary period such as observed for the satellites. Personally, at the risk of sticking my neck out, I think this is a great idea because, now that we have reached also in ground-based astronomy the age of digitization and most spectra are being taken with CCDs, it is entirely feasible to stock digital spectra in some kind of a library and to make them available to the public after a certain period. Only in this way can we start producing widely available atlases of spectra in the visual region, which anybody can use for classification.

Gerbaldi: Just one more thing: ground-based observations are not calibrated generally, but space observations are calibrated in flux. This is one more very important piece of information.

Garrison: Also, the IUE and other space observations are done all with one reduction procedure and the ground-based ones are simply not. The latter are done quite differently. However, the principle is there, and I agree completely with it.

6.5. Atlas details

Garrison: One item we need to get some opinions on is, if we were to produce a couple of electronically available atlases, – and recognizing that it takes an enormous effort to do that – what would be the most universally useful atlas to produce and what resolution would be useful?

I suggest that an R of 5000 or so, i.e., a resolution of 1 Å for a couple of pixels, might be a great way to start because then you could degrade that to 10 Å, if you want. If you were to get a tenth of an Ångstrom resolution, you could degrade that to one, and so on. You do need take a couple of standards to make sure that you are not seeing some instrumental profile. That would be an important thing.

Houk: As first priority, I suggest a library of the MK standards in digital form at a resolution of 1 to 3 Å.

Crawford: You should also put in samples of peculiar stars, not as classification standards, but as specimens.

Roman: I should like to propose that the digital atlas has a resolution slightly better than the MK System.

Garrison: Yes, in fact Richard Gray and I found that the 1 Å resolution of the 67 Å/mm plates was really essential to do the kind of careful work that he was tackling. The resolution is always a compromise. It is easier to degrade it than to put information in, but on the other hand you have to compromise, because you are not going to be able to reach very faint stars. One of the things about digital data is that, if you want to produce a standard atlas of half an Ångstrom resolution, you really should have that over three pixels at least. That means a high resolution spectrograph to start with, but then you can't get it all on the CCD. There are all kinds of compromises that we have to make. But I guess we have to keep in mind how it is going to be used and not just produce any old atlas, because that happens to be what we have got data for.

Lloyd-Evans: If one wants to include spectral high resolution, then of course one is not going to get a very wide wavelength coverage. I think that a lot of the need for an atlas is actually for rather low resolution observations of faint stars and other heavily reddened stars, so we are forced out of the normal blue region. We should consider, therefore, having at least two different atlases: one of high resolution, which you can reduce to a moderate resolution, but also one of much lower resolution with wide wavelength coverage, which will be useful for work on faint stars.

McCarthy: Phil Keenan, before leaving last night, said that he hoped someone would undertake the very important task of producing an atlas of carbon stars.

Garrison: Yes, he has mentioned that to me several times too, and to those people who have telescopes with infrared detectors, and to those who have atlases already. An atlas of carbon stars is different from anything else.

Liebert: You will have carbon dwarfs to incorporate now too.

Garrison: Yes, I've got one of those, or rather, a subgiant.

Liebert: They are all over the HR Diagram. In some cases just carbon is peculiarly strong; in other cases there are actual C_2 bands.

Lloyd-Evans: Our problem still is that we don't really have a classification system for the N stars. However, for the R stars there should be no great difficulty in producing an atlas with meaningful spectral types attached. So Keenan's idea of having parallel sequences for different subgroups within the carbon stars is a good one, certainly for the R type stars, because the overlapping of bands and band systems means that the background against which you see possible classification features differs very greatly between, for instance, stars of normal $^{12}C/^{13}C$ ratio and those which have enhanced ^{13}C. So any given feature really can't easily be compared between the two types of stars.

For the N type carbon stars we have at least the two types with the different isotopic ratios, but we still lack the basic temperature criteria and we still don't really have good luminosity criteria. So I'm not sure that we are ready to produce an atlas for these yet. Perhaps when I finish my current program, I'll be in a better position to comment on that.

7. Archiving

Corbally: I am much in agreement with what Nolan and Janet have been saying and which introduced the topic of atlases. Further, the commemorative medallion of the VATT that you were given this morning reminds me of the project we have to archive the data that comes from this new telescope. These data would particularly be of the CCD kind, which include the spectroscopic data. The archive will be kept at DAO under the same system that is being used for the CFHT's archiving and also for the Hubble Telescope data available there. The idea is that they will produce catalogues of the observed data, which will be available to anyone after whatever is decided to be the proprietary period for the observer. This archiving is intended as part of the new technology of this telescope, understanding that in this digital era it is something that observatories can do and should be doing. That archive will contain the raw data, but it will become much more useful when it also has processed data. That processed data might be put into subsets, grouping the objects, be they peculiar galaxies or types of spectra. All this is valuable but needs work, I'm well aware.

Garrison: I should like to underline that project. We discussed at the Argentina General Assembly of the IAU how archiving is an extremely important idea. Certainly ESO astronomers were pushing for archiving of both the raw data for permanent use and the reduced data. The data might be reduced in some standard way or by the original observer, who would understand the best way to make the reduction. A reduced spectrum of some sort must also be archived, because most people are not going to want to go into the archive and figure out how to reduce a spectrum taken with this instrument or that, but they might go in and use a reduced spectrum just to see what it is like.

von Hippel: There are other observatories that archive ground-based data. One, the Royal Greenwich Observatory, is in the La Palma consortium with the Dutch and Spanish. They archive the raw data; then they publish a catalogue a year later and make available on the computer everything that has been observed.

As a ground-based observer who takes a lot of CCD data, I think it would be very dangerous, at least initially, to have reduced data available because people would probably be using these data for slightly different purposes. Every observer wants to make sure they reduce the data in the best way possible, but unless you provide an awful lot of information about how they have reduced the data, you would not know whether it was usable for your purposes. That would be a continual worry to the rest of the community.

Corbally: Reduced data in archives is essentially a quick-look resource. It obviously does not absolve the investigator from learning how that data is reduced and whether it will address their particular problem. The original investigator will likely be accessible so that one can learn about the reduction. But the data do need to be reduced somewhat before they can be used even for quick-look purposes. What DAO is doing for the 2-D data, and eventually for the spectroscopic data, is to give these a standard reduction and then to compress the images. These compressed images are what you will be able to access over the networks. You would use these to answer some initial questions, but that is all.

Garrison: Yes, if I think about how I use the database of my plate file, I don't always want carefully to classify each star that I've taken. I mean, I've got 10,000 stars there, and every once in a while I will read in the *Ap.J.* some comment about some star and I will just want to go see what its spectrum looks like. I don't care how it has been reduced, or I may not even look at other standards to see how it has been classified. It would be a nightmare to have to pull out all the biases and all the flat fields and do the whole reduction just for a quick look. (*Responding to a suggestion by von Hippel ...*) Yes, why do we not use that nomenclature? Let us just call these "quick-look" data rather than reduced data, since this contains a caution.

Roman: I certainly agree that the raw data has to be archived and that there are times when you really do need to go back to the raw data and re-reduce it for your purposes. But I don't think that you should therefore not archive the reduced data or even not let people know that you have archived the reduced data. After all, for the most part, we use reduced data. If I want the trigonometric parallax of a star, I don't go back and ask to have copies of the hundred plates that have been used to deduce that parallax. I want the final result. I may want to know the final results from a half-a-dozen different observatories so that I can make my own guess as to how to weight the individual ones in the average, but I don't want raw data. I think that the same thing is true for spectral types in many cases.

Garrison: In photographic spectra, we have "reduced" the spectra by developing the darn thing. So it is already reduced when we see it.

von Hippel: Let us remember the difference between reduced and analyzed data. There *is* a difference, especially in the CCD domain, and a trigonometric parallax is an analyzed result; it is not reduced data. Reduced data are positions on plates.

Garrison: Yes, the spectral type is the equivalent of the parallax. I guess that is what you are saying.

Abt: This is a slightly negative evaluation of using archival data. At Kitt Peak we have taken over 100,000 photographic spectra. They were taken with three or four different telescopes and with different spectrographs. For each one, there are several cameras usually and a number of different gratings. This is a lot of heterogenous material. We spend a lot of time cataloguing this such that it is in beautiful form. We can look up in the card catalogue, for any star in the sky, all the plates taken from all the different spectrographs, and then it tells you when they were taken and all the other information. It has been very useful for myself and for nobody else. We get essentially no request from outside ever to look at this material, and the reason is very simple. The archive that *has* been very useful has been the IUE archive, because the IUE has two wavelength regions and two dispersions. It is a combination of only *four* different things. So if you look up something in IUE, you'd have a chance of one in four of finding something which matches what you want to match it against.

For those of us who do spectral classification, the one thing we do *not* want to change is all the parameters for our plates. The dispersions, the emulsions, the wavelength regions, and so on. We want to set up a single system in which we keep things fixed. This is particularly true for Nancy Houk. She does not want to change everything every time she goes to a new declination zone. We want to keep things fixed. So an archive which has a lot of heterogenous material is not terribly useful for people from the outside who want to use it for reference, though it might occasionally serve for a quick look. That is why the observatory materials in the past have not been terribly useful for people who want to look up things.

Now this can change because, when we get into the digital area, things like dispersions become less important. In other words, you just expand a spectrum or contract it to whatever scale you want. So standardization is much more likely to occur in the digital era than it did in the photographic era. Perhaps archiving for digital data, as for IUE now, will be very useful in the future but, frankly, we have wasted an awful lot of time at Kitt Peak worrying about and archiving material that was very heterogenous and so has not been requested.

Garrison: Well, I have to disagree because I've had a couple of students do projects in which they went to Mount Wilson and DAO and dug up in those old, old plate files and found just absolutely invaluable stuff. We have not used Kitt Peak because it is not old enough. So it is unfortunate that Kitt Peak's archives have not been used; but have they advertised it? I did not know that they existed and, frankly, I thought everybody kept them in their drawer.

Walborn: The Kitt Peak card catalogue is very interesting, and if that were on line, they might get a lot more requests.

White: I have a question about the archiving of the reduced and the original data. Along with the original data, are all the fields' calibration exposures taken and archived as well?

Garrison: If you want to open that can of worms, you can. We have been archiving for about four years on optical disk and we have now got about thirty or forty optical disks, but the reduced data only takes up about a tenth of one. So that gives you an idea of the proportion. Providing the reduced data is almost nothing; it is no storage problem.

Roman: I want to remind people that the astronomical data centers, and I don't mean just the NASA center, will welcome machine-readable data which is extensive enough to make a general machine-search useful. That may mean only a couple of hundred records; it does not mean necessarily two hundred thousand. These can be data such as derived data, e.g., spectral types, or we can also archive such things as the atlases that we have been talking about, provided that they are in each case in a form that other people can read and make use of. If the data are too extensive, say several times the size of the three Durchmusterung catalogues, then we are still interested in knowing about it. We will not try to archive the data, but we will put them in a master directory which lets people know where they are and how to obtain them. If anybody has any questions, they can contact me by e-mail.

Garrison: So we can pull it all together in the archives and not have to waste a lot of our large telescope time retaking spectra that somebody else has taken.

8. Aperture fever

Crawford: I want to follow up on some of what Nolan said by also trying to create some controversy through taking up the small telescope issue. I must be one of the few people – and some of the others are in this room – who have never caught the disease of "aperture fever" where, if it is not on the largest telescope that is at your observatory or somebody else's, in extragalactic, and nowadays in the infrared, it is not real astronomy. I like small telescopes a lot and I think they do a lot of good things along the lines of what we have talked about here: standardization, calibrations, and so on. But small telescopes are going away at national observatories and they are going away at other places. If you read the daily newspaper, you can predict that more of them are going to go away, if not the whole national observatories for that matter. So we have a real problem for observing time. The excitement of astronomy is not declining and it is not all just in physics. Even if it were, there still needs to be all this backup stuff. The astrophysicists, if they work without the calibrations, are eventually going to fall off the scaffolding, because the scaffolding is not going to be there. So I think small telescopes are tremendous, the complementary thing (that word again) to large telescopes, to radio, to the other wavelength regions and to space astronomy. We need them.

Furthermore, anybody who's ever looked at astro-economics, which is an interesting subject, full of lots of handwaving, can easily see that small telescopes are the only places where most people are going to get observing time. Predict, for example, how many graduate students are going to be able to use the 10-meter or how many of *you* are going to use the 10-meter. So small telescopes may be the only hope. Furthermore, with the new detectors, small telescopes are really very large telescopes and can do an awful lot, certainly for photometry and even for spectroscopy. Fortunately, there are a few excellent, small telescopes being used for these kind of things, but they are ones which are at an observatory or attached to somebody like Bob who loves this sort of stuff and keeps them going. But *they* could disappear rapidly too. In addition, the small telescopes are super-good for educational reasons, perhaps even more so than for research, because how do you capture the excitement of a student without their doing the actual research, the kind of stuff that we have talked about and which Richard was doing with real data?

Are small telescopes worth it? I would very much like to hear anyone comment on that, if not now, then at least to me by e-mail or mail. Some of us are trying on the side, slowly to produce a business plan that will find money for some small telescopes or at least link together those people interested in small telescopes who may have some time. I gave a couple of papers on this global network of small telescopes last summer in Ireland. The papers dialogue a bit about the value of small telescopes and give a spreadsheet about the astro-economics of both the cost and the value of small telescopes.

Garrison: One of the problems with small telescopes is that they are not usually equipped with good spectrographs: but there are a couple that do have these, including the little 24-inch in Chile which I operate from the University of Toronto. It is open to all, so if you want even a month or so of time on it, you just have to remember that the deadline is early November and write to me for an application form. Then there is a good chance that you will get your time.

Crawford: Well-equipped small telescopes, such as the 1-m on Kitt Peak, can be just as oversubscribed for darktime as the 4-m, so this is a matter of politics rather than reality.

Garrison: That is right. I am renewing my infrastructure grant for it this year, so I've done a lot of work in the last two months with statistics and so forth and it is rather interesting. CFHT produces more papers than we do because they have shorter runs, but one paper usually equals one star, and that one paper costs about $70,000. One paper with my little telescope down there, which sometimes includes thousands of hours of observing, costs $6,000. So that gives you a little perspective on that.

There are lots of other arguments one can use, because not all faint stars are interesting and not all interesting stars are faint. So you can design your project to use only a big telescope and you can say, "Oh, I can't do it without the biggest telescope in the world." Baloney! Design another project. There is a lot of science out there. With the 24-inch telescope, the MK spectrograph, and a digital detector I can reach close to twelfth magnitude. There are a couple of million stars brighter than that and I figure, if I can't find something to do that is interesting within a couple of million stars, I'm going to quit. We have even been monitoring a 23rd magnitude lens galaxy, unfiltered of course, but this is 23rd magnitude with a 24-inch telescope. Our seeing there is superb, of course, and we have got good optics.

Weaver: Yes, I did a calculation. Basically, a 36-inch telescope with a modern detector is equivalent to a 300-inch telescope when the 200-inch was built. So you get to see a lot of the universe with that 36-inch telescope. It seems to me that most of the problems we have been talking about concerning standards, and observing time, and large telescopes are taken care of, in part, by having an archive of the digital data that comes out from national observatories. I proposed about ten years ago that, if someone observes at a national observatory at national cost, it makes sense that he should, after some period of time, put this into the public domain. The observations obviously have to be reduced, and so I think that it should be part of the obligation of the observer to produce those reduced data for general public consumption. The idea I actually had then, especially with the digital data, was to put it on CD ROM and send it out to everybody once a year or after whatever the proprietary period was.

It also seems to me that an awful lot of work on large telescopes is repetitious. You have the case where someone goes and uses the spectrograph for radial velocities, then the next person uses it for classification, and the next person uses it for metal abundance or something. In the end, a lot of them could have done it with one or two of the standards instead of ten of the same type.

9. Future meetings

Abt: I should like to say something about future meetings. I think we all agree that this has been a very useful meeting for us because there are many aspects that will help us in our plans and in our application of the MK System. Spectral classification is what Chris calls "a powerful tool." But it is a *technique*, and those of us in this room, which includes a lot of people in the world who use this technique, enjoy talking with others about how we do things and what we can do with it. I have been thinking about what Jim Liebert said, and I find the reason we don't see students in this room is because students have not yet made the commitment to go into a certain subfield of astronomy, and therefore they are not particularly interested in techniques. They are still interested in astronomy and what they can learn in astronomy. Once they have made a commitment, going into cosmology or stellar astronomy or planetary astronomy or so on, then they will start learning more about the techniques that are used and what they can do with them.

The one thing I've felt some lack of this meeting – and this is very understandable because you can't talk about everything in one week – is applications of the MK System to astrophysical and stellar structure studies and other such things. In other words, how can we use the MK System to learn more about astrophysics, about stellar evolution, about evolution of galaxies, about cosmology perhaps even, and certainly other parts of astronomy? We need to emphasize in future meetings what can we do with the MK System that you can't do with the other systems such as photometric systems. For, if you have a technique which – let's take an extreme case – does not give you anything new, then people will say, "Well, why should I be interested in such a technique?" So I hope that in the future meetings we shall talk more about applications, about our own plans, and about suggestions of what other people can do uniquely with the MK System really to learn a lot more astrophysics. Thank you.

Garrison: I agree with you, Helmut. This has certainly not been principally an applications meeting. This has been more of a process meeting, and the scientific organizing committee set it up that way. But now that we are all using digital data, or most of us, we may not need another process meeting. In fact, the next meeting should be an applications meeting, and let's go and do some exciting stuff in preparation. For there are lots of people out there, that is cosmologists and extragalactic types, who are getting spectra of stars as spin-offs from their surveys and who are doing miserable things with them. So somehow, we have to get the word to them that here in the MK Process is a well-designed way of extracting exciting information from those spectra of stars that they might have considered garbage.

Well, that is probably a good note on which to end.

Editorial Note

This workshop's program, having the methodology of starting with a new technique which then leads into the science, can easily leave a final impression of concentrating on techniques rather than on science. It would be a pity to loose track of the fascinating science contained within these pages and so, as reminder, we leave the reader with some highlights:

Mihalas – the physical basis of classification;
Drilling – the physical parameters of OB stars;
Liebert and Sion – the physical basis of White Dwarf classification;
Gray – physical solutions for λ Bootis and Am star problems;
Walborn – parameters of O3 stars;
Hauck, Nicolet – physical data from spectral and photometric classification;
Boeshaar – physical information on M dwarfs from the infrared;
Parthasarathy – characteristics of post-AGB stars;
Chargeishvili – Galactic structure from OB stars.

New techniques were indeed examined – of precision classification via digital data, of extensions into the new wavelength regions of the IR and the UV, of automated classification, of interaction with developing photometric classifications – and we experienced something of the new science that these will bring. May the sense of excitement that all this generated in the workshop participants remain with readers of these proceedings.

INDEX